MALARIA VACCINE DEVELOPMENT

A Multi-Immune Response Approach

MALARIA VACCINE DEVELOPMENT

A Multi-Immune Response Approach

Edited by

STEPHEN L. HOFFMAN

Malaria Program
Naval Medical Research Institute
Rockville, Maryland

ASM PRESS
Washington, D.C.

1325 Massachusetts Ave., N.W.
Washington, DC 20005

Library of Congress Cataloging-in-Publication Data

Malaria vaccine development : a multi-immune response approach / edited by Stephen L. Hoffman

p. cm.

Includes index.

ISBN 1-55581-111-6

1. Malaria vaccines. I. Hoffman, Stephen L.

[DNLM: 1. Malaria Vaccines—immunology. 2. Malaria—immunology.

WC 750 M2372 1996]

QR189.5.M34M34 1996

616.9'362079—dc20

DNLM/DLC

for Library of Congress 96-4661

CIP

Printed in the United States of America

To my parents, Louise and Julian; my wife, Kim Lee; and my sons, Alexander, Seth, and Benjamin, whose love, support, encouragement, respect, and understanding provide the foundation for completion of this book and every other project I undertake.

Contents

Contributors

Roberto Amador • Instituto de Inmunología, Hospital San Juan de Dios, Universidad Nacional de Colombia, Bogotá, Colombia

Roy M. Anderson • Wellcome Centre for the Epidemiology of Infectious Disease and Department of Zoology, University of Oxford, Oxford OX13PS, United Kingdom

John J. Aponte • Instituto de Inmunología, Hospital San Juan de Dios, Universidad Nacional de Colombia, Bogotá, Colombia

Klavs Berzins • Department of Immunology, The Wenner-Gren Institute, Stockholm University, S-106 91 Stockholm, Sweden

Graham V. Brown • The Walter and Eliza Hall Institute of Medical Research, Post Office Royal Melbourne Hospital, Victoria 3050, and Infectious Diseases Service, Royal Melbourne Hospital, Parkville, Victoria 3052, Australia

Pierre Druilhe • Biomedical Parasitology, Pasteur Institute, Paris, France

Eileen D. Franke • Malaria Program, Naval Medical Research Institute, Rockville, MD 20852

Brian Greenwood • London School of Hygiene and Tropical Medicine, University of London, London WC1E 7HT, United Kingdom

Sunetra Gupta • Wellcome Centre for the Epidemiology of Infectious Disease and Department of Zoology, University of Oxford, Oxford OX13PS, United Kingdom

Stephen L. Hoffman • Malaria Program, Naval Medical Research Institute, Rockville, MD 20852

Anthony A. Holder • Division of Parasitology, National Institute for Medical Research, The Ridgeway, Mill Hill, London NW7 1AA, United Kingdom

Michael R. Hollingdale • Department of Biology, The University of Leeds, Leeds, United Kingdom

David C. Kaslow • Malaria Vaccines Section, Laboratory of Parasitic

Diseases, National Institute of Allergy and Infectious Diseases, National Institutes of Health, Bethesda, MD 20892

Louis H. Miller • Laboratory of Parasitic Diseases, National Institute of Allergy and Infectious Diseases, National Institutes of Health, Bethesda, MD 20892

Victor Nussenzweig • Michael Heidelberger Division of Immunology, Department of Pathology, New York University Medical Center, New York, NY 10010

Manuel E. Patarroyo • Instituto de Inmunología, Hospital San Juan de Dios, Universidad Nacional de Colombia, Bogotá, Colombia

Peter Perlmann • Department of Immunology, The Wenner-Gren Institute, Stockholm University, S-106 91 Stockholm, Sweden

J. H. L. Playfair • Department of Immunology, University College of London, London W1P 9PG, United Kingdom

Stephen J. Rogerson • The Walter and Eliza Hall Institute of Medical Research, Post Office Royal Melbourne Hospital, Victoria 3050, and Infectious Diseases Service, Royal Melbourne Hospital, Parkville, Victoria 3052, Australia

Photini Sinnis • Department of Medical and Molecular Parasitology and Department of Pathology, New York University Medical Center, New York, NY 10010

Malaria Vaccine Development: A Multi-Immune Response Approach
Edited by Stephen L. Hoffman

Chapter 1

Perspectives on Malaria Vaccine Development

Stephen L. Hoffman and Louis H. Miller

The multistage protozoan parasites *Plasmodium falciparum, Plasmodium vivax, Plasmodium malariae,* and *Plasmodium ovale* are the etiologic agents of malaria, a disease of almost unfathomable impact. *P. falciparum* stands out as the most important of these agents in its effect on human life. Every year, *P. falciparum* kills an estimated 1 to 3 million individuals and is responsible for hundreds of millions of clinical infections throughout the world (15, 16) (Fig. 1). The vast majority of the disease and death caused by *P. falciparum* occurs in sub-Saharan Africa. A century after Alphonse Laveran's Nobel Prize-winning discovery that malaria is caused by a protozoan in the blood (8) and Sir Ronald Ross's Nobel Prize-winning discovery that *Plasmodium* parasites are transmitted by mosquitoes (12), and 40 years after the publication of prominent malariologist Paul Russell's book *Man's Mastery of Malaria* (13), malaria remains one of the most important infectious diseases in the world.

Widespread and increasing resistance of the parasite to antimalarial drugs (Fig. 1 through 3), development of resistance of *Anopheles* mosquito vectors to commonly used insecticides, an inadequate infrastructure for delivery of control measures, population growth, and movement of nonimmune populations to malarious areas have all contributed to the persistence and in many cases worsening of the malaria problem (reviewed in reference 7). This deteriorating situation for malaria control has stimulated a search for new tools to control both transmission of the infection and the impact of the disease on populations. One such tool, vaccination, has led to the eradication of smallpox from the world and polio from the Western Hemisphere. In this book, distinguished scientists explain why such a control tool is not yet available for malaria, and they describe how they and other scientists throughout the world are working to develop

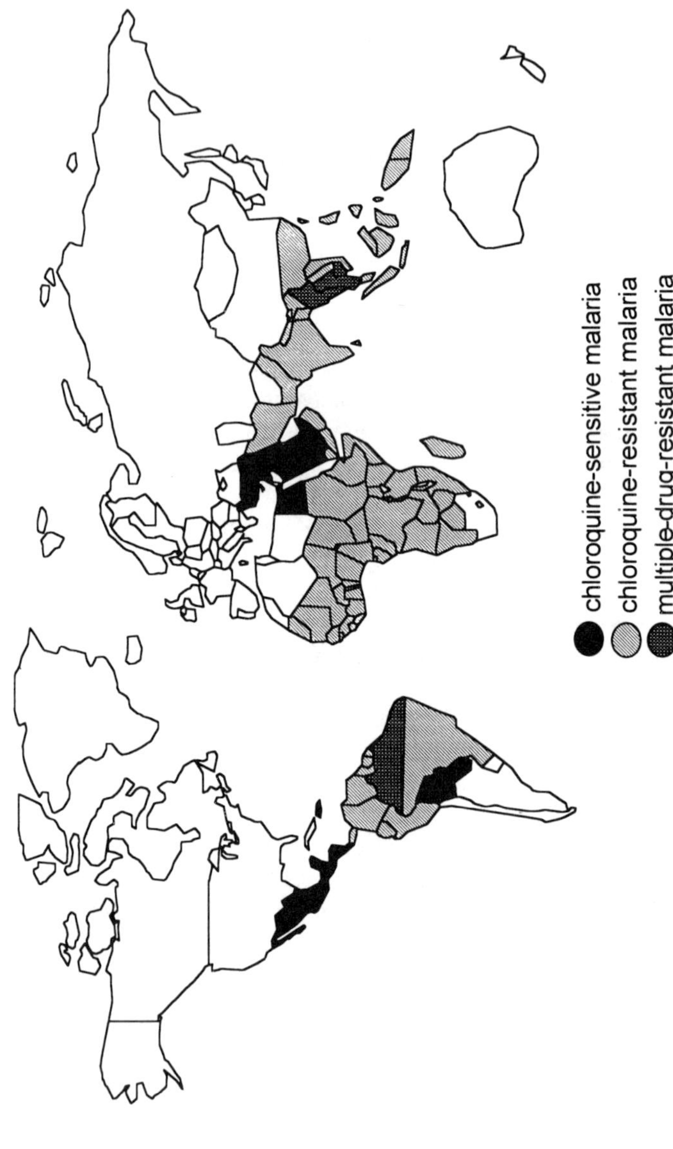

Figure 1. World distribution of malaria in 1995, with delineation of areas where *P. falciparum* is sensitive to chloroquine, resistant to chloroquine, and resistant to numerous antimalarial drugs. (Map provided by Jane Zucker, Centers for Disease Control and Prevention, Atlanta, Ga.)

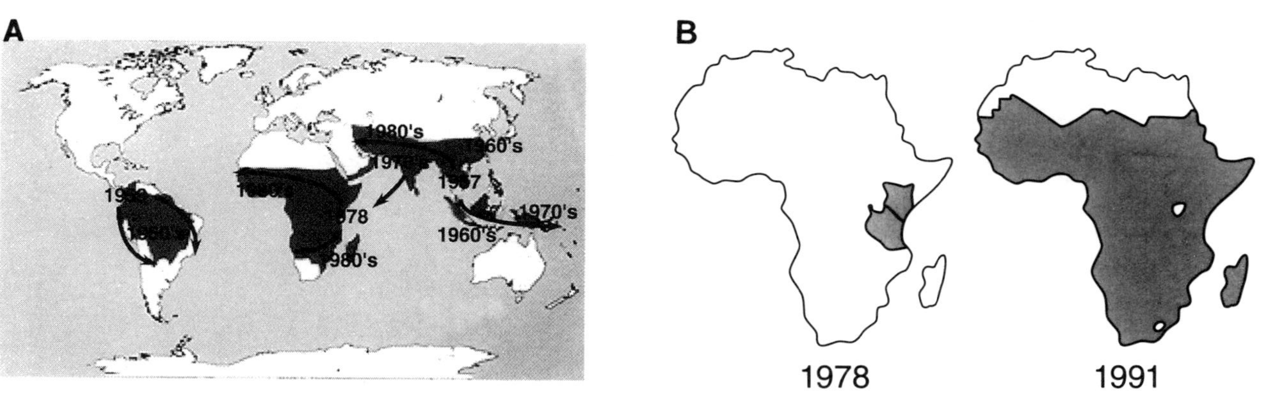

Cost of treatment for malaria

$0.01	Chloroquine
$0.13	Pyrimethamine/Sulfadoxine
$1.92	Mefloquine
$5.31	Halofantrine

■ Resistance Reported

Figure 2. The spread of chloroquine-resistant *P. falciparum* in the world (panel A; from Thomas Wellems) and in Africa (panel B; from Joel Breman).

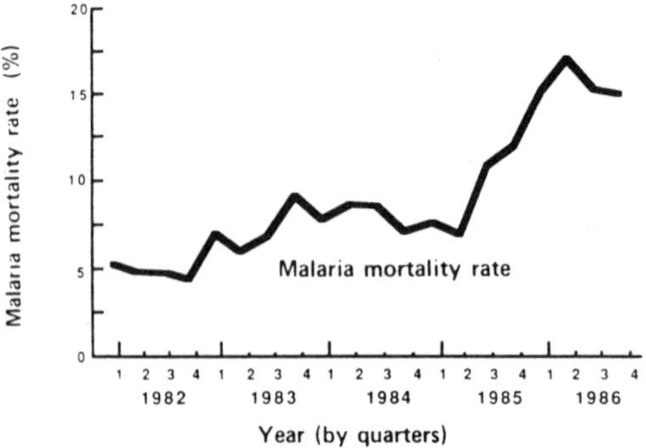

Figure 3. An increased malaria mortality rate in a hospital in Zaire in 1985 was associated with a high incidence of chloroquine-resistant *P. falciparum* in the community. Reprinted from reference 5a with permission of the World Health Organization.

vaccines that can be used in conjunction with drugs and antimosquito measures to control the impact of malaria and perhaps one day even eradicate from the world the *Plasmodium* parasites responsible for the disease.

For most infectious diseases for which we have effective vaccines, a single infection confers long-standing protective immunity. For example, a person who has had measles does not develop measles again. This type of sterile protective immunity does not exist for malaria. When a group of adult Kenyans who had lived in the same village for their entire lives and had been exposed to thousands of infected mosquitoes during their lives were radically cured of malaria, 85% developed a new asexual erythrocytic infection within 4 months (6). Notwithstanding this lack of naturally acquired sterile protective immunity, considerable data from studies in animals, from the field, and from work with volunteers clearly indicate that development of malaria vaccines is feasible.

Our first reason for believing that a vaccine can be made is that protective immunity has been induced in every animal model of malaria. Immunization with *Plasmodium knowlesi* asexual parasites in Freund's complete adjuvant induces a high level of immunity (10). Immune animals rapidly cleared their infections, despite the fact that *P. knowlesi* causes an infection that persists for more than 1 year. In this example, immunization led to immunity that was superior to immunity from natural infection.

Second, the density of parasitemias upon infection decreases with age (3), and the clinical manifestations of malaria are generally much milder in adults and older children than in infants and children under 5 years of age. If an individual survives beyond the age of 5 years in an area with intense transmission of malaria, then that individual's chance of dying from malaria drops drastically. These observations clearly indicate that repeated infection induces immune responses that reduce parasite burden in infected individuals and reduces the morbidity and mortality of malaria. If we could discover the immune mechanisms responsible for this protection and the antigenic targets of these protective immune responses, then we could develop vaccines that, like repeated natural infection, would reduce morbidity and mortality without eliminating infection.

Third, several studies have shown that immunoglobulin purified from the blood of adults who lived their entire lives in malarious areas of West Africa can passively transfer protection against *P. falciparum* (4). In the most recent study, administration of hyperimmune immunoglobulin G from West Africa to Thai children with multidrug-resistant *P. falciparum* infections reduced peripheral parasitemia by 99% (14) (see chapters 4 and 5). Such data indicate that antibodies against antigens expressed by erythrocytic-stage parasites can have a profound effect on eliminating the parasite. The challenges are to identify the antigenic targets of the antiparasitic immunoglobulin G and to develop vaccines that induce antibodies of the same specificities.

Fourth, immunization of mice and humans with radiation-attenuated sporozoites confers sterile protective immunity. This immunity, directed against the sporozoite in circulation or the parasite developing within hepatocytes, is antigen specific and does not appear to be strain specific (see chapters 2 and 3). These findings with radiation-attenuated sporozoites (11a) have provided the foundation for 20 years of work by numerous scientists who have been trying to elucidate the mechanisms and target antigens of the immune responses and to develop subunit vaccines that induce comparable immunity (see chapters 2 and 3).

Malaria vaccines are based on the parasite's complex life cycle (Fig. 4). Sporozoites are inoculated by *Anopheles* mosquitoes that feed on humans from dusk until dawn. The extracellular sporozoites are present in the bloodstream for less than 1 h. They invade hepatocytes, where a single, uninucleate sporozoite develops over 2 to 10 days, depending on the *Plasmodium* sp. (5 to 6 days for *P. falciparum*), to a mature liver-stage schizont. This liver-stage parasite ruptures, releasing up to 30,000 uninucleate merozoites, each of which can invade an erythrocyte and develop over 2 to 3 days (43 to 48 h for *P. falciparum*) into a mature, erythrocytic-stage schizont with 10 to 30 merozoites. When the infected erythrocyte ruptures, it releases the merozoites, each of which can invade an erythrocyte, and parasite-derived toxins. Alternatively, the merozoite can develop within the erythrocyte to the sexual stage called a gametocyte. When a mosquito ingests erythrocytes containing gametocytes, the parasite develops in the midgut of the mosquito into gametes that are released from erythrocytes. This release is followed by fertilization of the female by the male gamete, which leads to zygote formation and subsequently ookinete, oocyst, and sporozoite development (see chapter 8). Only the asexual erythrocytic stage of the life cycle is associated with pathology and disease.

Each stage has different antigens that lead to protective immunity, and in many cases, these antigens are not expressed at other stages of the life cycle. For example, extremely high levels of antibodies against the major surface protein of sporozoites, the circumsporozoite protein, do not recognize parasites from the erythrocytic stage of the life cycle. This fact, however, does not exclude the possibility that as we learn more about the genes expressed at different stages, antigens that induce protective immunity at all stages will be identified.

Vaccines to attack each stage of the life cycle are being developed (Fig. 4). The asexual erythrocytic stage of the parasite life cycle is responsible for all of the pathology and the clinical manifestations of malaria. The stages before the asexual erythrocytic parasites, the sporo-

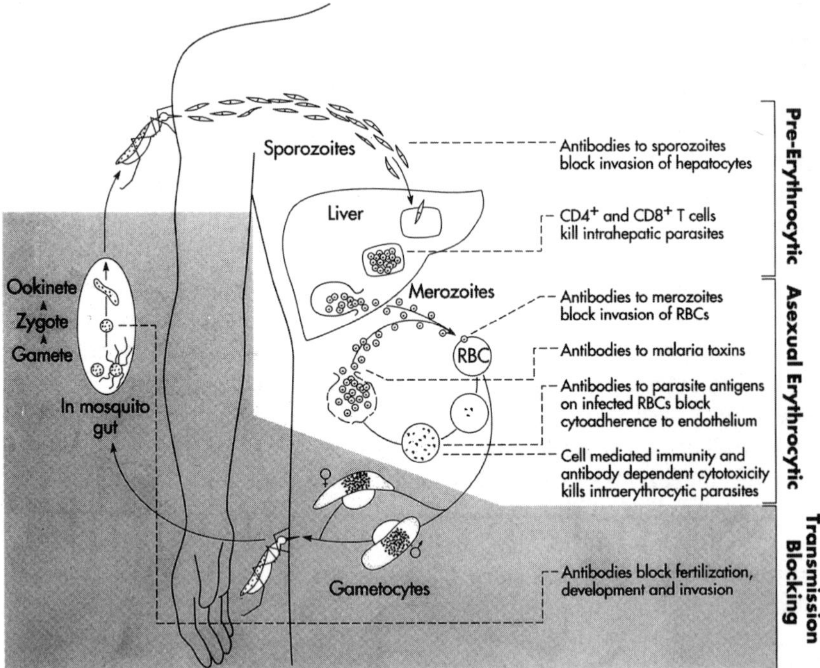

Figure 4. The life cycle of *Plasmodium* spp. highlights the immune mechanisms for pre-erythrocytic, asexual erythrocytic, and transmission-blocking vaccines. RBC, erythrocyte.

zoite and hepatic stages, are called pre-erythrocytic. Pre-erythrocytic vaccines induce immunity to antigens on sporozoites that are inoculated by the mosquito to initiate infection and against liver stages that develop from a uninucleate sporozoite into thousands of individual nucleated merozoites, each one able to invade an erythrocyte. Despite the differences in immune effector mechanisms and antigens for sporozoites and hepatic stages, the grouping of these vaccines under the single term pre-erythrocytic is based on the fact that a fully effective pre-erythrocytic vaccine will prevent development to the erythrocytic stages and thus prevent disease.

The second set of vaccines act against the asexual erythrocytic stages. The parasite invades erythrocytes and grows from a single nucleated trophozoite into a dividing, multinucleated schizont that produces approximately 10 to 30 merozoites, each able to invade another erythrocyte. At the time the infected erythrocytes rupture and release merozoites, they also release toxins that cause fever and possibly contribute to severe complications. Anti-asexual-erythrocytic-stage vaccines are being designed to prevent disease by reducing or completely inhibiting growth of the parasite in the blood (chapters 4 and 5), inhibiting the adherence of infected erythrocytes to endothelial cells and other erythrocytes (chapter 6), and blocking factors in the parasites, such as toxins, that cause disease (chapter 7).

Vaccines of the third type are directed against the sexual forms of the parasite in order to block infection of mosquitoes. Asexual erythrocytic parasites may differentiate into sexual stages (gametocytes) that are infectious to mosquitoes. Antibodies against antigens on the surfaces of male and female gametes, zygotes, and developing parasites within the mosquito midgut are the major targets of such vaccines (chapter 8). Such vaccines would not have any impact on the clinical manifestations of malaria in an individual. However, if administered successfully to a community, they could have a major impact on malaria transmission and thus, by reducing transmission, on malaria-associated morbidity and mortality in the individuals composing the community. In addition, a transmission-blocking vaccine would ideally be included in any field-deployed vaccine to prevent or slow the expansion of mutant parasite populations resistant to the vaccine.

In contrast to the approach to vaccines for smallpox, polio, and essentially all other infectious diseases for which vaccines are available, the approach to developing malaria vaccines, many malariologists believe, may require developing several functionally different types of malaria vaccines because of the differences in target groups. One approach is to develop vaccines designed to protect "individuals" by eliminating the parasite before it can cause any clinical manifestations. A pre-erythrocytic vaccine, a highly effective asexual erythrocytic vaccine that would completely block asexual multiplication, or a combination of the two would fulfill this objective. Such vaccines for preventing infection will be required for individuals who have never been or are infrequently exposed to malaria (nonimmunes) when they visit malarious areas. These nonimmunes include the estimated tens of millions of tourists from countries without malaria who annually visit countries where malaria is transmitted and the business, foreign service, and military personnel who visit these areas. The group of nonimmunes also includes individuals who live in nonmalarious areas of countries where malaria is transmitted but visit malarious regions of these countries. This rapidly expanding group of individuals may well be the largest nonimmune target group for a vaccine designed to prevent malaria.

Another approach is to develop vaccines designed to protect "communities." Effective vaccines being developed for completely preventing malaria among individuals by eliminating the parasite would have a major impact at the community level if the vaccines provided sustainable, long-term protection. However, there are concerns regarding both the level of protection achievable in the short term and the sustainability of the immunity induced by such vaccines in the field. In the long run, introduction of a vaccine that completely prevented infection in a community for only 5 years could be associated with more morbidity and mortality than not intervening at all. When the complete protection disappeared, a formerly semi-immune population that was protected against development of life-threatening malaria by continuous boosting of their immunity by low-grade infections might no longer be protected at all, and epidemics of severe malaria with high mortality rates could occur. Thus, there are efforts to develop vaccines that will reduce parasite burden or neutralize pathogenic properties without eliminating the parasite, with the goal of reducing the morbidity and mortality of malaria while maintaining immunity by continuous natural boosting. The primary target populations for such vaccines are the groups most severely affected by malaria, such as African infants and children, 1 to 3 million of whom are killed every year by *P. falciparum*, and pregnant women. The development of vaccines to reduce parasitemia is based on observations such as the high density of parasitemia in patients with cerebral malaria (11) and on reports of a clear association between parasite density at time of hospital admission and case fatality ratio (5). Parasitemia alone, however, is only one of the factors in disease, since children within an African village respond differently to high parasitemia. Some develop severe disease; others are minimally symptomatic. The suggestion that malaria toxins may be a factor in disease pathogenesis has fueled the search for toxins and the development of antitoxin vaccines (chapter 7).

Conceptualizing the deployment and impact of vaccines that induce long-term sterile immunity in a high percentage of recipients for protection either of individuals or of communities is relatively straightforward. Such vaccines will be easy to test in nonimmunes, who will develop no infection on challenge. Of greater difficulty will be the identification of vaccines that reduce disease or death without completely eliminating the parasites. For example, it is possible that a partially effective pre-erythrocytic, asexual, or transmission-blocking vaccine will reduce disease in a community. The surprising finding that permethrine-impregnated bed nets reduce mortality in children despite an unchanged infection rate in the children highlights the problem of predicting whether any particular intervention will have an impact (1, 2). To further complicate the picture of malaria, different levels of transmission and perhaps other epidemiologic factors (reviewed in reference 9) are associated with different patterns of disease. For example, in areas of intense transmission like western Kenya and Tanzania, severe anemia in infants and young children is the major severe manifestation of malaria, while in areas of less intense transmission, like The Gambia, cerebral malaria in 2- to 4-year-old children is the most common serious clinical manifestation of the disease (Fig. 5). The pathogenesis of these two severe complications or of death itself in the village is poorly defined. How, then, can we predict whether a particular vaccine will lead to reduced levels of disease other than by testing it within each setting? Results of vaccine trials may give clues to predicting the efficacy of a particular set of vaccine constructs. A partially effective pre-erythrocytic vaccine may decrease the incidence of disease, thus indicating that similar vaccines should be tested. Because of the diversity of the epidemiology, pathogenesis, and clinical manifestations of malaria, it may be necessary to develop multiple testing centers

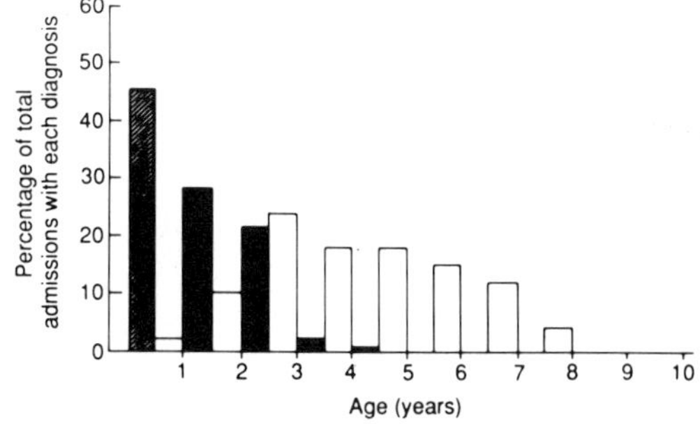

Figure 5. Age distribution of complications of malaria in African children. Severe anemia (dark bars) occurs earlier in childhood than cerebral malaria (open bars). Reprinted from reference 8a with permission of Cambridge University Press.

within Africa and other areas of the world to evaluate the efficacy of vaccines. The cost of such an empiric approach will be extremely high. Appropriately apportioning funds between this applied research and the basic research necessary to answer the questions critical to developing ideal malaria vaccines will challenge our wisdom in the years ahead.

In chapters 2 through 8 of this book, the authors provide an up-to-date review of efforts to develop vaccines that target each stage of the parasite life cycle. This work includes efforts to elucidate protective immune responses, identify the antigenic targets of the protective immune responses, and develop vaccine delivery systems that induce the required immune responses against the identified antigens. Vaccine-induced immune effector mechanisms include protective antibody, antibody-dependent cell-mediated immunity, and T-cell-dependent cellular immunity (Fig. 4). Protective antibodies may neutralize "toxic" molecules released by rupturing schizont-infected erythrocytes and prevent invasion of hepatocytes by sporozoites, invasion of erythrocytes by merozoites, invasion of mosquito midgut epithelium by ookinetes, adherence of infected erythrocytes to endothelial cells, and development of extracellular gametes, zygotes, and ookinetes in the mosquito midgut. Killing of parasites developing within hepatocytes is primarily mediated by effector T lymphocytes that recognize peptide-major histocompatibility complexes on the surfaces of infected hepatocytes or by molecules, such as cytokines, released from T cells or reticuloendothelial cells near the infected cell. The mediators enter the infected hepatocyte and destroy the parasite or induce the hepatocyte to produce substances that kill the parasite or itself. Attacking the infected erythrocyte, a cell without major histocompatibility complex molecules on its surface, can be done only by antibodies that recognize exposed parasite antigens or by cellular products like cytokines and free radicals that enter the infected cell and inhibit or kill the parasite.

Many important antigens and protective antibody and T-cell epitopes have been identified at each stage of the life cycle. Armed with these data on immune mechanisms and their antigenic targets and with recombinant DNA technology, modern methods of peptide synthesis, and new adjuvants, many thought that highly effective malaria vaccines would rapidly follow. This has not been the case, and in some respects, it is the inadequacy of these technologically advanced vaccines that is responsible for the delay. Purified recombinant proteins, synthetic peptides, and live recombinant viruses and bacteria have not been as immunogenic as expected, and major efforts to turn the promise of modern vaccinology into reality by improving their immunogenicity are under way.

An enormous amount of work is being done in an effort to develop these stage-specific vaccines. However, many malariologists believe that such vaccines represent only the first step and that an optimal malaria vaccine will have to induce protective immune responses against every stage. This perspective is based on the need to overcome several major problems facing malaria vaccinologists: (i) the unique antigenicities of the different stages of the life cycle, (ii) the sequence variabilities of important antigenic epitopes, (iii) the capacity of immune selection to enrich for parasites that are not susceptible to specific immune responses, and (iv) genetic restriction of T-cell responses to specific parasite epitopes. Dealing with these problems will require the development of multivalent, multi-immune response vaccines that optimally induce antibodies against extracellular and cell surface targets and induce protective $CD4^+$ and $CD8^+$ T cells that directly interact with infected hepatocytes or release cytokines that target infected hepatocytes and erythrocytes. Having overcome the daunting task of producing vaccines that induce protective immune responses against each stage of the life cycle, vaccinologists will be faced with the perhaps even more challenging objective of constructing such multivalent, multi-immune response vaccines. Currently, little experimental and less clinical data support this type of approach for any infectious agent, but it will be critical in coming years for vaccinologists to draw on the burgeoning knowledge in immune regulation and vaccine design to most efficiently develop such vaccines. Perhaps malaria vaccines will lead the way in this research.

In chapter 9, Manuel Patarroyo and his colleagues explain the steps they have used to develop, produce, and test the polymerized synthetic peptide SPf66, an immunogen with peptides derived from the sporozoite and merozoite stages of the life cycle. This candidate malaria vaccine is the only one to have been tested in large double-blind placebo-controlled trials in South America, Africa, and Asia. In chapter 10, Sunetra Gupta and Roy Anderson elucidate the critical role that mathematical modeling can play in predicting the effects of vaccination on the population dynamics of malaria, and in chapter 11, Brian Greenwood provides his perspective on what can be expected from malaria vaccines.

We still have a long way to go before safe, highly effective malaria vaccines will be ready for licensure and deployment to prevent infection or reduce malaria morbidity and mortality, either on their own or as part of an integrated program to diminish disease and control transmission of the infection. Despite the size and complexity of the problem, adequate resources for overcoming the many obstacles to developing effective vaccines have never been available (reviewed in reference 7). Nonetheless, enormous progress toward achieving these goals has been made during the past 10 to 20 years by numerous scientists throughout the world.

A systematic approach to developing malaria vaccines will require characterization of protective immune responses, identification of the antigenic targets of these protective immune responses, development of vaccine delivery systems that can induce the required immune responses against the identified target antigens, development of experimental challenge and field study methodologies to adequately assess the protective efficacy of the vaccines, and development of methodologies and surveillance tools with which to adequately determine the short- and long-term impacts of vaccines once they are developed. This work will require the study of naturally acquired immunity to malaria and the continued use, development, and refinement of model systems such as the rodent malaria, nonhuman-primate malaria, human malaria-nonhuman primate, *P. falciparum* irradiated-sporozoite human volunteer, and human malaria-human volunteer model systems. As we move toward the development and testing of many candidate malaria vaccines, we must not forget how critical it is to expand our understanding of the epidemiology, pathogenesis, and clinical manifestations of the complex, heterogeneous disease that we call malaria. Obtaining such information will be critical to the development of rational strategies for assessing candidate vaccines and deploying licensed malaria vaccines.

Acknowledgments. This work was supported by Naval Medical Research and Development Command Work Units 61102A00101-BFX.1431, 62787A00101-EFX.1432, and 63002A0010-HFX.1433.

Special thanks go to Sylvia I. Becker for editorial support.

REFERENCES

1. **Alonso, P. L., S. W. Lindsay, J. R. M. Armstrong, M. Conteh, A. G. Hill, P. H. David, G. Fegan, A. de Francisco, A. J. Hall, F.C. Shenton, K. Cham, and B. M. Greenwood.** 1991. The effect of insecticide-treated bed nets on mortality of Gambian children. *Lancet* **337:**1499–1502.
2. **Alonso, P. L., S. W. Lindsay, J. R. M. Armstrong Schellenberg, K. Keita, P. Gomez, F. C. Shenton, A. G. Hill, P. H. David, G. Fegan, K. Cham, and B. M. Greenwood.** 1993. A malaria control trial using insecticide-treated bed nets and targeted chemoprophylaxis in a rural area of the Gambia, West Africa. *Trans. R. Soc. Trop. Med. Hyg.* **87**(2):37–44.
3. **Beadle, C., P. D. McElroy, C. N. Oster, J. C. Beier, A. J. Oloo, F. K. Onyango, D. K. Chumo, J. D. Bales, J. A. Sherwood, S. L. Hoffman.** 1995. Impact of transmission intensity and age on *Plasmodium falciparum* density and associated fever: implications for malaria vaccine trial design. *J. Infect. Dis.* **172:**1047–1054.
4. **Cohen, S., I. A. McGregor, and S. Carrington.** 1961. Gamma-globulin and acquired immunity to human malaria. *Nature* (London) **192:**733–737.
5. **Field, J. W., and J. C. Niven.** 1937. A note on prognosis in relation to parasite counts in acute subtertian malaria. *Trans. R. Soc. Trop. Med. Hyg.* **30:**569–574.

5a. **Greenberg, A. E., et al.** 1989. Hospital-based surveillance of malaria-related paediatric morbidity and mortality in Kinshasa, Zaire. *Bull. W.H.O.* **67:**189–196.

6. **Hoffman, S. L., C. N. Oster, G. R. Woolett, J. C. Beier, J. D. Chulay, R. A. Wirtz, M. R. Hollingdale, and M. Mugambi.** 1987. Naturally acquired antibodies to sporozoites do not prevent malaria: vaccine development implications. *Science* **237:**639–642.

7. **Institute of Medicine.** 1991. *Malaria: Obstacles and Opportunities,* p. 1–6. National Academy Press, Washington, D.C.

8. **Laveran, A.** 1880. Note sur un nouveau parasite trové dans le sang de plusieurs malades atteints de fievre palustre. *Bull. Acad. Med. Ser. 2,* **9:**1235–1236.

8a. **Marsh, K.** 1992. Malaria—a neglected disease? *Parasitology* **104:**S53–S69.

9. **Miller, L. H., M. F. Good, and G. Milon.** 1995. Malaria pathogenesis. *Science* **264:**1878–1883.

10. **Mitchell, G. H., G. A. Butcher, and S. Cohen.** 1975. Merozoite vaccination against *Plasmodium knowlesi* malaria. *Immunology* **29:**397–407.

11. **Molyneux, M. E., T. E. Taylor, J. J. Wirima, and A. Borgstein.** 1989. Clinical features and prognostic indicators in paediatric cerebral malaria: a study of 131 comatose Malawian children. *Q. J. Med.* **71:**441–459.

11a. **Nussenzweig, R. S., and V. Nussenzweig.** 1984. Development of sporozoite vaccines. *Phil. Trans. R. Soc. London B* **307:**117–128.

12. **Ross, R.** 1897. On some peculiar pigmented cells found in two mosquitoes fed on malarial blood. *Br. Med. J.* **2:**1786–1788.

13. **Russell, P. F.** 1955. *Man's Mastery of Malaria.* Oxford University Press, London.

14. **Sabchareon, A., D. Outtara, P. Attanah, H. Bouharoun-Tayooun, P. Chantavanich, C. Foucault, T. Chongsuphajaisiddhi, and P. Druilhe.** 1991. Parasitologic and clinical human response to immunoglobulin administration in falciparum malaria. *Am. J. Trop. Med. Hyg.* **45:**297–308.

15. **Sturchler, D.** 1989. How much malaria is there worldwide? *Parasitol. Today* **5:**39.

16. **World Health Organization.** 1989. *Weekly Epidemiol. Rec.* **32:**241–247.

Malaria Vaccine Development: A Multi-Immune Response Approach
Edited by Stephen L. Hoffman

Chapter 2

Preventing Sporozoite Invasion of Hepatocytes

Photini Sinnis and Victor Nussenzweig

INTRODUCTION

In most areas where malaria is endemic, the total number of sporozoites in the salivary glands of an *Anopheles* mosquito is less than 10^4, and only a small proportion of these are injected into the host during the mosquito's blood meal. An estimate of the number of injected sporozoites was obtained by allowing *Anopheles stephensi* infected with *Plasmodium falciparum* to salivate into oil droplets. The median gland infection of the *A. stephensi* was 8,170, and the median number of ejected sporozoites was only 15 (range, 0 to 978) (55). When *A. stephensi* mosquitoes were allowed to feed through fresh mouse skin, the mean number of ejected sporozoites was 47 (median, 8), although the mean number of parasites in the salivary glands was more than 2×10^4 (median, 14,500) (49). After being injected into susceptible mammalian hosts, the parasites rapidly leave the blood circulation and enter hepatocytes, where they develop into the exoerythrocytic (EE) forms. Little is known about how sporozoites find their way to the liver and subsequently invade hepatocytes. In the first part of this chapter, we discuss the mechanisms involved in this rapid and specific sequestration of sporozoites by the liver. This is followed by a discussion on the development of malaria vaccines that aim to prevent sporozoite infection of hepatocytes.

MECHANISMS OF SPOROZOITE SEQUESTRATION IN THE LIVER

The mechanisms involved in the specific recognition of the liver by the parasites are a matter of controversy (reviewed in references 21, 29, and 70). Some studies suggest that sporozoites are arrested in the liver by Kupffer cells. This idea is appealing, because Kupffer cells line the liver sinusoids and are known to clear foreign particles from the blood, while hepatocytes are separated from the circulation by an endothelium. Although this endothelium is fenestrated, a fenestra is much smaller in diameter (0.1 μm) (73, 74) than a sporozoite is (1 μm).

The evidence supporting the Kupffer cell hypothesis is primarily ultrastructural. Meis et al. injected *Plasmodium berghei* sporozoites intraportally into Brown Norway rats and examined liver sections by electron microscopy (38). Ten minutes after injection, parasites were either free in the sinusoids or in endocytotic vacuoles of Kupffer cells. One hour later, sporozoites were found exclusively inside Kupffer cells, and there was evidence of fusion of lysosomes with intracellular vacuoles containing parasites. One Kupffer cell containing a parasite was found in the space of Disse adjacent to a hepatocyte. The free apical end of the sporozoite was penetrating the cytoplasm of the hepatocyte. On the basis of this observation, the suggestion was made that the Kupffer cell had transported the sporozoite from the sinusoidal lumen across the space of Disse toward the adjacent hepatocyte.

The Kupffer cell "gate hypothesis" was supported by subsequent studies that measured the clearance of sporozoites in rat livers by using an in vitro perfusion system (59, 62). In this system, between 67 and 80% of the sporozoites in the perfusate were removed during the first passage through the liver, and 95% were removed after 15 min of perfusion. When the liver donors were rats that had previously been injected with silica to destroy the Kupffer cells, only 52 and 57% of the sporozoite load was removed after 5 and 15 min of perfusion, respectively (59). In addition, when these silica-treated rats were injected with sporozoites and assayed for malaria infection, they had between 63 and 95% fewer EE forms in the liver when compared with controls (59, 69).

However, a similar study using liposome-encapsulated dichloromethylene diphosphonate (DCP) rather than silica to destroy the Kupffer cells reached the opposite conclusion (71). After injection into animals, these liposomes are taken up by macrophages, and the DCP is released in the lysosomal compartment and kills the cell. Rats that had been preinjected with DCP liposomes and injected with sporozoites had four to six times more EE forms than control rats. Furthermore, pretreatment with silica (but not with DCP liposomes) increased plasma levels of interleukin 6 (71), a known inhibitor of EE-form development (47). These studies suggest that the observed reduction in sporozoite clearance and EE-form development in silica-treated rats results from disruption of the sinusoidal architecture and/or the release of lymphokines by inflammatory cells.

Observations have also been made on the interaction between sporozoites and macrophages in vitro. The frequency of invasion of peritoneal macrophages or Kupffer cells by sporozoites was low, but the intracellular sporozoites were not destroyed (16, 57, 61, 67). When this interaction was studied in real time with video microscopy, sporozoites were found entering and exiting macrophages, which at times were destroyed by the parasites (67). When sporozoites were preincubated in immune serum, however, their association with Kupffer cells increased, and in contrast to normal sporozoites, the opsonized parasites were killed by the macrophages (16, 57). On the basis of these studies, it is difficult to reach a conclusion about the role of Kupffer cells and macrophages in malaria infection. Nevertheless, the destruction of opsonized parasites suggests that phagocytosis by macrophages may be an effector mechanism involved in protective immunity by antisporozoite antibodies.

An alternative to the theory that sporozoites must interact with Kupffer cells or endothelial cells before entering hepatocytes is the possibility that hepatocytes are invaded directly. When *P. berghei* sporozoites were injected into the tail veins of Sprague-Dawley rats and the animals were sacrificed 2 min after injection, sporozoites were found in hepatocytes (58). Sporozoites found in Kupffer cells were morphologically altered, undergoing destruction in endocytic vacuoles. In vitro studies demonstrated that sporozoites directly invade both hepatoma cell lines and primary hepatocyte cultures (30, 36). One group tried to identify hepatocyte receptors for sporozoites by incubating purified human liver membranes with the parasites and subsequently adding a cross-linking reagent. Two liver membrane proteins of 20 and 55 kDa bound to sporozoites (66), but these proteins have not been further characterized.

Others have reasoned that the liver ligand must be part of the major surface protein of sporozoites (Fig. 1), the circumsporozoite protein (CS protein; reviewed in reference 45). The central portion of the CS protein, which contains a series of amino acid repeats, is an unlikely candidate for containing the ligand, because its sequence varies in different species of malaria parasites infecting the same mammalian host. However, CS proteins have two stretches of amino acids that are highly conserved among all malaria parasites, region I and region II-plus (15, 60). Region I is composed of the amino acids KLKQP and is located on the 5′ boundary of

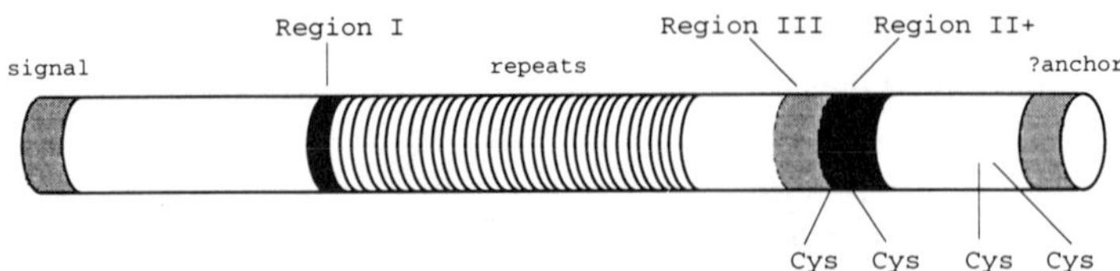

Figure 1. Schematic representation of the CS protein. The middle third of the protein is composed of repeats that vary in different species of *Plasmodium* parasites. N terminal and C terminal to the repeats are region I and region II-plus, respectively. These regions contain motifs that are highly conserved in CS proteins from all species of malaria parasites. Flanking region II-plus is region III, which includes the T-cell epitope Th2R. Region III is characterized by a strong propensity to maintain an α-helical structure. Also shown are the positions of the four cysteines found in all CS proteins. Hydrophobic amino acids at the C terminus may anchor the protein in the plasma membrane of the parasite or may be removed and replaced by a glycosylphosphatidylinositol anchor.

the repeats. Region II-plus is in the C terminus and includes two of the highly conserved pairs of cysteines.

When synthetic peptides representing region I were incubated with HepG2 cells (a hepatoma cell line), they bound specifically to the cells in a saturable manner. In addition, rabbit polyclonal antisera to this region partially inhibited *P. falciparum* invasion of HepG2 cells in vitro (1). In cross-linking studies with hepatoma cells, two proteins of 55 and 35 kDa were associated with the peptide; however, they have not been characterized.

Other investigators have focused on conserved region II-plus of CS protein (Table 1), which is also represented in a different membrane protein of malaria sporozoites called SSP2 or TRAP (26, 52, 54) and in Etp100, a protein from the coccidian parasite *Eimeria tenella* (10). Region II-plus encompasses a well known cell-adhesive motif of several host proteins such as thrombospondin, complement components (24, 52), and proteins involved in the development of the nervous system, F-spondin and Unc-5 (33, 34). Initial studies demonstrated that CS protein binds specifically to sulfated glycoconjugates (6, 48) and that this binding is region II-plus dependent (6). In addition, sulfated molecules such as fucoidan inhibit sporozoite invasion in vitro and sporozoite infectivity in mice (48). SSP2/TRAP also binds to sulfatides in a region II-plus-dependent manner, and, like CS protein, its interaction with HepG2 cells is inhibited by sulfated glycoconjugates (39).

In other studies, different recombinant constructs of the CS protein were incubated with frozen sections of various mouse organs. In liver sections, constructs containing region II-plus bound in a sinusoidal pattern

(4). Immunoelectron microscopy showed CS protein almost exclusively on the microvilli of the basolateral regions of hepatocytes in the space of Disse. CS protein did not bind to Kupffer cells, to liver endothelial cells, or to sections of spleen, brain, lungs, heart, or skeletal muscle. However, CS-protein binding was observed in mast cell granules and in the kidney, where it had a characteristic distribution in the tubular epithelia and in all basement membranes except that of the glomerulus (22). These extrahepatic sites are not accessible to the circulation and would not compete with the sites in the liver for binding to sporozoites in vivo.

CS protein also binds in a saturable manner to HepG2 cells. Synthetic peptides representing region II-plus inhibited both CS-protein binding to HepG2 cells and sporozoite invasion of these cells. When region II-plus peptides with various amino acid substitutions were used to competitively inhibit CS-protein binding to HepG2 cells, only those with downstream positively charged residues had inhibitory activity (60).

Like CS protein, SSP2/TRAP binds to the basolateral domain of hepatocytes in a region II-plus-dependent manner (51). In addition, anti-

Table 1. Region II-plus sequences of CS protein from several species of malaria parasites and homologous sequences from other proteins

Protein[a]	Sequence[b]
P. falciparum CS	E W S P **C** S V T **C** G N G I Q V **R** I **K**
P. vivax CS	E W T P **C** S V T **C** G V G V **R** V **R** **R** **R**
P. malariae CS	E W S P **C** S V T **C** G S G I **R** A **R** **R** **K**
P. knowlesi CS	E W T P **C** S V T **C** G N G V **R** I **R** **R** **K**
P. cynomolgi CS	E W S P **C** S V T **C** G K R V **R** M **R** **R** **K**
P. brasilianum CS	E W S P **C** S V T **C** G S G I **R** A **R** **R** **K**
P. berghei CS	E W S Q **C** N V T **C** G S G I **R** V **R** **K** **R**
P. yoelii CS	E W S Q **C** S V T **C** G S G V **R** V **R** **K** **R**
TRAP/SSP2	
P. falciparum	E W S P **C** S V T **C** G K G T **R** S **R** **K** **R**
P. yoelii	E W S E **C** S T T **C** D E G **R** **K** I **R** **R** **R**
Etp100	E W T E **C** S A T **C** G G G T **K** **H** **R** E **R**
Thrombospondin	P W S S **C** S V T **C** G D G V I T **R** I **R**
Properdin	P W G P **C** S V T **C** S K G T Q I **R** Q **R**
Complement protein 6	Q W T S **C** S K T **C** N S G T Q S **R** **H** **R**
Unc-5	D W S A **C** S S S **C** – – H **R** Y **R** T **R** A
F-spondin	E W S D **C** S V T **C** G K G M – – **R** T **R**

[a]SSP2/TRAP is another malaria sporozoite surface protein; Etp100 is a protein found in the micronemes of *E. tenella*; properdin is a complement protein; Unc-5 and F-spondin are neural adhesion molecules.
[b]Highly conserved cysteines and the downstream basic residues that are essential for binding to HSPGs are indicated by boldface type.

serum generated against recombinant SSP2/TRAP inhibits sporozoite invasion of HepG2 cells and primary hepatocytes in vitro (39, 53). It is not known, however, whether SSP2/TRAP is rapidly cleared by hepatocytes when it is injected into mice. Immunofluorescence studies with antiserum to SSP2/TRAP show that the way the protein is expressed on the surface of a sporozoite produces a mottled staining pattern. This is in contrast to the smooth, uniform pattern seen with antiserum to CS protein (9, 13, 51, 53). However, the absolute number of molecules of SSP2/TRAP and CS protein on the parasite surface is not known, so the relative contribution of region II-plus of each protein to hepatocyte binding cannot be determined.

The hepatocyte receptors for the CS protein are heparan sulfate proteoglycans (HSPGs), which perhaps belong to the syndecan family of proteoglycans (22). The binding of CS protein to liver sections and HepG2 cells was inhibited by treatment of the target cells with heparitinase but not with chondroitinase. Mutant CHO cells lacking HSPGs did not bind CS protein, whereas wild-type CHO cells and mutants lacking chondroitin sulfate did bind CS protein (22a, 48). Treatment of HepG2 cells with chlorate, an inhibitor of proteoglycan sulfation, decreased both CS protein binding and sporozoite invasion (59a). Since the basic residues in the C terminus of region II-plus are required for the binding of CS protein by HSPGs, it is most likely that the lysines and arginines of region II-plus form ionic bonds with the negatively charged sulfate molecules of the HSPGs.

Similarly to sporozoites, CS protein is rapidly cleared from the blood circulation and rapidly accumulates in the liver (5). Two minutes after intravenous injection, 70 to 80% of the CS protein was found in the liver. Light and electron microscopic studies showed that the protein accumulated in the same sinusoidal binding pattern that is observed in liver sections, suggesting that the injected CS protein was being cleared by the HSPGs of the hepatocyte microvilli. Although CS protein bound to kidney sections, it did not accumulate in the kidney after intravenous injection, probably because the fenestrae of the renal tubular epithelia are closed by a diaphragm that separates the underlying basement membranes from the blood circulation.

Whether or not sporozoites are captured in the liver by the same mechanism that captures CS protein is not known. The observation that fucoidan, a highly sulfated glycoconjugate, is a potent inhibitor of sporozoite infectivity in vivo supports this notion. One confounding issue is that the hepatocyte microvilli are within the space of Disse behind the fenestrae of the endothelial cells. Perhaps the long glycosaminoglycan chains (GAGs) of the HSPGs protrude through the fenestrae of the endothelial cells into the sinusoids and thus can retain the circulating sporozoites (21). Lyon et al. showed that the HSPGs of the liver are more highly sulfated than any other HSPGs studied to date, and most of this sulfation is along the distal third of the molecule (35), the part that might protrude through the fenestrae.

Recognition of the multiple protruding GAGs by CS and TRAP proteins on the parasite's surface would create a very high avidity binding interaction. After the parasites are arrested, the sporozoites must detach themselves from the GAGs to penetrate into the hepatocytes. We speculate that this is accomplished by shedding of the CS protein-HSPG complexes. In vitro, when CS protein on sporozoites is cross-linked by monoclonal antibodies (MAbs) to the repeats, the parasite's coat is shed as a sheath that it leaves behind (50). This phenomenon, called the circumsporozoite precipitin reaction (11, 68), may be an in vitro representation of a phenomenon that occurs during hepatocyte invasion.

HSPG molecules on cell surfaces are very abundant and potentially represent a high-capacity receptor system. Estimates range from 1×10^5 to 4×10^6 molecules per cell (32, 75). This may complicate attempts to develop reagents such as region II-plus peptidomimetics that would inhibit sporozoite invasion. The opposite approach of raising antibodies that specifically bind to region II-plus or to flanking regions may have a greater chance of success in preventing the initial sporozoite-hepatocyte interaction.

ANTIBODY-MEDIATED PROTECTION AGAINST SPOROZOITE INVASION

Because the number of sporozoites injected into the mammalian host is so small and because the sporozoites remain in the blood circulation for so short a time, it was generally thought that this parasite stage is not immunogenic and that protective immunity against it could not be achieved. These ideas were abandoned when it was shown that solid protection against malaria infection in rodents, monkeys, and humans is achieved by vaccination with irradiated sporozoites (reviewed in reference 45) and that humans in areas of endemic malaria have serum antibodies against sporozoites (43). The sterile immunity achieved by repeated immunization with irradiated sporozoites is mediated by neutralizing antibodies directed against sporozoite surface proteins and by cellular effector mechanisms that destroy the liver stages (reviewed in reference 42 and chapter 3). Nevertheless, studies in rodents, monkeys, and humans demonstrate that antibodies alone can protect against malaria infection. The initial observations were made with *P. berghei*, a rodent malaria parasite. When passively administered to mice, a (MAb 3D11) against the repeats of the CS protein prevented malar fection (77). Preincubation of *P. berghei* sporozoites with MAb, eve concentrations below 1 μg/ml, neutralized the infectivity of the parasit (19). Fab fragments of MAb 3D11 were also effective at higher conc trations, indicating that neither the class of antibody nor a secondary ev such as complement fixation, agglutination, or opsonization is requ for neutralization of sporozoite infectivity (19, 50). The infectivit chimpanzees of *P. falciparum* and *Plasmodium vivax* sporozoites, the most important malaria parasites of humans, is also neutralized by Fa fragments of a MAb against the repeats of the respective CS proteins (44) These data suggest that antibodies to the repeats can inhibit a function o the parasite that is essential for its survival in the mammalian host. Perhaps, as Stewart et al. suggest, the antirepeat antibodies immobilize th parasite (63).

The repeats of the CS protein of *P. falciparum*, $(NANP)_n$, are an tractive target for vaccine development. Screening a large number of *falciparum* sporozoites from different areas of the world showed that all isolates bear the same repeats (79), although there are variations in other regions of the protein (18). The protective effect of vaccines that elicit antibodies against sporozoites has already been documented in humans. In one study, volunteers were immunized with a conjugate of tetanus toxoid and the synthetic peptide $(NANP)_3$ in alum (27). Correlation betwee the titers of antipeptide and antisporozoite antibodies was excellent. Th three individuals with the highest titers and four controls were challenged by the bites of five heavily infected mosquitoes. One vaccinated individual did not develop a blood infection, and the other two had prolonged prepatent periods. In another study, the volunteers were injected with $R32tet_{32}$, a recombinant fusion protein containing 32 copies of the NANP repeat (2). Six individuals were challenged in the same manner. Parasitemia did not develop in the volunteer with the highest titer of antibodies and was delayed in two others. In a more recent vaccine trial, another recombinant fusion protein containing the NANP repeats, R32NS181, was formulated with a more potent adjuvant (28). Six of 11 volunteers had high titers of antibodies; of these 6, 2 did not develop an infection when bitten by five *P. falciparum*-infected mosquitoes, and 2 had delayed prepatent periods. Although not everyone with high titers of antibodies to the repeats was protected, those who were protected had high antibody titers.

All neutralizing antibodies to sporozoites are directed against repeats of the CS protein, but some antirepeat antibodies do not inhibit parasite infectivity (reviewed in reference 31). The level of protection in *Saimiri* z eys immunized with a recombinant *P. vivax* CS protein was low, though these monkeys had high serum antibody titers against the *vivax* repeats (12). However, passive transfer to *Saimiri* monkeys of 2 g of MAb NVS3, which recognizes the *P. vivax* CS protein repeat, pro- ted four of six monkeys against sporozoite challenge (8). Epitope map- ng demonstrated that MAb NVS3 recognized the AGDR sequence d within the DRAA/DGQPAG repeat, whereas the sera of monkeys inated with the recombinant protein did not specifically recognize sequence (8). Similar observations were made when the neutralizing apacities of different MAbs against the repeats of *P. berghei* CS protein were compared (80). The most likely explanation is that these antibodies differ in their binding affinities and/or fine specificities.

Since antibodies attack incoming sporozoites and do not recognize liver or blood stages, antibodies can achieve complete protection only hen all parasites are prevented from entering hepatocytes. It can be pre- ted a priori that the effectiveness of a vaccine that elicits only anti- odies will increase as the levels (and probably binding affinities) of the antibodies increase but will decline as the number of infective bites and the size of the inoculum escalate. This prediction was substantiated in various studies using the rodent malaria parasites *P. berghei* and *Plasmodium yoelii*. When mice were immunized with conjugates of repeat peptides and tetanus toxoid or with multiple antigen peptides (MAPs) containing CS-protein repeats, most of the animals that had high antibody iters against sporozoites were fully protected against challenge with sporozoites (64, 72, 78, 80). However, the degree of protection dropped precipitously when the challenge dose of sporozoites increased or when the antibody titers were lower.

Although low serum titers of antibodies to repeats do not prevent infection, a large proportion of sporozoites are inactivated. To obtain a more quantitative estimate of the effect of the antibodies, a specific DNA probe was used to compare parasite DNA in the livers of rats following injection with 10^5 *P. berghei* sporozoites preincubated with various concentrations of MAb 3D11 (19). About 90% of the parasites were neutralized when incubated with <0.8 μg of antibody per ml. When the dose was 8 μg/ml, no signal was detected in the liver, indicating that very few sporozoites had escaped neutralization.

Would a vaccine that eliminated most but not all pre-erythrocytic stages be useful? Would the severity and mortality of malaria be influenced by the dose of injected sporozoites and the number of merozoites released from the liver into the blood circulation? The answer to this question probably depends upon the individual's previous exposure to malaria and the degree of acquired resistance to the blood forms. The blood stages multiply rapidly, and protective immune responses against them develop slowly. Therefore, the pre-erythrocytic vaccines would be effective for travelers or for individuals living in areas of low endemicity only if these vaccines provided sterile immunity. On the other hand, in areas of high endemicity, some studies showed a correlation between levels of parasitemia and rates of exposure to *P. falciparum* sporozoites. In a recent study in western Kenya, where the average number of infective bites per day was 0.75 (range, 0 to 4.95), the correlation between exposure to infected mosquitoes in the previous 28 days and parasite density in the blood was highly significant (37). Importantly, the likelihood of fever or history of fever was much greater in children with parasite densities of 20,000/μl than in those with parasite densities of less than 5,000/μl. This finding is in agreement with several observations that show that severity of symptoms and fatality rates are associated with parasite density at the time of presentation (20, 56) and that malaria morbidity and mortality in areas of high endemicity decrease when insecticide-impregnated bed nets are used (14). It is therefore conceivable that in these areas, a reduction in the number of sporozoites reaching the liver will result in the alleviation of malaria symptomatology and perhaps mortality.

Suboptimal levels of antibodies to CS-protein repeats have been induced in most human volunteers in a limited number of trials of the first-generation vaccines. In human volunteers, the antibody response to the $(NANP)_3$-tetanus toxoid vaccine was dose dependent, but the vaccine dose could not be increased because of the toxicity of the carrier protein (27). Among the approaches being considered to remedy this situation is the utilization of MAP vaccines (reviewed in reference 41). MAPs contain a high density of selected B-cell and T-cell epitopes, are nontoxic to animals, and do not require carrier proteins (such as tetanus toxoid), which can cause local inflammatory responses and epitopic suppression to the attached peptide hapten. MAPs have been employed successfully to improve the immunogenicities of bacterial, viral, protozoan, and helminthic products. MAP vaccines containing *P. berghei* or *P. yoelii* CS-protein B-cell epitopes (repeats) and T-cell epitopes elicit high titers of antibodies to sporozoites and protection against homologous challenge (7, 64, 72, 78).

A MAP construct, $(T1B)_4$, which consists of four branches, each containing the B epitope $(NANP)_3$ and the T1 epitope DPNANPNVDP-NANPNV of *P. falciparum* CS protein, is under consideration for phase I human trials. The $(T1B)_4$ MAP was very immunogenic in three of four inbred strains of mice (17, 40) and in two species of *Aotus* monkeys (41). After reaction with viable sporozoites, the sera of the high-responder mice gave positive circumsporozoite precipitin reactions at dilutions of more than 1/800. These high antibody levels were not achieved in rodents when several other synthetic or recombinant CS-protein constructs were used. Also noteworthy is the good correlation in these sera between enzyme-linked immunosorbent assay titers against the synthetic peptide and immunofluorescence titers against sporozoites. This correlation indicates that a large proportion of the antipeptide antibodies recognize the native protein on the parasite surface. Responses in the monkeys were as large as those in high-responder mice. After three injections of $(T1B)_4$ MAPs formulated in Freund's adjuvant, the sera of these animals reached immunofluorescence titers of 10^5 or higher against *P. falciparum* sporozoites. When given to mice primed with *P. falciparum* sporozoites, the $(T1B)_4$ MAPs formulated in alum vigorously boosted the antibody response to CS protein (17). Thus, this MAP vaccine may increase the levels of anti-sporozoite antibodies in the sera of individuals from areas of malaria endemicity, with potential beneficial effects not only on the prevalence of malaria infection but also, as discussed above, on malaria morbidity.

One potential problem with this type of MAP construct is that its usefulness may be genetically restricted to only a portion of the human population, i.e., to individuals whose antigen-presenting cells bear class II major histocompatibility complex molecules that recognize the MAPs. Nevertheless, because two outbred species of *Aotus* monkeys and three inbred strains of mice ($H\text{-}2^b$, $H\text{-}2^a$, $H\text{-}2^k$) responded to the $(T1B)_4$ MAPs, a sizable proportion of humans may also be good responders.

The problem of genetic restriction can be circumvented by including in the MAP vaccines "universal" T-cell epitopes, which are recognized by a large proportion of humans, rather than T-cell epitopes from the CS protein. Two universal T-cell epitopes derived from tetanus toxoid (P2 and P30) were included into *P. falciparum* MAP vaccines together with the $(NANP)_6$ B-cell epitope, and they were very immunogenic in strains of mice that respond poorly to the repeats (25, 65). A similar construct was assayed successfully in a rodent model of malaria. A MAP vaccine containing P2 and P30 together with *P. yoelii* repeat sequence $(QGPGA)_4$ elicited high levels of antibodies to sporozoites and protected 78 to 100% of mice against challenge with this highly infectious species of malaria parasite (72). Protection was antibody mediated but independent of antibody subclass. Human phase I trials of *P. falciparum* MAPs containing universal epitopes are planned. However, this vaccine will generate T-cell memory for the P2 and P30 sequences rather than for CS-protein epitopes. Therefore, T cells from vaccinated individuals will not be boosted by sporozoites from infected mosquitoes, and the duration of protection will depend only upon the vigor of the response to the vaccine itself.

An attractive target for developing vaccines that prevent sporozoite penetration into hepatocytes is the C-terminal region of the CS protein and, in particular, region II-plus and its flanking regions. However, the conserved hepatocyte-adhesion motif is poorly immunogenic, most likely because it is represented in several host proteins (Table 1). The flanking sequence 5′ to region II-plus codes for a conserved motif of interspersed hydrophobic and hydrophilic amino acids (46), which would be predicted to form an amphipathic α-helix in the native protein. As shown in Fig. 2, this motif is found in all CS proteins of human, monkey, and rodent malaria parasites. This region, which we call region III, may have to preserve a common framework and function, perhaps associated with that of the neighboring cell adhesion motif, region II-plus. Region III contains the T-cell epitope Th2R and, despite overall conservation, is polymorphic (18, 76). Because all identified nucleotide substitutions in the codons for region III are nonsynonymous, the changes most likely result from strong

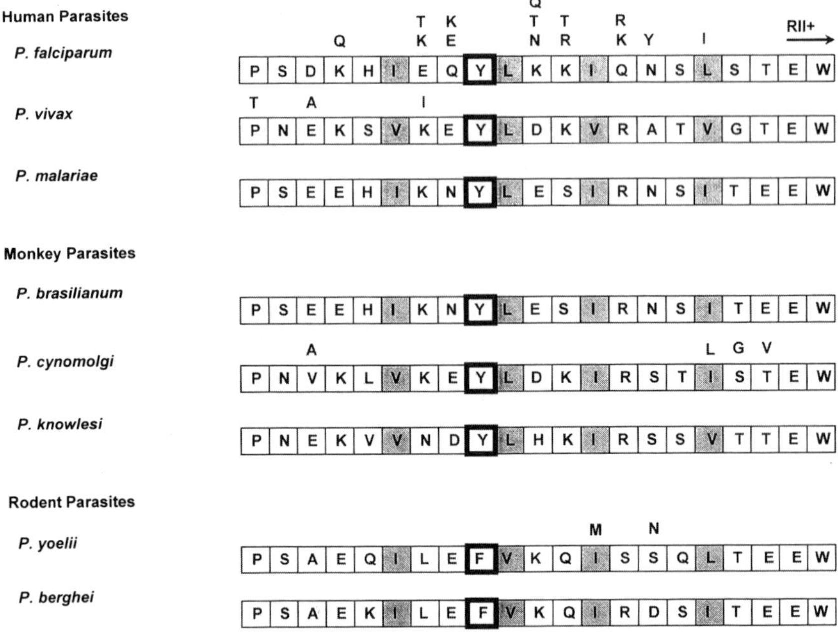

Figure 2. Region III from CS proteins of human, monkey, and rodent malaria parasites. All sequences are formed by interspersed segments of hydrophilic amino acids and conserved hydrophobic residues (gray boxes). In the center of the motif is a highly conserved tyrosine or phenylalanine residue (boldface boxes). The substitutions found to date are shown above the relevant sequence. The last two residues at the C terminus of region III are the first two residues of the neighboring region II-plus. Sequences were retrieved from GenBank, and alignments were performed by searching GenBank with the TFASTA algorithm from the University of Wisconsin Genetics Computer Group package.

selective pressure by the immune system of the host. Although other interpretations are possible (23), the selective pressure may be from antibodies that sterically hinder the interaction of region II-plus of the CS protein with the HSPGs of hepatocytes. Sporozoites bearing an antigenically distinct region III would have a selective advantage, thus providing an explanation for the accumulation of amino acid changes. This hypothesis would be strengthened if the sera from individuals living in areas of endemicity contain antibodies to region III variants. Although these sera contain antibodies to the C-terminal region of the CS protein (3), it has been difficult to ascertain their specificity by using synthetic peptides as antigens. In the native protein, region III most likely forms an amphipathic α-helix, and this structure is impossible to mimic with conventional synthetic peptides.

Antibodies that recognize the conformational epitopes encoded by region III may be elicited by recombinant antigens or by DNA vaccines that encode selected stretches of the C-terminal region of the CS protein. Although region III polymorphism complicates vaccine development, the problem is not insurmountable. Data accumulated from more than 100 CS-protein sequences of *P. falciparum* from diverse geographical areas of the world demonstrate that the variation in region III is limited to a few amino acid positions (Fig. 2; 18, 76). Even in the variant positions, only two amino acid changes have been reported, and most of these changes are conservative, a finding that supports the notion that maintenance of the α-helix in this region is required for parasite viability. Considering that a limited number of region III sequences can represent a large proportion of *P. falciparum* isolates worldwide, efforts to develop vaccines that target region III seem worthwhile.

CONCLUSIONS

The journey of the malaria sporozoite from the salivary gland of the mosquito to the liver of the vertebrate host remains one of the least understood portions of the parasite's life cycle. Although we know that hepatocytes bear receptors for the major surface protein of the sporozoite, we do not know whether this is the mechanism by which sporozoites are arrested in the liver. Sporozoites may initially interact with Kupffer cells or endothelial cells lining the liver sinusoids, although there is little evidence to support this hypothesis. Nonetheless, antibodies directed against the surface of the sporozoite can prevent malaria infection. Whether these antibodies prevent the parasite from "homing" to the liver or whether they interfere with parasite motility or viability is not known. The studies presented here and in the next chapter strongly suggest that a pre-erythrocytic malaria vaccine is feasible. Because antibody-mediated immunity must be as efficient as possible, efforts directed toward increasing the levels of antirepeat antibodies and/or generating antibodies to regions flanking region II-plus should markedly improve the next generation of malaria vaccines.

Acknowledgments. This work was supported by grants P01 AI35703 and K11 AI01175 from the National Institutes of Health.

We thank Elizabeth Nardin and Ruth S. Nussenzweig for their helpful comments and reviews of the manuscript and Marcelo Briones for his help with computer graphics and GenBank searches.

REFERENCES

1. **Aley, S. B., M. D. Bates, J. P. Tam, and M. R. Hollingdale.** 1986. Synthetic peptides from the circumsporozoite proteins of *Plasmodium falciparum* and *Plasmodium knowlesi* recognize the human hepatoma cell line HepG2-A16 in vitro. *J. Exp. Med.* **164:**1915–1921.
2. **Ballou, W. R., S. L. Hoffman, J. A. Sherwood, M. R. Hollingdale, F. A. Neva, W. T. Hockmeyer, D. Gordon, I. Schneider, R. A. Wirtz, J. F. Young, G. F. Wasserman, P. Reeve, C. L. Diggs, and J. D. Chulay.** 1987. Safety and efficacy of a recombinant DNA *Plasmodium falciparum* sporozoite vaccine. *Lancet* **i:**1277–1281.
3. **Calvo-Calle, J. M., E. H. Nardin, P. Clavijo, C. Boudin, D. Stuber, B. Takacs, R. S. Nussenzweig, and A. H. Cochrane.** 1992. Recognition of different domains of the *Plasmodium falciparum* CS protein by the sera of naturally infected individuals compared with those of sporozoite-immunized volunteers. *J. Immunol.* **149:**2695–2701.
4. **Cerami, C., U. Frevert, P. Sinnis, B. Takacs, P. Clavijo, M. J. Santosj, and V. Nussenzweig.** 1992. The basolateral domain of the hepatocyte plasma membrane bears receptors for the circumsporozoite protein of *Plasmodium falciparum* sporozoites. *Cell* **70:**1021–1035.
5. **Cerami, C., U. Frevert, P. Sinnis, B. Takacs, and V. Nussenzweig.** 1994. Rapid clearance of malaria circumsporozoite protein (CS) by hepatocytes. *J. Exp. Med.* **179:**695–701.
6. **Cerami, C., F. Kwakye-Berko, and V. Nussenzweig.** 1992. Binding of malarial CS protein to sulfatides and cholesterol sulfate: dependency on disulfide bond formation between cysteines in region II. *Mol. Biochem. Parasitol.* **54:**1–12.
7. **Chai, S. K., P. Clavijo, J. P. Tam, and F. Zavala.** 1992. Immunogenic properties of multiple antigen peptide systems containing defined T and B epitopes. *J. Immunol.* **149:**2385–2390.
8. **Charoenvit, Y., W. E. Collins, T. R. Jones, P. Millet, L. Yuan, G. H. Campbell, R. L. Beaudoin, R. J. Broderson, and S. L. Hoffman.** 1991. Inability of malaria vaccine to induce antibodies to a protective epitope within its sequence. *Science* **251:**668–671.
9. **Charoenvit, Y., M. F. Leef, L. F. Yuan, M. Sedegah, and R. L. Beaudoin.** 1987. Characterization of *Plasmodium yoelii* monoclonal antibodies directed against stage-specific sporozoite antigens. *Infect. Immun.* **55:**604–608.
10. **Clarke, L. E., F. M. Tomley, M. H. Wisher, I. J. Foulds, and M. E. G. Boursell.** 1990. Regions of an *Eimera tenella* antigen contain sequences which are observed in circumsporozoite proteins from *Plasmodium* spp. and which are related to the thrombospondin gene family. *Mol. Biochem. Parasitol.* **41:**269–280.

11. **Cochrane, A. H., M. Aikawa, M. Jeng, and R. S. Nussenzweig.** 1976. Antibody induced ultrastructural changes of malaria sporozoites. *J. Immunol.* **116:**859–861.
12. **Collins, W. E., R. S. Nussenzweig, T. K. Ruebush, I. C. Bathurst, E. H. Nardin, H. I. Gibson, G. H. Campbell, P. J. Barr, J. R. Broderson, J. C. Skinner, et al.** 1990. Further studies on the immunization of *Saimiri sciureus boliviensis* with recombinant vaccines based on the circumsporozoite protein of *Plasmodium vivax*. *Am. J. Trop. Med. Hyg.* **43:**576–583.
13. **Cowan G., S. Krishna, A. Crisanti, and K. Robson.** 1992. Expression of thrombospondin-related anonymous protein in *Plasmodium falciparum* sporozoites. *Lancet* **339:**1412–1413.
14. **D'Allessandro, U., B. O. Olaleye, W. McGuire, P. Langerock, S. Bennett, M. C. Aikins, M. C. Thomson, M. K. Cham, B. A. Cham, and B. M. Greenwood.** 1995. Mortality and morbidity from malaria in Gambian children after introduction of an impregnated bednet programme. *Lancet* **345:**479–483.
15. **Dame, J. B., J. L. Williams, T. F. McCutchan, J. L. Weber, R. A. Wirtz, W. T. Hockmeyer, W. L. Maloy, J. D. Haynes, I. Schneider, D. D. Roberts, G. S. Sanders, E. P. Reddy, C. L. Diggs, and L. H. Miller.** 1984. Structure of the gene encoding the immunodominant surface antigen on the sporozoite of the human malaria parasite *Plasmodium falciparum*. *Science* **225:**593–599.
16. **Danforth, H. D., M. Aikawa, A. Cochrane, and R. S. Nussenzweig.** 1980. Sporozoites of mammalian malaria: attachment to, interiorization and fate within macrophages. *J. Protozool.* **27:**193–202.
17. **De Oliveira, G. A., P. Clavijo, R. S. Nussenzweig, and E. H. Nardin.** 1994. Immunogenicity of an alum-adsorbed synthetic multiple-antigen peptide based on B and T cell epitopes of the *Plasmodium falciparum* CS protein: possible vaccine application. *Vaccine* **12:**1012–1017.
18. **Doolan, D. L., A. J. Saul, and M. F. Good.** 1992. Geographically restricted heterogeneity for the *Plasmodium falciparum* circumsporozoite protein: relevance for vaccine development. *Infect. Immun.* **60:**675–682.
19. **Ferreira, A., T. Morimoto, R. Altszuler, and V. Nussenzweig.** 1987. Use of a DNA probe to measure the neutralization of *Plasmodium berghei* sporozoites by a monoclonal antibody. *J. Immunol.* **138:**1256–1259.
20. **Field, J. W., and J. C. Niven.** 1937. A note on prognosis in relation to parasite counts in acute subtertian malaria. *Trans. R. Soc. Trop. Med. Hyg.* **30:**569–574.
21. **Frevert, U.** 1994. Malaria sporozoite-hepatocyte interactions. *Exp. Parasitol.* **79:**206–210.
22. **Frevert, U., P. Sinnis, C. Cerami, W. Shreffler, B. Takacs, and V. Nussenzweig.** 1993. Malaria circumsporozoite protein binds to heparan sulfate proteoglycans associated with the surface membrane of hepatocytes. *J. Exp. Med.* **177:**1287–1298.
22a. **Frevert, U., P. Sinnis, J. D. Esko, and V. Nussenzweig.** Cell surface glycosaminoglycans are not obligatory for *Plasmodium berghei* sporozoite invasion in vitro. *Mol. Biochem. Parasitol.*, in press.
23. **Good, M. F., J. A. Berzofsky, and L. H. Miller.** 1988. The T cell response to the malaria circumsporozoite protein. *Annu. Rev. Immunol.* **6:**663–688.
24. **Goundis, D., and K. B. M. Reid.** 1988. Properdin, the terminal complement components, thrombospondin and the circumsporozoite protein of malaria parasites contain similar sequence motifs. *Nature* (London) **335:**82–85.
25. **Grillot, D., D. Valmori, P. H. Lambert, G. Corradin, and G. Del Giudice.** 1993. Presentation of T-cell epitopes assembled as multiple-antigen peptides to murine and human T lymphocytes. *Infect. Immun.* **61:**3064–3067.
26. **Hedstrom, R. C., J. R. Campbell, M. L. Leef, Y. Charoenvit, M. Carter, M. Sedegah,**

R. L. Beaudoin, and S. L. Hoffman. 1990. A malaria sporozoite surface antigen distinct from the circumsporozoite protein. *Bull. W.H.O.* **68:**152–157.

27. **Herrington, D. A., D. F. Clyde, G. Losondky, M. Cortesia, F. R. Murphy, J. Davis, A. Baqar, A. M. Felix, E. P. Heimer, D. Gillessen, E. H. Nardin, R. S. Nussenzweig, V. Nussenzweig, M. R. Hollingdale, and M. M. Levine.** 1987. Safety and immunogenicity in man of a synthetic peptide malaria vaccine against *Plasmodium falciparum* sporozoites. *Nature* (London) **328:**257–259.
28. **Hoffman, S. L., R. Edelman, J. P. Bryan, I. Schneider, J. Davis, M. Sedegah, D. Gordon, P. Church, M. Gross, C. Silverman, M. Hollingdale, D. Clyde, M. Sztein, J. Losonsky, S. Paparello, et al.** 1994. Safety, immunogenicity, and efficacy of a malaria sporozoite vaccine administered with monophosphoryl lipid A, cell wall skeleton of mycobacteria, and squalane as adjuvant. *Am. J. Trop. Med. Hyg.* **51:**603–612.
29. **Hollingdale, M. R.** 1988. Biology and immunology of sporozoite invasion of liver cells and exoerythrocytic development of malaria parasites. *Prog. Allergy* **41:**15–48.
30. **Hollingdale, M. R., P. Leland, and A. L. Schwartz.** 1983. In vitro cultivation of the exoerythrocytic stage of *Plasmodium bergei* in a hepatoma cell line. *Am. J. Trop. Med. Hyg.* **32:**682–684.
31. **Jones, T. R., W. R. Ballou, and S. L. Hoffman.** 1993. Antibodies to the circumsporozoite protein and protective immunity to malaria sporozoites. *Prog. Clin. Parasitol.* **3:**103–117.
32. **Kjellen, L., A. Oldberg, and M. Hook.** 1980. Cell-surface heparan sulfate. *J. Biol. Chem.* **255:**10407–10413.
33. **Klar, A., M. Baldassare, and T. M. Jessell.** 1992. F-spondin: a gene expressed at high levels in the floor plate encodes a secreted protein that promotes neural cell adhesion and neurite extension. *Cell* **69:**95–110.
34. **Leung-Hagestejin, C., A. M. Spence, B. D. Stern, Y. Zhou, M. W. Su, E. M. Hedgecock, and J. G. Culotti**. 1992. UNC-5, a transmembrane protein with immunoglobulin and thrombospondin type 1 domains, guides cell and pioneer axon migrations in *C. elegans*. *Cell* **71:**289–299.
35. **Lyon, M., J. A. Deakin, and J. T. Gallagher.** 1994. Liver heparan sulfate structure. *J. Biol. Chem.* **269:**11208–11215.
36. **Mazier, D., R. L. Beaudoin, S. Mellouk, P. Druilhe, B. Texier, J. Trosper, F. Miltgen, I. Landau, C. Paul, O. Brandicort, C. Guguen-Guillouzo, and P. Langlois.** 1985. Complete development of hepatic stages of *Plasmodium falciparum* in vitro. *Science* **227:**440–442.
37. **McElroy, P. D., J. C. Beier, C. N. Oster, C. Beadle, J. A. Sherwood, A. J. Oloo, and S. L. Hoffman.** 1994. Predicting outcome in malaria: correlation between rate of exposure to infected mosquitoes and level of *Plasmodium falciparum* parasitemia. *Am. J. Trop. Med. Hyg.* **51:**523–532.
38. **Meis, J. F. G. M., J. P. Verhave, P. H. K. Jap, and J. H. E. T. Neuwissen.** 1983. An ultrastructural study on the role of Kupffer cells in the process of infection by *Plasmodium berghei* sporozoites in rats. *Parasitology* 86:231–242.
39. **Muller, H. M., I. Reckman, M. R. Hollingdale, H. Bujard, K. J. H. Robson, and A. Crisanti.** 1993. Thrombospondin related anonymous protein (TRAP) of *Plasmodium falciparum* binds specifically to sulfated glycoconjugates and to HepG2 hepatoma cells suggesting a role for this molecule in sporozoite invasion of hepatocytes. *EMBO J.* **12:**2881–2889.
40. **Munesinghe, D. Y., P. Clavijo, J. M. Calvo-Calle, R. S. Nussenzweig, and E. Nardin.** 1991. Immunogenicity of multiple antigen peptides (MAP) containing T and B cell epi-

topes of the repeat region of the *P. falciparum* circumsporozoite protein. *Eur. J. Immunol.* **21:**3015–3020.

41. **Nardin, E. H., G. A. De Oliveira, J. M. Calvo-Calle, and R. S. Nussenzweig.** 1995. The use of multiple antigen peptides (MAPs) in the analysis and induction of protective immune responses against infectious diseases. *Adv. Immunol.* **60:**105–149.
42. **Nardin, E. H., and R. S. Nussenzweig.** 1993. T cell responses to pre-erythrocytic stages of malaria: role in protection and vaccine development. *Annu. Rev. Immunol.* **11:**687–727.
43. **Nardin, E. H., R. S. Nussenzweig, I. A. McGregor, and J. H. Bryan.** 1979. Antibodies to sporozoites: their frequent occurrence in individuals living in an area of hyperendemic malaria. *Science* **206:**597–599.
44. **Nardin, E. H., V. Nussenzweig, R. S. Nussenzweig, W. E. Collins, K. T. Harinasuta, P. Tapchaisri, and Y. Chomcharn.** 1982. Circumsporozoite proteins of human malaria parasites *Plasmodium falciparum* and *Plasmodium vivax*. *J. Exp. Med.* **156:**20–30.
45. **Nussenzweig, V., and R. S. Nussenzweig.** 1989. Rationale for the development of an engineered sporozoite malaria vaccine. *Adv. Immunol.* **45:**283–334.
46. **Nussenzweig, V., and R. S. Nussenzweig.** 1991. Sporozoite malaria vaccine: where do we stand?, p. 1–14. *In* C. C. Wang (ed.), *Molecular and Immunological Aspects of Parasitism.* American Association for the Advancement of Science, Washington, D.C.
47. **Nussler, A., J. C. Drapier, L. Renia, S. Pied, F. Miltgen, M. Gentilini, and D. Mazier.** 1991. L-Arginine-dependent destruction of intrahepatic malaria parasites in response to tumor necrosis factor and/or interleukin 6 stimulation. *Eur. J. Immunol.* **21:**227–230.
48. **Pancake, S. J., G. H. Holt, S. Mellouk, and S. L. Hoffman.** 1992. Malaria sporozoites and circumsporozoites proteins bind specifically to sulfated glycoconjugates. *J. Cell Biol.* **6:**1351–1357.
49. **Ponnudurai, T., A. H. W. Lensen, G. J. A. van Gemert, M. G. Bolmer, and J. H. E. T. Meuwissen.** 1991. Feeding behavior and sporozoite ejection by infected *Anopheles stephensi. Trans. R. Soc. Trop. Med. Hyg.* **85:**175–180.
50. **Potocnjak, P., N. Yoshida, R. S. Nussenzweig, and V. Nussenzweig.** 1980. Monovalent fragments (Fab) of monoclonal antibodies to a sporozoite surface antigen (Pb44) protect mice against malarial infection. *J. Exp. Med.* **151:**1504–1513.
51. **Robson, K. J. H., U. Frevert, I. Reckmann, G. Cowan, J. Beier, I. G. Scragg, K. Takehara, D. H. L. Bishop, G. Pradel, R. Sinden, S. Saccheo, H. M. Muller, and A. Crisanti.** 1995. Thrombospondin-related adhesive protein (TRAP) of *Plasmodium falciparum:* expression during sporozoite ontogeny and binding to human hepatocytes. *EMBO J.* **14:**3883–3894.
52. **Robson, K. J. H., J. R. S. Hall, M. W. Jennings, T. J. R. Harris, K. Marsh, C. I. Newbold, V. E. Tate, and D. J. Weatherall.** 1988. A highly conserved amino-acid sequence in thrombospondin, properdin and in proteins from sporozoites and blood stages of a human malaria parasite. *Nature* (London) **335:**79–82.
53. **Rogers, W. O., A. Malik, S. Mellouk, K. Nakamura, M. D. Robers, A. Szarfman, D. M. Gordon, A. Nussler, M. Aikawa, and S. L. Hoffman.** 1992. Characterization of *Plasmodium falciparum* sporozoite surface protein 2. *Proc. Natl. Acad. Sci. USA* **89:**9176–9180.
54. **Rogers, W. O., M. D. Rogers, R. C. Hedstrom, and S. L. Hoffman.** 1992. Characterization of the gene encoding sporozoite surface protein 2, a protective *Plasmodium yoelii* sporozoite antigen. *Mol. Biochem. Parasitol.* **53:**45–52.
55. **Rosenberg, R., R. A. Wirtz, I. Schneider, and R. Burge.** 1990. An estimation of the number of malaria sporozoites ejected by a feeding mosquito. *Trans. R. Soc. Trop. Med. Hyg.* **84:**209–212.

56. **Rowe, A., J. Obeiro, C. I., Newbold, and K. Marsh.** 1995. *Plasmodium falciparum* rosetting is associated with malaria severity in Kenya. *Infect. Immun.* **63:**2323–2326.
57. **Seguin, M. C., W. R. Ballou, and C. A. Nacy.** 1989. Interactions of *Plasmodium bergei* sporozoites and murine Kupffer cells in vitro. *J. Immunol.* **143:**1716–1722.
58. **Shin, S. C. J., J. P. Vanderberg, and J. A. Terzakis.** 1982. Direct infection of hepatocytes by sporozoites of *Plasmodium bergei. J. Protozool.* **29:**448–454.
59. **Sinden, R. E., and J. E. Smith.** 1982. The role of the Kupffer cell in the infection of rodents by sporozoites of *Plasmodium:* uptake of sporozoites by perfused liver and establishment of infection in vivo. *Acta Trop.* **39:**11–27.
59a. **Sinnis, P.** Unpublished results.
60. **Sinnis, P., P. Clavijo, D. Fenyo, B. T. Chait, C. Cerami, and V. Nussenzweig.** 1994. Structural and functional properties of region II-plus of the malaria circumsporozoite protein. *J. Exp. Med.* **180:**297–306.
61. **Smith, J. E., and J. Alexander.** 1986. Evasion of macrophage microbicidal mechanisms by mature sporozoites of *Plasmodium yoelii yoelii. Parasitology* **93:**33–38.
62. **Smith, J. E., and R. E. Sinden.** 1981. Studies on the uptake of sporozoites of *P. yoelii nigeriensis* by perfused rat liver. *Trans. R. Soc. Trop. Med. Hyg.* **75:**188–189.
63. **Stewart, M. J., R. J. Nawrot, S. Schulman, and J. P. Vanderberg.** 1986. *Plasmodium berghei* sporozoite invasion is blocked in vitro by sporozoite-immobilizing antibodies. *Infect. Immun.* **51:**859–864.
64. **Tam, J. P., P. Clavijo, Y. Lu, V. Nussenzweig, R. S. Nussenzweig, and F. Zavala.** 1990. Incorporation of T and B epitopes of the circumsporozoite protein in a chemically defined synthetic vaccine against malaria. *J. Exp. Med.* **171:**299–306.
65. **Valmori, D., A. Pessi, E. Bianchi, and G. Corradin.** 1992. Use of human universally antigenic tetanus toxin T cell epitopes as carriers for human vaccination. *J. Immunol.* **149:**717–721.
66. **van Pelt, J. F., J. Kleuskens, M. Hollingdale, J. P. Verhave, T. Ponnudurai, J. H. E. T. Meuwissen, and S. H. Yap.** 1991. Identification of plasma membrane proteins involved in the hepatocyte invasion of *Plasmodium falciparum* sporozoites. *Mol. Biochem. Parasitol.* **44:**225–232.
67. **Vanderberg, J., S. Chew, and M. J. Stewart.** 1990. *Plasmodium* sporozoite interactions with macrophages in vitro: a videomicroscopic analysis. *J. Protozool.* **37:**528–536.
68. **Vanderberg, J., R. S. Nussenzweig, and H. Most.** 1969. Protective immunity produced by injection of x-irradiated sporozoites of *P. berghei.* V. In vitro effects of immune serum on sporozoites. *Mil. Med.* **134:**1183–1187.
69. **Verhave, J. P., J. H. E. T. Meuwissen, and J. Golenser.** 1980. The dual rate of macrophages in the sporozoite-induced malaria infection. A hypothesis. *Int. J. Nucl. Med. Biol.* **7:**149–156.
70. **Vreden, S. G. S.** 1994. The role of Kupffer cells in the clearance of malaria sporozoites from the circulation. *Parasitol. Today* 10:304–308.
71. **Vreden, S. G. S., R. W. Sauerwein, J. P. Verhave, N. Van Rooijen, J. H. E. T. Meuwissen, and M. F. Van Den Broek.** 1993. Kupffer cell elimination enhances development of liver schizonts of *Plasmodium bergei* in rats. *Infect. Immun.* **61:**1936–1939.
72. **Wang, R., Y. Charoenvit, G. Corradin, R. Porrozzi, R. L. Hunter, G. Glenn, C. R. Alving, P. Church, and S. L. Hoffman.** 1995. Induction of protective polyclonal antibodies by immunization with a *Plasmodium yoelii* circumsporozoite protein multiple antigen peptide vaccine. *J. Immunol.* **154:**2784–2793.
73. **Wisse, E.** 1970. An electron microscopic study of the fenestrated endothelial lining of rat liver sinusoids. *J. Ultrastruct. Res.* **31:**125–150.

74. **Wisse, E., R. B. De Zanger, P. Van der Smissen, and R. S. McCuskey.** 1985. The liver sieve: considerations concerning the structure and function of endothelial fenestrae, the sinusoidal wall and the space of Disse. *Hepatology* **5:**683–692.
75. **Yanagisha, M., and V. Hascall.** 1992. Cell surface heparan sulfate proteolgycans. *J. Biol. Chem.* **267:**9451–9454.
76. **Yoshida, N., S. M. Di Santi, A. P. Dutra, R. S. Nussenzweig, V. Nussenzweig, and V. Enea.** 1990. *Plasmodium falciparum:* restricted polymorphism of T cell epitopes of the circumsporozoite protein in Brazil. *Exp. Parasitol.* **71:**386–392.
77. **Yoshida, N., R. S. Nussenzweig, P. Potocnjak, V. Nussenzweig, and M. Aikawa.** 1980. Hybridoma produces protective antibodies directed against the sporozoite stage of malaria parasite. *Science* **207:**71–73.
78. **Zavala, F., and S. Chai.** 1990. Protective anti-sporozoite antibodies induced by a chemically defined synthetic vaccine. *Immunol. Lett.* **25:**271–274.
79. **Zavala, F., A. Masuda, P. M. Graves, V. Nussenzweig, and R. S. Nussenzweig.** 1985. Ubiquity of the repetitive epitope of the CS protein in different isolates of human malaria parasites. *J. Immunol.* **135:**2790–2793.
80. **Zavala, F., J. P. Tam, P. J. Barr, P. J. Romero, V. Ley, R. S. Nussenzweig, and V. Nussenzweig.** 1987. Synthetic peptide vaccine confers protection against murine malaria. *J. Exp. Med.* **166:**1591–1596.

Malaria Vaccine Development: A Multi-Immune Response Approach
Edited by Stephen L. Hoffman

Chapter 3

Attacking the Infected Hepatocyte

Stephen L. Hoffman, Eileen D. Franke, Michael R. Hollingdale, and Pierre Druilhe

INTRODUCTION

The observation that sterile immunity to malaria can be induced by inoculation with irradiated sporozoites (25–27, 87, 98, 99) has focused attention during the last 20 years on the sporozoite and liver stages of the parasite life cycle. Initially, the focus was almost exclusively on antibody-mediated mechanisms directed against the circumsporozoite (CS) protein that would prevent effective sporozoite invasion of hepatocytes. However, during the more than a decade since the cloning and sequencing of the gene encoding the *Plasmodium falciparum* CS protein in 1984 (29, 33), there has been a steady increase in the emphasis on attacking the *Plasmodium*-infected hepatocyte (Fig. 1). Progress in identifying immune mechanisms capable of eliminating infected hepatocytes, in identifying the targets of these immune mechanisms, and in developing methods of inducing immune responses against these defined targets has been dramatic. Gamma interferon (34, 66, 78, 115) and, subsequently, other cytokines (78, 91) were the first molecules shown to have protective activity directed against infected hepatocytes. $CD8^+$ T lymphocytes (115, 139) were the next ones emphasized, and several years later, $CD4^+$ T lymphocytes received attention (30, 96) (Fig. 2). Antibodies (23, 97) and $\gamma\delta$ T cells (130) have significant activity against infected hepatocytes. Initial studies emphasized the CS protein as the target of protective immune responses in infected hepatocytes. Subsequently, sporozoite surface protein 2 (SSP2) (45, 56, 105), liver-stage antigen-1 (LSA-1) (43, 46, 147), and other pre-erythrocytic-stage (sporozoite and liver-stage)-specific antigens have been emphasized as targets recognized by these immune mechanisms.

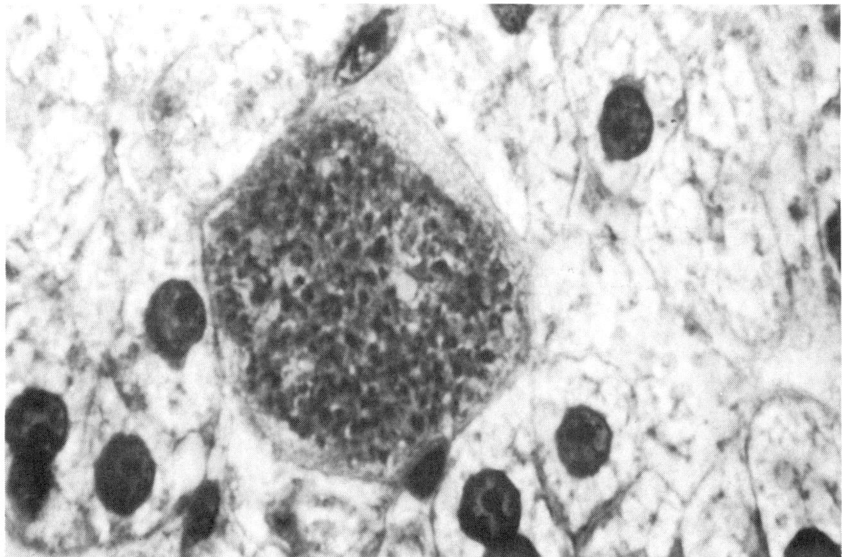

Figure 1. Giemsa-stained mature *P. falciparum* liver schizont obtained from *Saimiri sciureus* 6 days after sporozoite injection. (Provided by P. Druilhe.)

More recently, antigens first expressed in the liver stage but also expressed in the blood stage of the life cycle have been considered as potential targets (5, 23, 110, 128). Researchers thought at first that induction of $CD8^+$ protective cytotoxic lymphocyte (CTL) activity would require immunization with live vectors expressing the CS protein or other target proteins. Now we know that immunization with synthetic or recombinant peptides in an appropriate delivery system (68, 133, 140, 143) or with DNA vaccines (31a, 120) can also induce these types of response against *Plasmodium* peptides. In this chapter, we summarize the development of our current knowledge and the directions of future research on identifying the immune mechanisms that attack infected hepatocytes, identifying the targets of these protective immune responses, and developing vaccines designed to induce protective immunity against these targets.

WHY IS THE INFECTED HEPATOCYTE A GOOD TARGET?

The liver stage of the malaria parasite life cycle lasts for 2 days in the rodent malaria parasites *Plasmodium yoelii* and *Plasmodium berghei* and generally for 5 to 14 days in the human malaria parasites. However, the primary liver stage may last for 6 to 12 months, and in the case of the re-

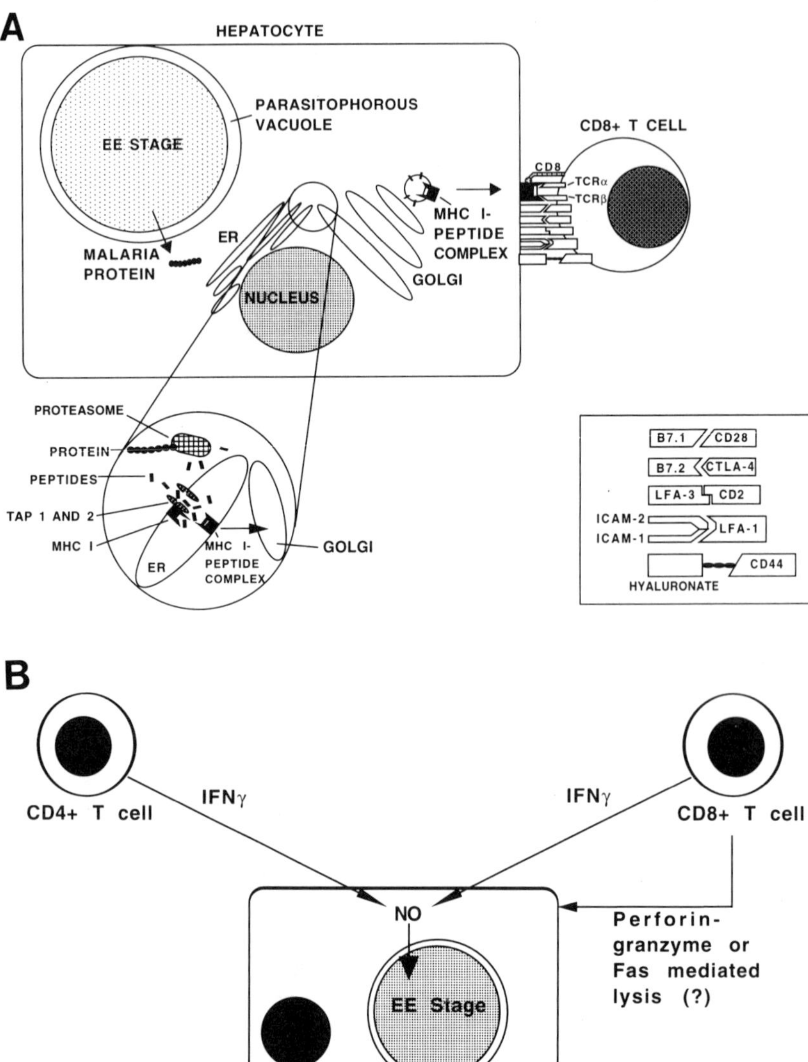

Figure 2. (A) Schematic representation of the proposed processing of *Plasmodium*-derived proteins within hepatocytes and the presentation of peptide-MHC complexes to $CD8^+$ T lymphocytes. Interaction between the infected hepatocyte and $CD8^+$ T lymphocytes probably involves a set of receptor ligands as shown; however, these interactions have not been established for *Plasmodium*-infected hepatocytes. (B) Schematic representation of proposed mechanisms for the antiparasitic activity of $CD4^+$ and $CD8^+$ T lymphocytes against infected hepatocytes. Abbreviations: EE stage, exoerythrocytic stage; ER, endoplasmic reticulum; IFNγ, gamma interferon; NO, nitric oxide.

lapsing malarias caused by *Plasmodium vivax* and *Plasmodium ovale*, a secondary erythrocytic-stage infection caused by hypnozoites, which are forms of the parasite latent in the liver for up to 2 to 3 years after sporozoite inoculation, may occur. Some antigens expressed by sporozoites or by asexual blood stages of the parasite life cycle are also expressed in hepatocytes, and thus, it is possible that infected hepatocytes act as antigen depots that induce immune responses not only against infected hepatocytes but also against other stages of the life cycle. During the liver stage of the life cycle, no symptoms or signs of malaria and no evidence of pathology are manifested. Thus, despite the fact that the parasite is intracellular, the liver stage represents an ideal target for vaccine-induced immune responses because the duration of this stage and its lack of association with disease. Furthermore, as is pointed out below, the infected hepatocyte is the target of multiple well-defined T-lymphocyte responses, each of which can provide sterile protective immunity in the absence of other parasite-specific immune responses. Finally, accumulating data suggest that immune responses that only reduce the numbers of parasites that emerge from infected hepatocytes may actually have an impact on malaria-associated morbidity and mortality without providing sterile protective immunity (6, 13, 45a).

THE LIVER STAGE: BIOLOGY, IDENTIFIED PROTEINS, AND POTENTIAL FOR IMMUNE ATTACK

Biology

Sporozoites are injected by the bite of female *Anopheles* mosquitoes and specifically localize in the liver sinusoids. Contrary to earlier assumptions, it is now believed, but not definitely proven, that the numbers of inoculated sporozoites are small, about 10 to 100 per bite (93, 108). Once within the sinusoid, sporozoites must cross the endothelium into the space of Disse before invading hepatocytes, unless a segment of the hepatocyte membrane is actually exposed in the sinusoid (see chapter 2). Regardless, the mechanism whereby the sporozoite makes its way to the hepatocyte membrane remains controversial. Passage through endothelial or Kupffer cells has been proposed (51, 76). Since sporozoites interact poorly with Kupffer cells in vitro (122) and since Kupffer cell depletion has little effect on sporozoite infectivity in vivo (136), it is more likely that sporozoites interact with endothelial cells (51a). Molecular specificity appears to be crucial for the localization of sporozoites in the liver and is probably mediated by interactions between region II of the CS protein and TRAP/PfSSP2 and sulfatides (18, 19, 38, 85, 90). Subsequently, molecular interactions are also likely to be required for the actual invasion of hepatocytes (52b). Infectivity is probably mediated by a cascade of interactions between sporozoites and host molecules (55). *P. falciparum* sporozoites develop within human hepatocytes in vivo and in vitro, and two hepatocyte-specific receptors associated with development have been identified (134). Sporozoites appear to invade apically by invagination of the hepatocyte cell membrane, which forms a parasitophorous vacuole membrane (PVM) that surrounds the developing liver stage or exoerythrocytic (EE) parasite (4). As during merozoite invasion of erythrocytes (81), material appears to be secreted from sporozoite rhoptries during invasion (4).

Shortly after invasion, the inner membrane and subpellicular microtubules of the thin (1.5 by 10 to 20 μm) sporozoite break down, and this area bulges out, creating a uninucleated trophozoite 3 to 5 μm in diameter bounded by a plasma membrane and situated in a vacuole surrounded by a PVM (4). The trophozoite develops into a mature liver-stage schizont that occupies most of the volume of the hepatocyte (Fig. 1) (51, 74). During schizogony, parasite antigens are inserted into the PVM, which appears to form deep invaginations into the infected hepatocyte (10), especially adjacent to the hepatocyte nucleus (51). Some parasite proteins are actually found in the cytoplasm of the hepatocyte (23, 51). Apart from one report (97), no parasite antigens have been detected on the hepatocyte surface by microscopy.

Murine and human malaria parasites within the liver show significant biological differences. Development of mature EE schizonts containing at least 2,000 uninucleated merozoites takes 42 to 48 h for the rodent malaria parasites *P. yoelii* and *P. berghei* (39). *P. falciparum* EE parasites more closely resemble avian parasites (74) and develop over 5 to 6 days, forming up to 30,000 merozoites within a schizont 80 to 100 μm in diameter (73). During *P. vivax* EE development, some trophozoites do not develop further and may persist for several years as small hypnozoites (5 μm in diameter) within hepatocytes (59). Relapses of *P. vivax* malaria may arise from hypnozoites that are triggered to develop by unknown mechanisms. Following rupture of infected hepatocytes, EE merozoites are released and invade erythrocytes, or they may also be phagocytized by Kupffer cells (129). The signals and parasite molecules responsible for the development of a uninucleate trophozoite to a fully mature liver-stage schizont and for the rupture of this schizont are unknown.

Identified Proteins

We now have considerable evidence that the potent, sterile protective immunity induced by immunization with radiation-attenuated sporozoites is primarily mediated by immune T lymphocytes that attack infected hepatocytes (see below) (Fig. 2). The facts that this protective immunity is not strain specific, is efficacious in essentially all humans who receive optimal immunization and challenge, and withstands enormous sporozoite challenge in mice may indicate that the immune response is directed against multiple *Plasmodium* proteins in infected hepatocytes. Thus, in view of the limited success obtained with single-molecule-based vaccines (i.e., CS protein), considerable effort has been put into identifying additional novel parasite proteins specifically expressed in infected hepatocytes. In this section, we introduce these proteins and describe their structures and localization within infected hepatocytes (Table 1). Later, we describe the evidence that they are the targets of protective immune responses.

CS protein

The first protein identified as a target of protective immune responses directed against the infected hepatocyte was the CS protein (146). This protein is a major surface protein of sporozoites but is also found in micronemes (3, 37). The CS protein is brought into the hepatocyte when the sporozoite invades and resides on the EE parasite membrane and the PVM, where it is detectable by microscopy, throughout the liver stages of rodent malarias (9, 53), *P. vivax* (51, 52a), and *P. falciparum* (79). There is no evidence for transcription, translation, or expression of this protein within hepatocytes; some data indicate that these events do not take place (9). This protein is probably carried into the hepatocyte by the invading sporozoite (9). The CS protein ranges in size from 44 to 67 kDa depending on the species of parasite. The CS protein genes of different *Plasmodium* spp. (*Plasmodium knowlesi, P. falciparum, P. vivax, P. berghei, P. yoelii, Plasmodium cynomolgi,* etc.) encode proteins with similar structures but little homology at the nucleotide and amino acid levels (111). All CS proteins have a central region of tandemly repeated amino acids, but the sequences of these repeat regions are all quite distinct. Amino and carboxy terminal to the repeat regions, all CS proteins have domains designated region I and II, respectively, that are highly conserved at the amino acid level among different *Plasmodium* spp. The other regions flanking the repeat region contain most of the identified T-cell epitopes, and there is little sequence homology among different species. In the case of *P. falciparum,* these flanking regions contain sequences with a great deal of variability among different strains.

SSP2/TRAP

Another protein identified in infected hepatocytes and shown to be the target of protective immune responses is the *P. yoelii* SSP2 (PySSP2) (45, 105). Charoenvit and colleagues immunized mice with irradiated *P. yoelii*

Table 1. Antigens expressed in infected hepatocytes

Antigen[a]	Mol wt (10^3)	Species cloned	Localization in:		
			Sporozoite	EE	Blood stage
CS protein	42–67	*P. falciparum, P. yoelii, P. reichenowi, P. berghei, P. vivax, P. knowlesi, P. malariae, P. brasilianum, P. cynomolgi, P. simium*	Surface, microneme	Parasite membrane, PVM	No?
SSP2/TRAP	140 (*P. yoelii*) 90 (*P. falciparum*)	*P. falciparum, P. yoelii*	Surface, microneme	Yes	Yes (*P. falciparum*)
LSA-1	200	*P. falciparum*	No	Vacuole	No
PyHEP17–PfEXP-1	17/23	*P. yoelii/P. falciparum*	No	PVM, host cell cytoplasm	Yes
STARP	78	*P. falciparum, P. reichenowi*	Surface	Yes	Yes
SALSA	70	*P. falciparum*	Surface	Vacuole	?
LSA-3	205	*P. falciparum, P. yoelii*	Surface, internal organelles	Yes	No
LSA-2	230	*P. berghei*	No	PVM	No
Pbl.1	35	*P. berghei*	No	PVM	No
CSP-2	42/54	*P. falciparum*	Surface, microneme	PVM, endoplasmic reticulum	No
hsp70	70	*P. falciparum*	No	Nuclei, cytoplasm	Yes
GRP78	78	*P. falciparum*	No	Nuclear membrane, endoplasmic reticulum	Yes

[a]Other blood-stage antigens first expressed in infected hepatocytes include *P. falciparum* merozoite surface protein 1 (128), the 220-kDa glutamate-rich protein (17), the serine-rich antigen (128), *P. falciparum* erythrocyte membrane 2 (128), the acidic-basic repeat antigen (128), and the rhoptry antigen (128).

sporozoites and produced a monoclonal antibody (MAb) directed at a 140-kDa sporozoite protein (21). The gene encoding this protein was cloned and sequenced (45, 105), and the protein was named SSP2. PySSP2 is present in the micronemes of sporozoites, on the sporozoite surface, and throughout the liver stage in *P. yoelii*-infected hepatocytes (3). The gene encoding *P. falciparum* SSP2 (PfSSP2) has now been identified and characterized (104); it was shown to be the previously described, approximately 90-kDa (by immunoblot), thrombospondin-related anonymous protein (TRAP) (101). Pf-SSP2/TRAP is present on the surfaces (28, 104) and in the micronemes of sporozoites (104) and for the first 4 days of the *P. falciparum* liver stage, but it has not been shown to be present in fully mature, 3D7 late-liver-stage schizonts (104). Both PySSP2 and PfSSP2/TRAP have domains homologous to region II of the CS protein. PySSP2 has a central repeat region (16 copies of NPNEPS), and some strains of *P. falciparum* have a 3-amino-acid repeat, PNN. PfSSP2/TRAP shows considerable polymorphism (100). PySSP2 has not been shown to be expressed in infected erythrocytes. However, Pf-SSP2/TRAP is expressed at low levels by the erythrocytic stages of some strains of *P. falciparum* (101).

LSA-1

The second protein identified in infected hepatocytes was *P. falciparum* LSA-1 (43). Its gene was identified, together with other genes encoding *P. falciparum* protein expressed at the pre-erythrocytic stage, in genomic DNA by differential immunoscreening of expressed proteins. Antibodies from humans who had been on antimalaria chemoprophylaxis were used to identify sporozoite and liver-stage antigens (43). Expression of LSA-1 is detected throughout liver schizogony and increases as maturation progresses, demonstrating active synthesis (35). LSA-1 is localized to flocculent material in the parasitophorous vacuole, which forms a stroma in which hepatic merozoites are released, and may adhere to the merozoites (35). There is no evidence for expression of *P. falciparum* LSA-1 (PfLSA-1) in sporozoites or infected erythrocytes. PfLSA-1 encodes a protein predicted to be 230 kDa (147) and found to be 200 kDa (35). It is to date the only *P. falciparum* liver-stage antigen with a molecular mass that has been determined with *P. falciparum* liver-stage extracts obtained from culture. The protein contains a large central repeat region that is polymorphic in length and is flanked by relatively invariant nonrepeat regions (35, 136a, 144).

PyHEP17/PfEXP-1

A MAb (NYLS3) was produced from mice immunized with a suspension of mouse hepatocytes infected with mature *P. yoelii* liver-stage parasites. This MAb recognizes infected hepatocytes 6 h after sporozoite invasion, the parasite throughout the liver stage, and rings and trophozoites of infected erythrocytes (23). Immunoelectron microscopy demonstrates abundant protein on the surfaces of the parasitophorous vacuoles of infected hepatocytes and erythrocytes and in the hepatocyte and erythrocyte cytoplasms. In a Western blot (immunoblot), the MAb recognizes a 17-kDa band in extracts of infected erythrocytes. This MAb eliminates *P. yoelii*-infected but not *P. berghei*-infected hepatocytes from in vitro culture, and when given in passive transfer, it reduces the density of *P. yoelii* blood-stage parasitemia after blood-stage challenge (23). The gene has been cloned (30a), and the striking sequence homology and antigenic cross-reactivity indicate that PyHEP17 is the *P. yoelii* homolog of *P. falciparum* exported protein-1 (PfEXP-1), which has the same expression pattern as PyHEP17 (110). The identity between PyHEP17 and PfEXP-1 is 37% (60 of 161 residues) at the amino acid level (30a).

STARP

The sporozoite threonine- and asparagine-rich protein (STARP) is a 78-kDa protein that was chosen because it was consistently detected on the surfaces of a series of sporozoites from laboratory strains and wild isolates of *P. falciparum* (36). It has been detected on each single sporozoite tested in an unusual nonhomogeneous distribution. The 2-kb gene was identified as described below for the sporozoite and liver-stage antigen (SALSA) gene. It contains a 5′ miniexon and a large central exon with a complex repetitive structure that encodes a mosaic of multiple disperse motifs and tandem 45- and 10-amino-acid repeats. In contrast to the perfectly conserved 45-amino-acid repeats, the 10-amino-acid repeats show both length and sequence diversity. Northern (RNA) blots and reverse transcriptase PCR conclusively demonstrate the expression of STARP in the sporozoite stage. Presence of the protein in the liver stage was supported by results from an immunofluorescence assay using antibodies directed to the repetitive as well as the nonrepetitive regions.

The extent of structural conservation was analyzed in *P. falciparum* and heterologous species. In various field isolates of *P. falciparum*, size variation was observed in the 45- and 10-amino-acid repeats; however, the 5′ and 3′ nonrepetitive regions appeared to be well conserved (36). The sequence of the homologous STARP gene in the primate malaria species *Plasmodium reichenowi* was determined; overall homologies at the DNA and amino acid levels were 94 and 88%, respectively, with the greatest divergence in the 10-amino-acid repeat region. A similar conserved gene may exist in the rodent malaria species *P. yoelii*, *P. berghei*, and *Plasmodium chabaudi*, as shown by DNA hybridization and nonrepetitive probes, though not with the repeat region. In the same assays, no homology was detected in *P. vivax* and *Plasmodium gallinaceum.*

SALSA

In order to identify SALSA and other non-CS pre-erythrocytic proteins, a strategy similar to that used for LSA-1 was employed. A set of 120 DNA clones expressing pre-erythrocytic antigens was screened by using human sera that strongly recognized native parasite proteins on the sporozoite surface and in liver stages but did not recognize CS protein and LSA-1 (69). In contrast to LSA-1, SALSA is one of the many molecules shared between sporozoites and liver stages. Its molecular weight is 70 kDa in sporozoite extracts, and it is actively synthesized during the liver phase. It was detected in all liver forms screened from three isolates.

Like many other liver-stage molecules, SALSA appears to be associated with the flocculent material that in rodent malaria (*P. berghei*) is preferentially ingested by infiltrating macrophages and neutrophils (72). Immunoelectron microscopy localizes this material to small vacuoles in the peripheries of young liver stages and large vacuoles within mature EE parasites (9, 75). Upon maturation and formation first of the pseudocytomeres and thereafter of individual merozoites, this flocculent material forms a stroma into which the hepatic merozoites are released. In the sporozoite stage, electron microscopy located SALSA mostly, if not only, at the surface. No analog of SALSA could be detected either by DNA hybridization or by antibody cross-reactivity in any other human or rodent species (17a).

LSA-3

LSA-3 is an abundant 205-kDa protein expressed in both sporozoite and liver stage but not blood stages. Epitopes contained in the LSA-3 repeat region cross-react with several of the malarial glutamic acid-rich proteins in blood stages. The corresponding 5- to 6-kb gene shows polymorphism in the number though not the sequence of repeats. It has two exons, which encode a 5′ nonrepetitive region (a block of eight amino acid repeats arranged in four main distinct motifs) and a large 3′ nonrepetitive region also containing a short block, of four amino acid repeats. Immunoelectron microscopy shows that the antigen is associated with both the internal organelles and the surfaces of sporozoites and has the same distribution as LSA-1 in liver stages. An analog of LSA-3 was identified in *P. yoelii* but not *P. berghei* by antigenic similarities. Antibodies to LSA-3 recombinant proteins and peptides derived from the *P. falciparum* repeat and 5′ nonrepeat regions fully cross-react with *P. yoelii* sporozoites and liver stages, though not with asexual blood stages, and detect a 205-kDa protein in *P. yoelii* sporozoite extracts.

Heat shock proteins

Two members of the heat shock protein 70 family, the 70-kDa Pfhsp and a glucose-regulated protein, Pfgrp (78 kDa), have been cloned from *P. falciparum* (8, 15, 62, 63, 145). Although undetectable in *P. falciparum* sporozoites, expression of both proteins was induced after sporozoite invasion of hepatocytes (61). Pfhsp was localized in the nuclei and cytoplasm of EE parasites and Pfgrp was found in their nuclear membranes and endoplasmic reticula (61). Recently, *P. falciparum* hsp60 with significant homology to human hsp60 and implicated in protective $\gamma\delta$ T-cell responses (130) was cloned and sequenced (127a).

LSA-2

Although expression of LSA-1 was detected throughout liver schizogony, no cross-reaction between human antibodies recognizing LSA-1 repeat and nonrepeat regions and *P. berghei* and *P. yoelii* liver stages occurred. Mice immunized with a repeat LSA-1 peptide developed antibodies that recognized a 230-kDa *P. berghei* EE-specific liver-stage antigen (LSA-2) (52) that was localized on the PVM (10) and may be an analog of LSA-1.

CSP-2

Mice immunized with *P. falciparum* sporozoites were completely protected against challenge with *P. berghei* but not *P. yoelii* sporozoites (124). Sera and a MAb from these mice recognized a 42/54-kDa antigen designated CS protein 2 (CSP-2) on both *P. falciparum* and *P. berghei* EE parasites (125). The MAb was protective in passive transfer.

Pbl.1

The *P. berghei* protein Pbl.1 was identified by a MAb produced by immunizing mice with *P. berghei* EE parasites. Pbl.1 is not found on *P. berghei* sporozoites and has been demonstrated 3 h after sporozoite invasion of hepatocytes. It is localized in the parasitophorous vacuole and on the PVM (127) and is not present in infected erythrocytes. The gene encoding Pbl.1 has not been reported.

Blood-stage antigens that are first expressed in infected hepatocytes

Several blood-stage antigens, including merozoite surface protein 1 (PfMSP-1) (128), the 220-kDa glutamate-rich protein (17), the serine-rich antigen (128), erythrocyte membrane protein 2 (PfEMP2) (128), the acidic-base repeat antigen (128), and rhoptry antigen 1 (128), are first expressed in infected hepatocytes. Their significance as targets of protective immune responses has not been established.

Immune Responses That Could Target Infected Hepatocytes

The major emphasis of work on immune responses that target infected hepatocytes has been on $CD8^+$ T lymphocytes. Such $CD8^+$ T cells could be activated by recognition of peptides from any of the parasite proteins described above that are associated with class I major histocompatibility complex (MHC) molecules on the surfaces of infected hepatocytes or on antigen-presenting cells. $CD8^+$ T cells could release perforins and granzymes, which destroy the infected hepatocytes, as well as gamma interferon and other cytokines, which induce the infected hepatocyte to produce substances like nitric oxide that initiate killing of the parasite. Alternatively, CTLs may destroy infected hepatocytes by crosslinking of CTL membrane ligands with apoptosis-inducing target cell receptors such as APO-1/Fas (reviewed in reference 14). Our concept of this interaction is shown in Fig. 2. The standard chromium release cytotoxicity assay has been used by most researchers to indicate the appropriate induction of $CD8^+$ T cells. However, if gamma interferon and/or other cytokines are responsible for protection, then the standard assay, which is more likely to measure the effect of classic cytotoxic T lymphocytes (CTLs), may not be the appropriate in vitro assay to use to predict protection. An assay that measures the number of gamma-interferon-producing $CD8^+$ T cells may be more appropriate (82).

Notwithstanding the emphasis that has been placed on $CD8^+$ T cells, it is now quite clear that $CD4^+$ T cells can also recognize parasite peptides with class II MHC molecules and eliminate infected hepatocytes by similar mechanisms. Furthermore, $CD4^+$ T-cell-mediated immunity may be just as potent as $CD8^+$ T-cell-mediated protection. Alternatively, T cells remote from the specific infected hepatocyte could be activated by presentation of antigens found in sporozoites or secreted from hepatocytes; the activated T cells may secrete cytokines that induce the infected hepatocyte to kill the parasite. Recently, $\gamma\delta$ T cells were shown to have activity against infected hepatocytes in the absence of $\alpha\beta$ T cells (130). Antibodies could recognize parasite proteins on infected hepatocytes and either directly or with complement or cells (antibody-dependent cellular cytotoxicity) eliminate the infected hepatocyte (23, 97). Finally, antibodies that recognize proteins on EE merozoites could block the initial invasion of erythrocytes.

GOALS OF ATTACKING INFECTED HEPATOCYTES

The irradiated-sporozoite vaccine induces sterile protective immunity; it produces immune responses that prevent sporozoites from effectively invading hepatocytes or prevent liver-stage parasites from developing to maturity and releasing infective merozoites or both. A vaccine that duplicates the sterile immunity afforded by the irradiated-sporozoite vaccine (see below) would be ideal. By preventing infection of erythrocytes, it would prevent the development of disease and stop transmission, because no sexual-stage parasites (gametocytes) would develop. Some investigators now believe that even if such a vaccine did not completely prevent blood-stage infection but only reduced the numbers of parasites that emerged from the liver, it could have a profound effect on the morbidity and mortality associated with malaria. The evidence for this belief is indirect. First, widespread use of insecticide-impregnated bed nets had a dramatic effect on mortality in sub-Saharan Africa (6, 12) even though the bed nets did not prevent all cases of malaria and in some areas they did not even reduce the incidence of malaria. In effect, they functioned much like a partially effective sporozoite or liver-stage vaccine. The bed nets mechanically reduced the numbers of bites by sporozoite-infected mosquitoes, and thus fewer merozoites were released from the liver, resulting in a lower blood-stage-parasite density, fewer clinical manifestations, and reduced mortality. Another piece of information that supports this contention is an observation made in a region of western Kenya where virtually all children aged 6 months to 1 year have *P. falciparum* parasitemia every day of the year. Blood-stage-parasite density correlated directly with the numbers of infective mosquitoes to which the children had been exposed during the past month (13, 71). These data suggest that reduced exposure to infective mosquitoes corresponds to a decreased frequency of exposure to sporozoites, which results in a lower blood-stage parasite density and, ultimately, fewer clinical symptoms. Finally, there has been a report of an association between the presence of the class I HLA allele, HLA Bw53, and a reduced risk of severe malaria (45a). Since hepatocytes are the only cells expressing class I HLA molecules that *P. falciparum* resides in, this observation suggests that incomplete anti-liver stage immunity may be responsible for a reduction in severe malaria. A liver-stage vaccine could in essence have a biological effect similar to that of bed nets or reduced transmission: all three act to reduce the numbers of parasites that emerge from the liver. Furthermore, immune responses to infected hepatocytes should be boosted by repeated exposure to mosquito-inoculated sporozoites and therefore have an excellent chance of being sustained, even in individuals who never develop erythrocytic-stage infections. Perhaps there will be several liver-stage vaccines: one to provide sterile protective immunity for nonimmunes visiting malarious areas and a second, probably combined with a blood-stage vaccine, for individuals living their entire lives in highly malarious regions of the world.

DATA INDICATING THAT INFECTED HEPATOCYTES ARE TARGETS OF PROTECTIVE IMMUNE RESPONSES

Irradiated-Sporozoite Model

Immunization with radiation-attenuated sporozoites induces a stage-specific protective immune response in rodents, humans, and monkeys (25–27, 44, 87, 98, 99). Protection is against sporozoite challenge but not against blood-stage infection. The protective immunity must be directed against the sporozoite in circulation or against infected hepatocytes or against both. This finding is consistent with the observation that irradiated sporozoites develop only partially within hepatocytes (123) and thus never become mature liver-stage parasites that initiate blood-stage infection. Incubation of sporozoites with MAbs against the CS protein repeat region in culture neutralizes sporozoite infectivity in vitro (54) and in vivo (86). More important, MAbs passively protect mice (22, 94) and monkeys (20) against sporozoite-induced malaria (see chapter 2). Yet a large body of evidence now indicates that the immunity induced by irradiated-sporozoite vaccine is mediated in large part by T cells that recognize malaria antigens on the surfaces of infected hepatocytes. The potential role of T cells in this immunity was first suggested by data showing that μ-suppressed mice immunized with irradiated sporozoites are protected against challenge (24) and that spleen cells transferred from sporozoite-immunized mice protect naive mice (135). In 1987, adoptive transfer of immune T cells into naive mice was shown to protect against malaria in the absence of antibodies (32). Two groups, one working with the A/J-*P. berghei* system and one working with the BALB/c-*P. yoelii* system, showed that protective immunity induced by immunization with irradiated sporozoites is abrogated by in vivo depletion of CD8$^+$ T cells; the antibodies induced by immunization with irradiated sporozoites are not adequate to protect against sporozoite-induced malaria, and depletion of CD4$^+$ T cells has no effect on protection (115, 139). These data strongly suggest that this immunity is dependent on CD8$^+$ T lymphocytes recognizing malaria peptides on the surfaces of infected hepatocytes in association with class I MHC molecules.

These conclusions are supported by studies showing that mice immunized with irradiated *P. berghei* sporozoites and challenged with large numbers of live sporozoites develop parasite-specific, $CD8^+$ T-cell-dependent inflammatory infiltrates in their livers (48). In addition, spleen cells from mice immunized with irradiated sporozoites eliminate infected hepatocytes from Kupffer cell-free, hepatocyte cultures in an MHC- and species-specific manner (48, 50), indicating that immune T cells recognize *Plasmodium* antigens on the surfaces of infected hepatocytes and eliminate these cells. The initial assumption was that this protection is mediated by $CD8^+$ CTLs. However, considerable in vivo data now indicate that "classic" CTLs may not be involved. The protective immunity induced by irradiated *P. berghei* sporozoites in A/J mice and *P. berghei* sporozoites in BALB/c mice is abrogated by in vivo treatment of the mice with anti-gamma interferon (115, 122a). This abrogation is not found in *P. berghei* (NK65) (48)-immunized BALB/c mice. Adoptive transfer of a $CD8^+$ T-cell clone against the *P. yoelii* CS protein derived from a mouse immunized with radiation-attenuated sporozoites, a clone that endogenously produces large quantities of gamma interferon, protects against *P. yoelii.* This protective immunity is eliminated by in vivo treatment of the mice with anti-gamma interferon (137). Recently, irradiated sporozoite-induced immunity in the *P. berghei*-BALB/c model system was reversed by treatment with agents such as aminoguanidine that prevent the production of nitric oxide (122a).

Inducible nitric oxide synthase (iNOS) expression in livers following sporozoite challenge is restricted to infected hepatocytes and is dependent on the persistence of the irradiated parasites in the livers of immunized animals (58, 113). By accelerating the removal of preexisting irradiated parasites from the hepatocytes of immunized animals by using the antimalarial drug primaquine, the abilities of immunized animals to express iNOS in response to sporozoite challenge corresponded to a loss of liver-stage protection (58, 113).

Finally, data now indicate that $CD4^+$ T cells and $\gamma\delta$ T cells may also contribute to irradiated-sporozoite-induced protection (130, 131).

Thus, we now have considerable data indicating that the protective immunity induced by the irradiated-sporozoite vaccine is primarily directed against the infected hepatocyte. There are still, however, many incompletely answered major questions. For example, there is no clear demonstration of how T cells pass from the Kupffer cell-lined sinusoids into the space of Disse so as to attack infected hepatocytes. With the exception of the work on heat shock proteins (97), no *Plasmodium* antigens have been identified on the surface of infected hepatocytes using antibodies as probes, and uninfected hepatocytes express MHC molecules at low levels, if at all.

In Vivo and In Vitro Evidence of the Role of Cytokines

Studies that involve the addition of cytokines to hepatocyte cultures and the systemic administration of cytokines have provided important insights into the mechanisms by which vaccine-induced cellular immunity may actually eliminate infected hepatocytes from culture and afford protection in vivo.

Addition of cytokines to hepatocyte cultures

Reports published nearly a decade ago showed that the addition of gamma interferon to hepatocyte cultures reduces or eliminates *P. berghei* (34) and *P. falciparum* (78). Subsequent reports showed that in the murine hepatocyte-*P. berghei* (77) and the human hepatocyte-*P. falciparum* (79) systems, this activity is reversed by inhibiting iNOS, indicating that gamma interferon induces the infected hepatocyte to produce L-arginine-derived nitrogen oxides that are toxic to the intracellular parasite (77).

Analysis of the pattern of secretion of certain $CD4^+$ T-cell clones suggests that other cytokines may also be involved (30, 88, 95, 96). The inhibitory effect of interleukin 1 (IL-1) and IL-6 on intrahepatic development of human and murine parasites has been reported (78, 92). Tumor necrosis factor (TNF) inhibits development of *P. berghei* in vitro in a hepatoma cell line (114), but TNF alone is not effective in primary cultures of *P. yoelii*-infected hepatocytes (77, 89). In cocultures of hepatocytes and nonparenchymal cells, TNF induces nonparenchymal cells to inhibit parasites by releasing IL-6 (88, 89).

Systemic administration of cytokines

Systemic administration of gamma interferon partially protects mice and monkeys against *P. berghei* (34) and *P. cynomolgi* (66), respectively. In addition, administration of recombinant IL-12 to mice (119) and rhesus monkeys (28a) provides 100% protection against sporozoite challenge with *P. yoelii* in mice and *P. cynomolgi* in monkeys. The protection in mice is entirely eliminated by administration of a MAb against gamma interferon and is eliminated in 50% of mice by administration of N^G-monomethyl-L-arginine to inhibit nitric oxide synthesis. Current data suggest that IL-12 induces T cells and natural killer (NK) cells to produce gamma interferon, which induces infected hepatocytes to produce the nitric oxide that kills the developing parasite (119).

Studies with Subunit Vaccines against Specific Parasite Proteins

Studies involving subunit vaccines against specific parasite proteins are discussed in the next section.

EVIDENCE THAT SPECIFIC PROTEINS ARE THE TARGET OF PROTECTIVE IMMUNE RESPONSES AGAINST INFECTED HEPATOCYTES

CS Protein

In the late 1980s, the only target antigen available for study was the CS protein expressed on the surfaces of sporozoites (37, 146) and on the parasite membrane and PVM of developing EE parasites (9). A single region of the *P. falciparum* CS protein was identified as including a CTL epitope (64), and analogous regions were later identified in *P. berghei* and *P. yoelii* CS proteins (106, 107, 138). A CTL clone recognizing the CTL epitope of the *P. berghei* CS protein adoptively transferred complete protection against challenge with *P. berghei* sporozoites (106, 107), and CTL against the *P. yoelii* epitope eliminated infected hepatocytes from culture in an antigen-specific, MHC-restricted manner (138). In subsequent studies, the transfer of a similar $CD8^+$ CTL clone against the *P. yoelii* CS protein transferred protection (103, 137), and if this clone was transferred 3 h after sporozoite inoculation, it still provided protection (103). Since sporozoites are thought to enter hepatocytes within an hour of inoculation, this experiment indicated that the CTL clones were recognizing a CS protein peptide on the surfaces of infected hepatocytes and were either destroying the infected hepatocyte or rendering the parasite nonfunctional. This concept was further supported by data demonstrating that radiolabeled protective but not nonprotective CTL clones could be found in apposition to infected hepatocytes after adoptive transfer in vivo (102). The mechanism whereby these $CD8^+$ T cells prevent further development of the parasites is unknown. They may act through the release of pore-forming proteins or through cytokines. However, as described above, work in the rodent irradiated-sporozoite model systems strongly suggests that the immunity induced by the sporozoite vaccine is primarily mediated by the release of cytokines that induce the infected hepatocyte to produce nitric oxide. It has now been shown that the protective immunity induced by immunization of mice with a *P. yoelii* CS protein DNA vaccine (120) is completely eliminated by treatment of the mice with a MAb against gamma interferon or by administration of aminoguanidine, an inhibitor of iNOS (30a).

There is also evidence that the protection may require specific adhesion molecules such as CD44 on the surfaces of effector $CD8^+$ T lymphocytes (102). CD44 may be required for optimal interaction with the infected hepatocyte, or activation of CD44 may induce expression of other molecules, such as LFA-1, that facilitate the interaction of the effector cell with the parasitized cell.

A number of studies have demonstrated that $CD4^+$ T cells also have protective activity. $CD4^+$ T cells directed against amino acids 59 through 79 from the amino terminus of the *P. yoelii* CS protein can recognize CS protein peptides on the surfaces of infected hepatocytes, eliminate infected hepatocytes from culture, and adoptively transfer protection against malaria (70, 95, 96). $CD4^+$ T-cell clones from a human volunteer immunized with *P. falciparum* sporozoites specifically lysed autologous B cells pulsed with a synthetic peptide representing the C-terminal region of the CS protein (83).

Thus, $CD8^+$ T cells against a 9-amino-acid peptide sequence on the *P. yoelii* and *P. berghei* CS proteins and $CD4^+$ T cells against a 21-amino-acid epitope (amino acids 59 to 79) on the *P. yoelii* CS protein can protect mice against challenge with sporozoites in the absence of other parasite-specific immune responses.

SSP2/TRAP

Immunization with irradiated sporozoites induces not only antibody production but also CTLs against PySSP2 (56). Spleen cells from mice immunized with irradiated sporozoites produced CTLs against P815 mouse mastocytoma cells transfected with a 1.5-kb fragment of the gene encoding PySSP2 (56). Adoptive transfer of a $CD8^+$ CTL clone against PySSP2 protected 100% of mice against challenge (57). Like the T-cell clones against *P. yoelii* CS protein, transfer of the protective T-cell clone 3 h after sporozoite inoculation, at a time when the parasites had already entered hepatocytes, still prevented development of blood-stage infection. These experiments clearly establish that anti-PySSP2 T cells can protect against malaria in the absence of other parasite-specific immune responses and that these cells must be affecting the parasite developing within the hepatocytes. The PySSP2 epitope recognized by the protective clone has not yet been definitively identified.

$CD4^+$ T-cell-dependent protection against a PySSP2 epitope can be induced by active immunization (see below). Thus, in addition to the $CD8^+$ and $CD4^+$ protective T-cell epitopes on the *P. yoelii* CS protein, there are completely protective $CD8^+$ and $CD4^+$ T-cell epitopes on PySSP2.

LSA-1

Mice immunized with an LSA-1 peptide produced antibodies that recognize *P. berghei* LSA-2 (10, 52) and were protected against challenge with *P. berghei* sporozoites. Immune spleen cells from these mice killed *P. berghei*-infected hepatocytes in vitro.

PyHEP17/PfEXP-1

MAb NYLS3 against PyHEP17 MAb eliminates *P. yoelii*-infected hepatocytes from culture but does not affect *P. berghei*-infected hepatocytes (23). The effect on hepatocytes does not require complement. When passively transferred into mice, it reduces the level of liver-stage schizonts but does not provide complete protection against blood-stage infection. These data indicate that antibodies against a single epitope on PyHEP17 can have a profound effect on the parasite developing within the liver. The gene encoding PyHEP17 has been cloned and sequenced (30a), and a PyHEP17 DNA vaccine has been constructed. This vaccine protects against *P. yoelii* sporozoite challenge but not against blood-stage challenge (31a). Since PyHEP17 is not expressed in sporozoites, these data demonstrate that the PyHEP17 vaccine protects by attacking infected hepatocytes.

STARP, SALSA, and LSA-3

Experiments similar to those described above for mice are very difficult to perform with the novel molecules recently identified in *P. falciparum* when the homologous gene in rodent species has not yet been described. Nevertheless, cell lines directed to SALSA were derived by stimulating peripheral blood lymphocytes from immune chimpanzees with peptides. These cell lines exerted strong $CD8^+$-dependent, class I-restricted CTL activity on autologous B cells pulsed with the same peptide (17a). Similar results were obtained with LSA-3 (13a).

Data from studies on influenza, human immunodeficiency virus, and malaria make it likely that a series of CTL epitopes will be identified on many proteins regardless of whether these epitopes can indeed be targets for CTL in vivo. In vitro assays aimed at investigating the activities of cell lines derived from chimpanzees or humans on autologous chimp or MHC-matched human *P. falciparum*-infected hepatocytes are likely to be essential in predicting the relevance of this CTL activity to protection.

Heat Shock Proteins

Immunization of $\alpha\beta$ T-cell-deficient mice with irradiated *P. yoelii* sporozoites significantly inhibits liver-stage development but does not provide immunity against development of parasitemia. This activity against infected hepatocytes is abolished by depletion of $\gamma\delta$ cells (130), and adoptive transfer of a $\gamma\delta$ T-cell clone reduces liver-stage development by 50% compared to that of controls. The protective $\gamma\delta$ T-cell clone proliferated with recombinant *Mycobacterium bovis* hsp65, a protein with significant sequence homology to human hsp60. These results suggest that $\gamma\delta$ T cells may eliminate infected hepatocytes by recognition of either host or *Plasmodium* hsp60. Cloning of *P. falciparum* hsp60 (127a) should further these investigations.

CSP-2

We have no data on CSP-2.

EVIDENCE THAT IMMUNIZATION OF HUMANS WITH IRRADIATED SPOROZOITES OR NATURAL EXPOSURE TO MALARIA INDUCES IMMUNE RESPONSES THAT COULD ATTACK INFECTED HEPATOCYTES

CD8$^+$ T-Cell Responses

CS protein

The first step in the process of determining CD8$^+$ T-cell responses was the identification of CTL epitopes on the *P. falciparum* CS protein. These responses were first identified in 1988 by Kumar and coworkers (64), who identified a single CD8$^+$CTL epitope on the *P. falciparum* CS protein in B10.BR mice. Peripheral blood mononuclear cells (PBMC) from volunteers immunized with irradiated sporozoites of *P. falciparum* had genetically restricted, CD8$^+$ T-cell-dependent, peptide-specific cytolytic activity against the same region, Pf 7G8 CS 368-390 (KPKDELDYENDIEKKICKMEKCS) (67). When studied in a standard restimulation assay (67), these PBMC also lysed target cells transiently transfected with the gene encoding the *P. falciparum* protein. Three of four volunteers studied had CTLs against this peptide and transiently transfected targets, but one of these three was not protected when challenged with live sporozoites. This result demonstrated in humans what had been already demonstrated in the *P. yoelii* rodent model system: the presence of CTLs against a single epitope or protein as revealed by a chromium release assay does not indicate that an individual will be protected. Subsequently, Sedegah and colleagues (121) used the same assay to show that Kenyans with life-long natural exposure to malaria had genetically restricted, peptide-specific CD8$^+$ T-cell-dependent CTL activity against the same epitope. Doolan and colleagues (31) demonstrated that PBMC from Australians who had lived in malarious areas also had cytolytic activity against a similar region of the CS protein, but they did not demonstrate genetic restriction of the response or T-cell subset dependence of the activity. It was next shown that the first eight amino acids of PfCSP 368–390 peptide, KPKDELDY, bound to HLA-B35 and that this peptide was the target of CTLs among Gambians naturally exposed to malaria (46). This peptide (KPKDELDY), a variant of it (KSKDELDY), and two other *P. falciparum* CS protein-derived petides (MPNDPNRNV [PfCSP 300–308, a peptide that binds to HLA-B7] and LRKPKHKKL [PfCSP 105–113, a peptide that binds to HLA-B8]) are the targets of CTLs derived from the peripheral blood of Gambians with natural exposure to malaria (2). $CD8^+$ T-cell dependence and genetic restriction of CTL activity were not demonstrated. An additional HLA-A2.1-restricted CTL epitope on the *P. falciparum* CS protein, PfCSP 7G8 334-342 (YLKKIKNSL), was identified by stimulating cells from individuals from Burkina Faso and deriving CTL lines and clones (16). Thus, HLA-A2, -B7, -B8, and -B35 $CD8^+$ CTL epitopes have now been identified on the *P. falciparum* CS protein.

SSP2/TRAP

In the same study described above for *P. falciparum* CS protein, of Gambians naturally exposed to malaria, Aidoo and colleagues (2) described two HLA-A2.1-restricted epitopes (PfSSP2/TRAP 3-11 and 500-508 [HLGNVKYLV and GIAGGLALL, respectively]), and two HLA-B8-restricted epitopes (107-115 and 109-117 [ASKNKEKAL and KNKEKALI, respectively]) on this protein. Wizel and colleagues studied volunteers immunized with radiation-attenuated *P. falciparum* sporozoites and identified 12 peptides that included HLA-A2-restricted epitopes that sensitized target cells for HLA-restricted, peptide-specific, $CD8^+$ T-cell-dependent cytolytic activity (141). One of these peptides included the sequence of peptide PfSSP2/TRAP 3-11 (HLGNVKYLV) identified by Aidoo and colleagues (2). Wizel and colleagues showed that peripheral blood lymphocytes from volunteers immunized with radiation-attenuated sporozoites 14 days earlier have direct cytotoxic activity without in vitro stimulation, demonstrating for the first time circulating active CTLs against a *Plasmodium* protein. Such data raise the possibility that these activated cells may mobilize to the liver to kill infected hepatocytes. All previous work had used restimulation assays. Wizel and colleagues conducted similar studies with peptides conforming to HLA-B8 binding motifs and identified the same two peptides identified by Aidoo and colleagues as targets of $CD8^+$ T-cell-dependent, HLA-restricted, peptide-specific CTL activity (142). Furthermore, Wizel and colleagues, as above, showed circulating activated CTLs against one of the peptides.

PfLSA-1

CD8$^+$ T-cell-dependent, HLA-B53-restricted cytolytic activity against a conserved CTL epitope (PfLSA-1 1786-94 [KPIVQYDNF]), in the C-terminal region has been identified in individuals exposed to natural infection with *P. falciparum* sporozoites in The Gambia (46). In addition, an HLA-B35 epitope (PfLSA-1850-1857 [KPNDKSLY]) and an HLA-B17 epitope (PfLSA-1 1854-61 [KSLYDEHI]) have been identified by using cells from The Gambia (2).

Other pre-erythrocytic proteins

As yet, little has been published to demonstrate CD8$^+$ CTLs against other pre-erythrocytic-stage proteins of *P. falciparum*. However, an HLA-A2.2-restricted CTL epitope was recently described in STARP (2), and unpublished data indicate that humans naturally exposed to malaria have CTL activity against several other epitopes from STARP, SALSA, and LSA-3. In contrast to CTL activity found in immunized chimpanzees (17a), the activity found in subjects exposed under field conditions is usually low; this indicates that if the level of activity can be artificially raised by vaccination with better antigen presentation systems, perhaps it can be raised to protective levels.

CD4$^+$ T-Cell Responses

Many papers detail CD4$^+$ T-cell responses to the *P. falciparum* CS protein and most of the other liver-stage proteins described here. These papers show that humans naturally exposed to malaria, and in some cases to irradiated sporozoites, have CD4$^+$ T cells sensitized to liver-stage proteins (35, 41, 49, 60). This finding is expected from our current understanding of the interaction of short peptides, class II MHC molecules, and the CD4$^+$ T-cell receptor. A CD4$^+$ CTL clone that recognizes a peptide containing amino acids 337 through 346 (KIQNSLSTEW) of the NF54 *P. falciparum* CS protein in the context of HLA-DR7 was derived from a human immunized with irradiated sporozoites (83). Subsequently, DR1-, -4, -7, and -9-restricted T-cell clones from three sporozoite-immunized volunteers were shown to recognize overlapping but distinct epitopes within amino acids 326 to 345 of the *P. falciparum* CS protein (84). With the recent increased interest in the role of CD4$^+$ T-cell effectors against infected hepatocytes, such studies have taken on increased importance. Studies that determine the prevalence of CD4$^+$

T-cell responses to invariant peptides from liver-stage antigens in individuals of defined HLA haplotypes from areas of endemicity will be critical in developing this approach to vaccine development. Preliminary data for a number of liver-stage proteins, including LSA-1 (35, 55a, 60, 144), LSA-3 (13a), and SALSA (17a), indicate a high prevalence of response and a low prevalence of variation in epitopes. Since good priming of $CD4^+$ T cells may be essential to the expansion of $CD8^+$ CTLs and since $CD4^+$ cells are also effective on their own, these antigens should stimulate further studies of their use in vaccines.

Summary

It now seems quite clear that immunization with radiation-attenuated sporozoites and natural exposure to malaria induce a wide range of T-cell responses against the liver-stage proteins being considered for vaccine development. These studies are critical in identifying T-cell epitopes to be included in vaccines and for use in the refining assays that will determine whether vaccines induce the immune responses expected. Unfortunately, determining whether a specific immune response measured by these assays translates into protection against malaria is virtually impossible. However, these studies can be used to exclude vaccines that produce an inadequate immune response.

PROGRESS TOWARD DEVELOPING VACCINES

A vaccine includes the target of the desired antibody or T-lymphocyte-mediated protective immune response plus a delivery system designed to optimize the immune responses to these B and T epitopes. Malaria vaccine developers are in the forefront of modern vaccinology, utilizing synthetic peptides, purified recombinant proteins, live recombinant viruses, bacteria, protozoa, and plasmid DNA as immunogens and a wide variety of adjuvants and vehicles as delivery systems to optimize the induction of protective immunity.

Vaccines Designed To Produce Protective Antibodies

No vaccine has been shown to produce antibodies that attack infected hepatocytes and to produce sterile protective immunity in rodent model systems, nonhuman primates, or humans. The only data available that indicate that antibodies attack the antigens in infected hepatocytes come from experiments with antibodies against heat shock proteins (97) and the antigen PyHEP17 described above (23).

Vaccines Designed To Induce Protective T-Cell Responses in Animal Models

The interest in developing vaccines to induce protective T-cell responses is enormous. Most of the work has focused on inducing protective $CD8^+$ T-cell responses, but there is now great interest in inducing protective $CD4^+$ T-cell responses.

CS protein

$CD8^+$ T cells in rodent plasmodia. Considerable efforts have been made to produce vaccines that actively induce protective $CD8^+$ T-cell responses against the *P. berghei* and *P. yoelii* CS proteins. In the *P. berghei* system, oral immunization of mice with a recombinant *Salmonella typhimurium* expressing *P. berghei* CS protein induces CTLs against this protein and protects 50 to 75% of the mice against challenge with *P. berghei* sporozoites (109). Like the immunity found after immunization with irradiated-sporozoite vaccine, this immunity was abrogated by in vivo depletion of $CD8^+$ T cells (1). When mice were immunized with a recombinant vaccinia virus expressing the *P. berghei* CS protein, they produced CTLs against the *P. berghei* CS protein and were not protected (112). Recent work with another recombinant vaccinia indicates that this can be accomplished (64a). However, mice immunized with recombinant vaccinia (116), *S. typhimurium* (117), or pseudorabies virus (118) expressing the *P. yoelii* CS protein produced excellent immune responses, but they were not protected against the highly infectious *P. yoelii* sporozoites. However, when BALB/c mice were immunized with irradiated P815 mastocytoma cells transfected with the gene encoding the *P. yoelii* CS protein, 50 to 85% of them were protected against challenge (56). This immunity was also eliminated by in vivo depletion of $CD8^+$ T cells. Recently, priming with a recombinant influenza virus expressing the only known *P. yoelii* CS protein $CD8^+$ epitope and boosting with a recombinant vaccinia expressing the entire *P. yoelii* CS protein provided $CD8^+$ T-cell-dependent protection to 60% of mice (65).

One of the most exciting recent developments in vaccinology is the demonstration that immunization with "naked" DNA induces protective antibody and cellular immune responses. This technology has now been used to construct an experimental *P. yoelii* CS protein vaccine. Immunization with a *P. yoelii* CS protein plasmid DNA vaccine induces extremely high levels of anti-*P. yoelii* CS protein antibodies and CTLs (120). This immunity protects up to 83% of mice against challenge with *P. yoelii* sporozoites and is eliminated by in vivo depletion of $CD8^+$ T lymphocytes (120), in vivo treatment with an anti-gamma-interferon MAb, and treatment with aminoguanidine, an inhibitor of iNOS. This vaccine protects only BALB/c mice and provides little or no protection in four other strains of mice. The protection is severely genetically restricted (30a).

CD4$^+$ T lymphocytes in rodent plasmodia. BALB/c mice immunized with a multiple antigen peptide containing a known protective CD4$^+$ T-cell epitope from the *P. berghei* CS protein were protected against challenge with sporozoites (80).

Perspective. As described in section III, the data indicating that mice immunized with subunit *P. yoelii* CS protein subunit vaccines have excellent CTL responses by the chromium release assay but are often not protected call into question the appropriateness of using the CTL assay to indicate adequate immunization. Furthermore, it is important to note that the irradiated-sporozoite vaccine protects 100% of mice against challenge with up to 100,000 sporozoites. In contrast, these vaccines against the *P. yoelii* CS protein protect only 50 to 75% of mice against challenge with 100 to 200 sporozoites. The irradiated-sporozoite vaccine either is inducing better immune responses against the CS protein or is also inducing immune responses against other sporozoite and liver-stage antigens. Antibody and T-cell assays do not provide any indication that responses against the CS protein are better after immunization with the irradiated-sporozoite vaccine than after immunization with the recombinant CS-protein vaccines. Thus, investigators have spent considerable time trying to discover additional parasite proteins that are the targets of irradiated sporozoite-induced protective immunity.

SSP2

CD8$^+$ T lymphocytes in rodents. SSP2 was the first non-CS-protein target of irradiated-sporozoite-induced protective immunity described (see above). After demonstrating that mice immunized with irradiated *P. yoelii* sporozoites produce CTLs against PySSP2, Khusmith and colleagues showed that immunization with PySSP2 would protect against challenge (56). Mice were immunized with P815 cells transfected with a 1.5-kb gene fragment encoding part of PySSP2; 50 to 70% of the mice were protected, and the protection was eliminated by in vivo depletion of CD8$^+$ T cells. A plasmid DNA vaccine encoding the entire PySSP2 induced approximately 50% protection in A/J mice (136b).

CD4$^+$ T lymphocytes in rodents. Recently, it was demonstrated that immunization with an 18-amino-acid peptide from PySSP2 in the anionic, block copolymer adjuvant TiterMax provides 100% protection in A/J mice but little protection in two other strains of mice (136c). This immunity is eliminated by in vivo elimination of antibodies to $CD4^+$ T cells but not of antibodies to $CD8^+$ T cells. Microscopic analysis of livers from immunized and challenged mice studies indicate that the immunity is directed against infected hepatocytes and eliminated by treatment with anti-gamma interferon. Furthermore, in vitro studies indicate that this peptide activates only a Th1 T-cell response. The protection against this short, simple peptide is more consistent and profound than that found with any univalent vaccine designed to induce protective $CD8^+$ T-cell responses.

PyHEP17–PfEXP-1

A plasmid DNA vaccine including the cDNA for the part of the gene encoding PyHEP17 induces 80 to 90% protection in A/J mice, approximately 50% protection in B10.BR mice, and less than 30% protection in BALB/c, B10.Q, and C57BL/6 mice. This protection is entirely dependent on $CD8^+$ T lymphocytes, gamma interferon, or nitric oxide (30a).

Combination vaccines

When it was shown that immunization with recombinant mastocytoma cells expressing the *P. yoelii* CS protein or PySSP2 vaccines gave only partial protection against malaria (50 to 75%), Khusmith and colleagues immunized BALB/c mice with transfected P815 cells expressing *P. yoelii* CS protein and PySSP2 and achieved 100% protection (56). Furthermore, as is observed after immunization with irradiated sporozoites, this immunity was completely reversed by in vivo depletion of $CD8^+$ T cells. In the same strain of mice, additive protection could be achieved by combining immunogens.

When it was demonstrated that the *P. yoelii* CS protein DNA vaccine protected BALB/c mice well but not other strains of mice and that the PyHEP17 DNA vaccine protected A/J mice well but not BALB/c mice, experiments were conducted to determine whether this genetic restriction of protection could be circumvented by combining the two vaccines. The combination vaccine worked. It provided 80 to 90% protection for BALB/c and A/J mice and additive protection (80 to 90%) in B10.BR mice. Unfortunately, the combination did not protect C57BL/6 mice or B10.Q mice; additional antigens are required (30a).

Perspective

Steady work over the years has established that mice can be consistently protected against challenge with rodent malaria sporozoites by inducing T-cell responses against infected hepatocytes. Many of the failures in the human experimental challenge system have been predicted by experiments in the rodent model. Likewise, the successful protection of humans by irradiated-sporozoite vaccine was predicted by work in the rodent model system. The rodent subunit vaccines are now good but not optimal. They do not withstand a large sporozoite challenge, do not induce lifelong protection, and have not yet been shown to provide high-level, consistent protection in outbred mice and across many strains of inbred mice. It is our contention that such findings suggest that similar vaccines will not provide humans the long-lasting, consistent, sterile protective immunity expected of such vaccines. It is hoped that refinement of vaccination regimens and vaccines in rodents will lead to better protection. The same specific refinements may not prove to be appropriate for human vaccines, but the demonstration that critical experimentation can lead to higher levels of protection will provide an important foundation for similar studies in humans.

Vaccines Designed To Produce Protective Cellular Immune Responses in Experimentally Infected Volunteers

Univalent *P. falciparum* CS-protein vaccines

Although numerous *P. falciparum* CS-protein vaccines have been tested in humans (reviewed in reference 47; see chapter 2), most have been designed to induce protective antibody responses against the *P. falciparum* CS-protein repeat region. The strong body of data from the rodent malaria system and the demonstration that humans immunized with irradiated sporozoites and naturally exposed to malaria produce CTLs against the *P. falciparum* CS protein have turned the attention of vaccine developers toward producing human vaccines that induce CTL against the *P. falciparum* CS protein. Humans immunized with radiation-attenuated sporozoites and with subunit *P. falciparum* CS-protein repeat region vaccines develop $CD4^+$ T-cell responses against the *P. falciparum* CS-protein repeat region (11, 83, 84), and these vaccines may have induced protective anti-repeat-region T-cell responses against infected hepatocytes. However, the major emphasis of *P. falciparum* CS-protein vaccines designed to induce protective T-cell responses is on the regions of the protein flanking the central repeat region. A number of formulations of *P. falciparum* CS protein produce CTLs in rodents (64, 68, 140). Several vaccines that include the flanking regions of the *P. falciparum* CS protein have been tested in humans. Humans were immunized with a single oral dose of a recombinant *Salmonella typhi* expressing the *P. falciparum* CS protein, and one individual showed $CD8^+$ T-cell-dependent, genetically restricted, peptide-specific cytolytic activity against peptide 368-390 from the *P. falciparum* CS protein (40). Recently, a recombinant protein including the repeat region and entire carboxy terminus of the *P. falciparum* CS protein on hepatitis B particles was shown to induce CTLs in humans and to protect 2 of 8 volunteers who received one formulation (42).

Multivalent recombinant vaccinia virus

A recombinant attenuated vaccinia virus called NYVAC has been constructed (129a). This recombinant vaccinia virus expresses seven *P. falciparum* proteins: PfCSP, PfSSP2/TRAP, PfLSA-1, PfMSP-1, PfAMA-1, PfSERA, and Pfs25. The first three are primary targets for liver-stage vaccine development, and the fourth antigen, PfMSP-1, is known to be expressed in infected hepatocytes. Human studies have been initiated with this multistage vaccine.

Perspective

To date, few vaccines designed to produce protective immune responses against infected hepatocytes have been studied in humans. However, in the next few years, abundant information regarding induction of protective T-cell responses against infected hepatocytes in humans should be forthcoming.

Current Status of Field Trials

No field trials have been done with vaccines designed to induce protective immune responses against infected hepatocytes.

OBSTACLES TO DEVELOPING VACCINES THAT TARGET INFECTED HEPATOCYTES

Reproducible data now demonstrate that antibodies can eliminate infected hepatocytes from in vitro culture (23). Thus, it may be possible to produce protective antibodies against the infected hepatocyte. If the protective epitopes are conserved among strains of *P. falciparum*, such epitopes could become important components of all vaccines. However, the major emphasis of work on attacking infected hepatocytes is aimed at inducing protective T-cell responses. Even if a protective epitope is identified and is conserved among all strains of a species of *Plasmodium* that infects humans, it will probably not be able to induce protective T-cell responses in all vaccinees, because T cells recognize short peptides associated with MHC molecules: 8- to 10-amino-acid peptides for class I-restricted $CD8^+$ T cells and 12- to 24-amino-acid peptides for class II-restricted $CD4^+$ T cells. There may be "universal" or "degenerate" $CD8^+$ and $CD4^+$ *P. falciparum* T epitopes that are presented with all class I or class II HLA molecules on the surfaces of infected hepatocytes, but this is unlikely. Thus, vaccine developers will have to identify different protective epitopes for individuals with different HLA backgrounds. One approach will be to immunize with multiple full-length proteins without identifying specific epitopes and then hope that the diversity within the protein is great enough to sensitize all individuals. Another approach will be to systematically identify T-cell epitopes among individuals who represent the majority of HLA phenotypes and to construct vaccines that include all of these epitopes (2, 141, 142). However, there is also evidence that in nonmalaria molecules, T-cell epitopes associating with class I antigens are clustered so that relatively small defined regions may contain the range of nonamers that could associate with a wide range of diverse class I MHC molecules (7). Either approach may be severely limited by variation of these epitopes within the parasite. The genetic restriction of T-cell responses and the parasite polymorphism of potentially important T-cell epitopes are serious obstacles to the development of effective vaccines designed to attack infected hepatocytes.

SUMMARY AND CONCLUSIONS

In 1982, no gene encoding a liver-stage protein had been cloned, and there was no direct evidence that the infected hepatocyte is the target of protective immune responses. In fact, many thought that infected hepatocytes were sequestered from the immune system. There was considerable doubt as to whether T cells could actually attack *Plasmodium*-infected hepatocytes and whether such hepatocytes expressed the major histocompatibility antigens required for specific T-lymphocyte recognition. As of 1996, the genes encoding more than 10 *Plasmodium* proteins that are expressed in infected hepatocytes have been cloned and characterized. $CD8^+$ and $CD4^+$ T-cell clones have been shown to provide sterile protection of mice in adoptive transfer experiments, and there is no doubt that parasite-specific T cells can recognize infected hepatocytes following recognition of parasite peptides in combination with class I and II MHC molecules. The obstacles to be overcome before the ideal liver-stage malaria vaccine is produced are still considerable, but enormous progress has been made in the last decade. The first generation of vaccines designed to induce immune responses that attack infected hepatocytes has entered clinical trials. Even if these first-generation vaccines do not provide protection, they will provide an important foundation for the next generation of vaccines. We expect dramatic progress in the next decade as more liver-stage targets of protective immune responses are incorporated into vaccines and as vaccine construction and delivery systems are improved. It is unlikely that the ideal malaria vaccine will be designed to attack only the infected hepatocyte, but we believe that liver-stage protection will be an essential part of the ideal malaria vaccination.

Acknowledgments. This work was supported in part by Naval Medical Research and Development Command work units 61102A.S13.00101.BFX.1431, 612787A.870.00101.EFX.1432, and 623002A.810.00101.HFX.1433.

Special thanks go to Benjamin Wizel for helpful comments and Sylvia I. Becker for editorial support.

REFERENCES

1. **Aggarwal, A., S. Kumar, R. Jaffe, D. Hone, M. Gross, and J. Sadoff.** 1991. Oral Salmonella: malaria circumsporozoite recombinants induce specific CD8+ cytotoxic T cells. *J. Exp. Med.* **172:**1083–1090.
2. **Aidoo, M., A. Lalvani, C. E., Allsopp, M. Plebanski, S. J. Meisner, P. Krausa, M. Browning, S. Morris Jones, F. Gotch, D. A. Fidock, M. Takiguchi, K. J. H. Robson, B. M. Greenwood, P. Druilhe, H. C. Whittle, and A. V. S. Hill.** 1995. Identification of conserved antigenic components for a cytotoxic T lymphocyte-inducing vaccine against malaria. *Lancet* **345:**1003–1007.
3. **Aikawa, M., C. T. Atkinson, L. M. Beaudoin, M. Sedegah, Y. Charoenvit, and R. Beaudoin.** 1990. Localization of CS and non-CS antigens in the sporogonic stages of Plasmodium yoelii. *Bull. W.H.O.* **68**(Suppl.)**:**165–171.
4. **Aikawa, M., A. Schwartz, S. Uni, R. Nussenzweig, and M. Hollingdale.** 1984. Ultrastructure of in vitro cultured exoerythrocytic stage of Plasmodium berghei in a hepatoma cell line. *Am. J. Trop. Med. Hyg.* **33:**792–799.
5. **Aley, S. B., J. W. Barnwell, M. D. Bates, W. E. Collins, and M. R. Hollingdale.** 1987. Plasmodium vivax: exoerythrocytic schizonts recognized by monoclonal antibodies against blood-stage schizonts. *Exp. Parasitol.* **64:**188–194.
6. **Alonso, P. L., S. W. Lindsay, J. R. M. Armstrong Schellenberg, K. Keita, P. Gomez, F. C. Shenton, A. G. Hill, P. H. David, G. Fegan, K. Cham, and B. M. Greenwood.** 1993. A malaria control trial using insecticide-treated bed nets and targeted chemoprophylaxis in a rural area of The Gambia, West Africa. *Trans. R. Soc. Trop. Med. Hyg.* **87**(Suppl. 2):37–44.
7. **Androlewicz, M. J., and P. Cresswell.** 1994. Human transporters associated with antigen processing possess a promiscuous peptide-binding site. *Immunity* **1:**7–14.
8. **Ardeshir, F., J. E. Flint, S. J. Richman, and R. T. Reese.** 1987. A 75 kd merozoite surface protein of Plasmodium falciparum which is related to the 70 kd heat-shock proteins. *EMBO J.* **6:**493–499.
9. **Atkinson, C. T., M. Aikawa, S. B. Aley, and M. R. Hollingdale.** 1989. Expression of Plasmodium berghei circumsporozoite antigen on the surface of exoerythrocytic schizonts and merozoites. *Am. J. Trop. Med. Hyg.* **41:**9–17.
10. **Atkinson, C. T., M. R. Hollingdale, and M. Aikawa.** 1992. Localization of a 230-kd parasitophorous vacuole membrane antigen of Plasmodium berghei exoerythrocytic schizonts (LSA-2) by immunoelectron and confocal laser scanning microscopy. *Am. J. Trop. Med. Hyg.* **46:**533–537.
11. **Ballou, W. R., S. L. Hoffman, J. A. Sherwood, M. R. Hollingdale, F. A. Neva, W. T. Hockmeyer, D. M. Gordon, I. Schneider, R. A. Wirtz, J. F. Young, G. F. Wasserman,**

P. Reeve, C. L. Diggs, and J. D. Chulay. 1987. Safety and efficacy of a recombinant DNA Plasmodium falciparum sporozoite vaccine. *Lancet* **i:**1277–1281.

12. **Beach, R. F., T. K. Ruebush, J. D. Sexton, P. L. Bright, A. W. Hightower, J. G. Breman, D. L. Mount, and A. J. Oloo.** 1993. Effectiveness of permethrin-impregnated bed nets and curtains for malaria control in holoendemic area of western Kenya. *Am. J. Trop. Med. Hyg.* **49:**290–300.
13. **Beadle, C., P. D. McElroy, C. N. Oster, J. C. Beier, A. J. Oloo, F. K. Onyango, D. K. Chumo, J. D. Bales, J. A. Sherwood, and S. L. Hoffman.** 1995. Impact of transmission intensity and age on Plasmodium falciparum density and associated fever: implications for malaria vaccine design. *J. Infect. Dis.* **172:**1047–1054.

13a. **BenMohammed, L., et al.** Submitted for publication.

14. **Bertoletti, A., A. Sette, A. Penna, M. Levrero, F. V. Chisari, F. Fiaccadori, and C. Ferrari.** Unpublished data.
15. **Bianco, A. E., J. M. Favaloro, T. R. Burkot, J. G. Culvenor, P. E. Crewther, G. V. Brown, R. F. Anders, R. L. Copel, and D. J. Kemp.** 1986. A repetitive antigen of Plasmodium falciparum that is homologous to heat shock protein 70 of Drosophila melanogaster. *Proc. Natl. Acad. Sci. USA* **83:**8713–8717.
16. **Blum Tirouvanziam, U., C. Servis, A. Habluetzel, D. Valmori, Y. Men, F. Esposito, L. Del Nero, N. Holmes, N. Fasel, and G. Corradin.** 1995. Localization of HLA-A2.1-restricted T cell epitopes in the circumsporozoite protein of Plasmodium falciparum. *J. Immunol.* **154:**3922–3931.
17. **Borre, M. B., M. Dziegiel, B. Hogh, E. Petersen, K. Rieneck, E. Riley, J. F. Meis, M. Aikawa, K. Nakamura, M. Harada, A. Wind, P. H. Jakobsen, J. Cowland, S. Jepsen, N. H. Axelsen, and J. Vuust.** 1991. Primary structure and localization of a conserved immunogenic Plasmodium falciparum glutamate rich protein (GLURP) expressed in both the preerythrocytic and erythrocytic stages of the vertebrate life cycle. *Mol. Biochem. Parasitol.* **49:**119–131.

17a. **Bottius, E., et al.** Submitted for publication.

18. **Cerami, C., U. Frevert, P. Sinnis, B. Takacs, P. Clavijo, M. J. Santos, and V. Nussenzweig.** 1992. The basolateral domain of the hepatocyte plasma membrane bears receptors for the circumsporozoite protein of Plasmodium falciparum sporozoites. *Cell* **70:**1021–1033.
19. **Cerami, C., F. Kwakye Berko, and V. Nussenzweig.** 1992. Binding of malarial circumsporozoite protein to sulfatides [Gal(3-SO_4)β1-Cer] and cholesterol-3-sulfate and its dependence on disulfide bond formation between cysteines in region II. *Mol. Biochem. Parasitol.* **54:**1–12.
20. **Charonenvit, Y., W. E. Collins, T. R. Jones, P. Millet, L. Yuan, G. H. Campbell, R. L. Beaudoin, J. R. Broderson, and S. L. Hoffman.** 1991. Inability of malaria vaccine to induce antibodies to a protective epitope within its sequence. *Science* **251:**668–671.
21. **Charoenvit, Y., M. L. Leef, L. F. Yuan, M. Sedegah, and R. L. Beaudoin.** 1987. Characterization of *Plasmodium yoelii* monoclonal antibodies directed against stage-specific sporozoite antigens. *Infect. Immun.* **55:**604–608.
22. **Charoenvit, Y., S. Mellouk, C. Cole, R. Bechara, M. F. Leef, M. Sedegah, L. F. Yuan, F. A. Robey, R. L. Beaudoin, and S. L. Hoffman.** 1991. Monoclonal, but not polyclonal antibodies protect against Plasmodium yoelii sporozoites. *J. Immunol.* **146:**1020–1025.
23. **Charoenvit, Y., S. Mellouk, M. Sedegah, T. Toyoshima, M. F. Leef, P. De la Vega, R. L. Beaudoin, M. Aikawa, V. Fallarme, and S. L. Hoffman.** 1995. Plasmodium yoelii: 17-kDa hepatic and erythrocytic stage protein is the target of an inhibitory monoclonal antibody. *Exp. Parasitol.* **80:**419–429.

24. **Chen, D. H., R. E. Tigelaar, and F. I. Weinbaum.** 1977. Immunity to sporozoite-induced malaria infection in mice. I. The effect of immunization of T and B cell-deficient mice. *J. Immunol.* **118:**1322–1327.
25. **Clyde, D. F., V. C. McCarthy, R. M. Miller, and R. B. Hornick.** 1973. Specificity of protection of man immunized against sporozoite-induced falciparum malaria. *Am. J. Med. Sci.* **266:**398–401.
26. **Clyde, D. F., V. C. McCarthy, R. M. Miller, and W. E. Woodward.** 1975. Immunization of man against falciparum and vivax malaria by use of attenuated sporozoites. *Am. J. Trop. Med. Hyg.* **24:**397–401.
27. **Clyde, D. F., H. Most, V. C. McCarthy, and J. P. Vanderberg.** 1973. Immunization of man against sporozoite-induced falciparum malaria. *Am. J. Med. Sci.* **266:**169–177.
28. **Cowan, G., S. Krishna, A. Crisanti, and K. Robson.** 1992. Expression of thrombospondin-related anonymous protein in Plasmodium falciparum sporozoites. *Lancet* **339:**1412–1413.
28a. **Crutcher, J. M., et al.** Submitted for publication.
29. **Dame, J. B., J. L. Williams, T. F. McCutchan, J. L. Weber, R. A. Wirtz, W. T. Hockmeyer, W. L. Maloy, J. D. Haynes, I. Schneider, D. Roberts, G. S. Sanders, E. P. Reddy, C. L. Diggs, and L. H. Miller.** 1984. Structure of the gene encoding the immunodominant surface antigen on the sporozoite of the human malaria parasite Plasmodium falciparum. *Science* **225:**593–599.
30. **Del Giudice, G., D. Grillot, L. Renia, I. Muller, G. Corradin, J. A. Louis, D. Mazier, and P. H. Lambert.** 1990. Peptide-primed CD4+ cells and malaria sporozoites. *Immunol. Lett.* **25:**59–64.
30a. **Doolan, D. L., et al.** Submitted for publication.
31. **Doolan, D. L., R. A. Houghten, and M. F. Good.** 1991. Location of human cytotoxic T cell epitopes within a polymorphic domain of the Plasmodium falciparum circumsporozoite protein. *Int. Immunol.* **3:**511–516.
31a. **Doolan, D. L., M. Sedegah, R. C. Hedstrom, P. Hobart, Y. Charoenvit, and S.L. Hoffman.** Circumventing genetic restriction of protection against malaria with multi-gene DNA immunization: $CD8^+$ T cell, interferon γ, and nitric oxide dependent immunity. *J. Exp. Med.*, in press.
32. **Egan, J. E., J. L. Weber, W. R. Ballou, M. R. Hollingdale, W. R. Majarian, D. M. Gordon, W. L. Maloy, S. L. Hoffman, R. A. Wirtz, I. Schneider, G. R. Woollett, J. F. Young, and W. T. Hockmeyer.** 1987. Efficacy of murine malaria sporozoite vaccines: implications for human vaccine development. *Science* **236:**453–456.
33. **Enea, V., J. Ellis, F. Zavala, D. E. Arnot, A. Asavanich, A. Masuda, I. Quakyi, and R. S. Nussenzweig.** 1984. DNA cloning of Plasmodium falciparum circumsporozoite gene: amino acid sequence of repetitive epitope. *Science* **225:**628–630.
34. **Ferreira, A., L. Schofield, V. Enea, H. Schellekens, P. van der Meide, W. E. Collins, R. S. Nussenzweig, and V. Nussenzweig.** 1986. Inhibition of development of exoerythrocytic forms of malaria parasites by gamma-interferon. *Science* **232:**881–884.
35. **Fidock, D. A., H. Gras Masse, J. P. Lepers, K. Brahimi, L. Benmohamed, S. Mellouk, C. Guerin Marchand, A. Londono, L. Raharimalala, J. F. Meis, G. Langsley, C. Roussilhon, A. Tartar, and P. Druilhe.** 1994. Plasmodium falciparum liver stage antigen-1 is well conserved and contains potent B and T cell determinants. *J. Immunol.* **153:**190–204.
36. **Fidock, D. A., S. Sallenave Sales, J. A. Sherwood, G. S. Gachihi, M. F. Ferreira da Cruz, A. W. Thomas, and P. Druilhe.** 1994. Conservation of the Plasmodium falciparum sporozoite surface protein gene, STARP, in field isolates and distinct species of Plasmodium. *Mol. Biochem. Parasitol.* **67:**255–267.

37. **Fine, E., M. Aikawa, A. H. Cochrane, and R. S. Nussenzweig.** 1984. Immuno-electron microscopic observation on Plasmodium knowlesi sporozoites: localization of protective antigen and its precursors. *Am. J. Trop. Med. Hyg.* **33:**220–226.
38. **Frevert, U., P. Sinnis, C. Cerami, W. Shreffler, B. Takacs, and V. Nussenzweig.** 1993. Malaria circumsporozoite protein binds to heparan sulfate proteoglycans associated with the surface membrane of hepatocytes. *J. Exp. Med.* **177:**1287–1298.
39. **Garnham, P. C. C.** 1980. Malaria in its various vertebrate hosts, p. 96–144. *In* J. P. Kreier (ed.), *Malaria*, vol. 1. *Epidemiology, Chemotherapy, Morphology, and Metabolism.* Academic Press, Inc., New York.
40. **Gonzalez, C., D. Hone, F. R. Noriega, C. O. Tacket, J. R. Davis, G. Losonsky, J. P. Nataro, S. Hoffman, A. Malik, E. Nardin, M. B. Sztein, D. G. Heppner, T. R. Fouts, A. Isibasi, and M. M. Levine.** 1994. Salmonella typhi vaccine strain CVD 908 expressing the circumsporozoite protein of Plasmodium falciparum: strain construction and safety and immunogenicity in humans. *J. Infect. Dis.* **169:**927–931.
41. **Good, M. F., D. Pombo, I. A. Quakyi, E. M. Riley, R. A. Houghten, A. Menon, D. W. Alling, J. A. Berzofsky, and L. H. Miller.** 1987. Human T-cell recognition of the circumsporozoite protein of Plasmodium falciparum: immunodominant T-cell domains map to the polymorphic regions of the molecule. *Proc. Natl. Acad. Sci. USA* **85:**1199–1203.
42. **Gordon, D. M., T. W. McGovern, U. Krzych, J. C. Cohen, I. Schneider, R. LaChance, D. G. Heppner, G. Yuan, M. Hollingdale, M. Slaoui, P. Hauser, P. Voet, J. C. Sadoff, and W. R. Ballou.** 1995. Safety, immunogenicity, and efficacy of a recombinantly produced Plasmodium falciparum circumsporozoite protein-hepatitis B surface antigen subunit vaccine. *J. Infect. Dis.* **171:**1576–1585.
43. **Guerin-Marchand, C., P. Druilhe, B. Galey, A. Londono, J. Patarapotikul, R. L. Beaudoin, C. Dubeaux, A. Tartar, O. Mercereau Puijalon, and G. Langsley.** 1987. A liver-stage-specific antigen of Plasmodium falciparum characterized by gene cloning. *Nature* (London) **329:**164–167.
44. **Gwadz, R. W., A. H. Cochrane, V. Nussenzweig, and R. S. Nussenzweig.** 1979. Preliminary studies on vaccination of rhesus monkeys with irradiated sporozoites of Plasmodium knowlesi and characterization of surface antigens of these parasites. *Bull. W.H.O.* **57**(Suppl. 1):165–173.
45. **Hedstrom, R. C., J. R. Campbell, M. L. Leef, Y. Charoenvit, M. Carter, M. Sedegah, R. L. Beaudoin, and S. L. Hoffman.** 1990. A malaria sporozoite surface antigen distinct from the circumsporozoite protein. *Bull. W.H.O.* **68**(Suppl.):152–157.
45a. **Hill, A. V. S., C. E. M. Allsopp, D. Kwiatkowski, N. M. Anstey, P. Twumasi, P. A. Rowe, S. Bennett, D. Brewster, A. J. McMichael, and B. M. Greenwood.** 1991. Common West African HLA antigens are associated with protection from severe malaria. *Nature* (London) **352:**595–600.
46. **Hill, A. V. S., J. Elvin, A. C. Willis, M. Aidoo, C. E. Allsopp, F. M. Gotch, X. M. Gao, M. Takiguchi, B. M. Greenwood, A. R. Townsend, A. J. McMichael, and H. C. Whittle.** 1992. Molecular analysis of the association of HLA-B53 and resistance to severe malaria. *Nature* (London) **360:**434–439.
47. **Hoffman, S. L., E. D. Franke, W. O. Rogers, and S. Mellouk.** 1993. Preerythrocytic malaria vaccine development, p. 149–167. *In* M. F. Good and A. J. Saul (ed.), *Molecular Immunological Considerations in Malaria Vaccine Development.* CRC Press, Inc., Boca Raton, Fla.
48. **Hoffman, S. L., D. Isenbarger, G. W. Long, M. Sedegah, A. Szarfman, L. Waters, M. R. Hollingdale, P. H. van der Miede, D. S. Finbloom, and W. R. Ballou.** 1989. Sporo-

zoite vaccine induces genetically restricted T cell elimination of malaria from hepatocytes. *Science* **244:**1078–1081.

49. **Hoffman, S. L., C. N. Oster, C. Mason, J. C. Beier, J. A. Sherwood, W. R. Ballou, M. Mugambi, and J. D. Chulay.** 1989. Human lymphocyte proliferative response to a sporozoite T cell epitope correlates with resistance to falciparum malaria. *J. Immunol.* **142:**1299–1303.
50. **Hoffman, S. L., W. Weiss, S. Mellouk, and M. Sedegah.** 1990. Irradiated sporozoite vaccine induces cytotoxic T lymphocytes that recognize malaria antigens on the surface of infected hepatocytes. *Immunol. Lett.* **25:**33–38.
51. **Hollingdale, M. R.** 1988. Malaria and the liver, p. 1195–1211. *In* I. M. Arias (ed.), *The Liver: Biology and Pathobiology.* Raven Press, New York.

51a. **Hollingdale, M. R., et al.** Submitted for publication.

52. **Hollingdale, M. R., M. Aikawa, C. T. Atkinson, W. R. Ballou, G. Chen, J. Li, J. F. Meis, B. Sina, C. Wright, and J. Zhu.** 1990. Non-CS pre-erythrocytic protective antigens. *Immunol. Lett.* **25:**71–76.

52a. **Hollingdale, M. R., W. E. Collins, C. C. Campbell, and A. L. Schwartz.** 1985. In vitro culture of two populations (dividing and non-dividing) of exoerythrocytic parasites of Plasmodium vivax. *Am. J. Trop. Med. Hyg.* **34:**216–222.

52b. **Hollingdale, M. R., J. M. Foster, adn P. Reeves.** Submitted for publication.

53. **Hollingdale, M. R., P. Leland, J. L. Leef, and A. L. Schwartz.** 1983. Entry of Plasmodium berghei sporozoites into cultured cells, and their transformation into trophozoites. *Am. J. Trop. Med. Hyg.* **32:**685–690.
54. **Hollingdale, M. R., E. H. Nardin, S. Tharavanij, A. L. Schwartz, and R. S. Nussenzweig.** 1984. Inhibition of entry of Plasmodium falciparum and P. vivax sporozoites into cultured cells: an in vitro assay of protective antibodies. *J. Immunol.* **132:**909–913.
55. **Hollingdale, M. R., R. E. Sinden, and J. F. van Pelt.** 1993. Sporozoite invasion. *Nature* (London) **362:**26.

55a. **Kazura, J. W., et al.** Submitted for publication.

56. **Khusmith, S., Y. Charoenvit, S. Kumar, M. Sedegah, R. L. Beaudoin, and S. L. Hoffman.** 1991. Protection against malaria by vaccination with sporozoite surface protein 2 plus CS protein. *Science* **252:**715–718.
57. **Khusmith, S., M. Sedegah, and S. L. Hoffman.** 1994. Complete protection against *Plasmodium yoelii* by adoptive transfer of a $CD8^+$ cytotoxic T-cell clone recognizing sporozoite surface protein 2. *Infect. Immun.* **62:**2979–2983.
58. **Klotz, F. W., L. F. Scheller, M. C. Seguin, N. Kumar, M. A. Marletta, S. J. Green, and A. F. Azad.** 1995. Co-localization of inducible-nitric oxide synthase and Plasmodium berghei in hepatocytes from rats immunized with irradiated sporozoites. *J. Immunol.* **154:**3391–3395.
59. **Krotoski, W. A., P. C. Garnham, R. S. Bray, D. M. Krotoski, R. Killick Kendrick, C. C. Draper, G. A. Targett, and M. W. Guy.** 1982. Observations on early and late post-sporozoite tissue stages in primate malaria. I. Discovery of a new latent form of Plasmodium cynomolgi (the hypnozoite), and failure to detect hepatic forms within the first 24 hours after infection. *Am. J. Trop. Med. Hyg.* **31:**24–35.
60. **Krzych, U., J. A. Lyon, T. Jareed, I. Schneider, M. R. Hollingdale, D. M. Gordon, and W. R. Ballou.** 1995. T lymphocytes from volunteers immunized with irradiated Plasmodium falciparum sporozoites recognize liver and blood stage malaria antigens. *J. Immunol.* **155:**4072–4077.
61. **Kumar, N., H. Nagasawa, J. B. Sacci, Jr., B. J. Sina, M. Aikawa, C. Atkinson, P. Uparanukraw, L. B. Kubiak, A. F. Azad, and M. R. Hollingdale.** 1993. Expression of

members of the heat-shock protein 70 family in the exoerythrocytic stages of Plasmodium berghei and Plasmodium falciparum. *Parasitol. Res.* **79:**109–113.

62. **Kumar, N., C. A. Syin, R. Carter, I. Quakyi, and L. H. Miller.** 1988. Plasmodium falciparum gene encoding a protein similar to the 78-kDa rat glucose-regulated stress protein. *Proc. Natl. Acad. Sci. USA* **85:**6277–6281.
63. **Kumar, N., and H. Zheng.** 1992. Nucleotide sequence of a Plasmodium falciparum stress protein with similarity to mammalian 78-kDa glucose-regulated protein. *Mol. Biochem. Parasitol.* **56:**353–356.
64. **Kumar, S., L. H. Miller, I. A. Quakyi, D. B. Keister, R. A. Houghten, W. L. Maloy, B. Moss, J. A. Berzofsky, and M. F. Good.** 1988. Cytotoxic T cells specific for the circumsporozoite protein of Plasmodium falciparum. *Nature* (London) **334:**258–260.

64a. **Lanar, D., et al.** Submitted for publication.

65. **Li, S., M. Rodrigues, D. Rodriguez, J. R. Rodriguez, M. Esteban, P. Palese, R. S. Nussenzweig, and F. Zavala.** 1993. Priming with recombinant influenza virus followed by administration of recombinant vaccinia virus induces CD8+ T-cell-mediated protective immunity against malaria. *Proc. Natl. Acad. Sci. USA* **90:**5214–5218.
66. **Maheshwari, R. K., C. W. Czarniecki, G. P. Dutta, S. K. Puri, B. N. Dhawan, and R. M. Friedman.** 1986. Recombinant human gamma interferon inhibits simian malaria. *Infect. Immun.* **53:**628–630.
67. **Malik, A., J. E. Egan, R. A. Houghton, J. C. Sadoff, and S. L. Hoffman.** 1991. Human cytotoxic T lymphocytes against Plasmodium falciparum circumsporozoite protein. *Proc. Natl. Acad. Sci. USA* **88:**3300–3304.
68. **Malik, A., M. Gross, T. Ulrich, and S. L. Hoffman.** 1993. Induction of cytotoxic T lymphocytes against the *Plasmodium falciparum* circumsporozoite protein by immunization with soluble recombinant protein without adjuvant. *Infect. Immun.* **61:**5062–5066.
69. **Marchand, C., and P. Druilhe.** 1990. How to select Plasmodium falciparum pre-erythrocytic antigens in an expression library without defined probe. *Bull. W.H.O.* **68**(Suppl.):158–164.
70. **Mazier, D., L. Renia, A. Nussler, S. Pied, M. Marussig, J. Goma, D. Grillot, F. Miltgen, J. C. Drapier, G. Corradin, G. Del Giudice, and G. E. Grau.** 1990. Hepatic phase of malaria is the target of cellular mechanisms induced by the previous and the subsequent stages. A crucial role for liver nonparenchymal cells. *Immunol. Lett.* **25:**65–70.
71. **McElroy, P. D., J. C. Beier, C. N. Oster, C. Beadle, J. A. Sherwood, A. J. Oloo, and S. L. Hoffman.** 1994. Predicting outcome in malaria: correlation between rate of exposure to infected mosquitoes and level of Plasmodium falciparum parasitemia. *Am. J. Trop. Med. Hyg.* **51:**523–532.
72. **Meis, J. F., P. H. Jap, M. R. Hollingdale, and J. P. Verhave.** 1987. Cellular response against exoerythrocytic forms of Plasmodium berghei in rats. *Am. J. Trop. Med. Hyg.* **37:**506–510.
73. **Meis, J. F., T. Ponnudurai, B. Mons, A. Van Belkum, P. M. van Eerd, P. Druilhe, and H. Schellekens.** 1990. Plasmodium falciparum: studies on mature exoerythrocytic forms in the liver of the chimpanzee, Pan troglodytes. *Exp. Parasitol.* **70:**1–11.
74. **Meis, J. F., P. J. Rijntjes, J. P. Verhave, T. Ponnudurai, M. R. Hollingdale, J. E. Smith, R. E. Sinden, P. H. Jap, J. H. Meuwissen, and S. H. Yap.** 1986. Fine structure of the malaria parasite Plasmodium falciparum in human hepatocytes in vitro. *Cell Tissue Res.* **244:**345–350.
75. **Meis, J. F., J. P. Verhave, P. H. Jap, F. Hess, and J. H. Meuwissen.** 1981. An ultrastructural study of developing stages of exo-erythrocytic Plasmodium berghei in rat hepatocytes. *Parasitology* **82:**195–204.

76. **Meis, J. F., J. P. Verhave, P. H. Jap, and J. H. Meuwissen.** 1983. An ultrastructural study on the role of Kupffer cells in the process of infection by Plasmodium berghei sporozoites in rats. *Parasitology* **86:**231–242.
77. **Mellouk, S., S. J. Green, C. A. Nacy, and S. L. Hoffman.** 1991. IFN-gamma inhibits development of Plasmodium berghei exoerythrocytic stages in hepatocytes by an L-arginine-dependent effector mechanism. *J. Immunol.* **146:**3971–3976.
78. **Mellouk, S., R. K. Maheshwari, A. Rhodes-Feuillette, R. L. Beaudoin, N. Berbiguier, H. Matile, F. Miltgen, I. Landau, S. Pied, J. P. Chigot, R. M. Friedman, and D. Mazier.** 1987. Inhibitory activity of interferons and interleukin 1 on the development of Plasmodium falciparum in human hepatocyte cultures. *J. Immunol.* **139:**4192–4195.
79. **Mellouk, S. O., S. L. Hoffman, Z. Z. Liu, P. De la Vega, T. R. Billar, and A. K. Nussler.** 1994. Nitric oxide-mediated antiplasmodial activity in human and murine hepatocytes induced by gamma interferon and the parasite itself: enhancement by exogenous tetrahydrobiopterin. *Infect. Immun.* **62:**4043–4046.
80. **Migliorini, P., B. Betschart, and G. Corradin.** 1993. Malaria vaccine: immunization of mice with a synthetic T cell helper epitope alone leads to protective immunity. *Eur. J. Immunol.* **23:**582–585.
81. **Miller, L. H., M. Aikawa, J. G. Johnson, and T. Shiroishi.** 1979. Interaction between cytochalasin B-treated malarian parasites and erythrocytes: attachment and junction formation. *J. Exp. Med.* **149:**172–184.
82. **Miyahira, Y., K. Murata, D. Rodriguez, J. R. Rodriguez, M. Esteban, M. M. Rodrigues, and F. Zavala.** 1995. Quantification of antigen specific CD8+ T cells using an ELISPOT assay. *J. Immunol. Methods* **181:**45–54.
83. **Moreno, A., P. Clavijo, R. Edelman, J. Davis, M. Sztein, D. Herrington, and E. Nardin.** 1991. Cytotoxic CD4+ T cells from a sporozoite-immunized volunteer recognize the Plasmodium falciparum CS protein. *Int. Immunol.* **3:**997–1003.
84. **Moreno, A., P. Clavijo, R. Edelman, J. Davis, M. Sztein, F. Sinigaglia, and E. Nardin.** 1993. CD4+ T cell clones obtained from Plasmodium falciparum sporozoite-immunized volunteers recognize polymorphic sequences of the circumsporozoite protein. *J. Immunol.* **151:**489–499.
85. **Muller, H. M., I. Reckmann, M. R. Hollingdale, H. Bujard, K. J. Robson, and A. Crisanti.** 1993. Thrombospondin related anonymous protein (TRAP) of Plasmodium falciparum binds specifically to sulfated glycoconjugates and to HepG2 hepatoma cells, suggesting a role for this molecule in sporozoite invasion of hepatocytes. *EMBO J.* **12:**2881–2889.
86. **Nardin, E. H., V. Nussenzweig, R. S. Nussenzweig, W. E. Collins, K. T. Harinasuta, P. Tapchaisri, and Y. Chomcharn.** 1982. Circumsporozoite proteins of human malaria parasites Plasmodium falciparum and Plasmodium vivax. *J. Exp. Med.* **156:**20–30.
87. **Nussenzweig, R. S., J. Vanderberg, H. Most, and C. Orton.** 1967. Protective immunity produced by the injection of X-irradiated sporozoites of Plasmodium berghei. *Nature* (London) **216:**160–162.
88. **Nussler, A., J.-C. Drapier, L. Renia, S. Pied, F. Miltgen, M. Gentilini, and D. Mazier.** 1991. L-Arginine-dependent destruction of intrahepatic malaria parasites in response to tumor necrosis factor and/or interleukin 6 stimulation. *Eur. J. Immunol.* **21:**227–230.
89. **Nussler, A., S. Pied, J. Goma, L. Renia, F. Miltgen, G. E. Grau, and D. Mazier.** 1991. TNF inhibits malaria hepatic stages in vitro via synthesis of IL-6. *Int. J. Immunol.* **3:**317–321.
90. **Pancake, S. J., G. D. Holt, S. Mellouk, and S. L. Hoffman.** 1992. Malaria sporozoites and circumsporozoite proteins bind specifically to sulfated glycoconjugates. *J. Cell Biol.* **117:**1351–1357.

91. **Pied, S., A. Nussler, M. Pontet, F. Miltgen, H. Matile, P. H. Lambert, and D. Mazier.** 1989. C-reactive protein protects against preerythrocytic stages of malaria. *Infect. Immun.* **57:**278–282.
92. **Pied, S., L. Renia, A. Nussler, F. Miltgen, and D. Mazier.** 1991. Inhibitory activity of IL-6 on malaria hepatic stages. *Parasite Immunol.* **13:**211–217.
93. **Ponnudurai, T., A. H. W. Lensen, G. J. Van Gemert, M. G. Bolmer, and J. H. Meuwissen.** 1991. Feeding behaviour and sporozoite ejection by infected Anopheles stephensi. *Trans. R. Soc. Trop. Med. Hyg.* **85:**175–180.
94. **Potocnjak, P., N. Yoshida, R. S. Nussenzweig, and V. Nussenzweig.** 1980. Monovalent fragments (Fab) of monoclonal antibodies to a sporozoite surface antigen (Pb44) protect mice against malaria infection. *J. Exp. Med.* **151:**1504–1513.
95. **Renia, L., D. Grillot, M. Marussig, G. Corradin, F. Miltgen, P. H. Lambert, D. Mazier, and G. Del Giudice.** 1993. Effector functions of circumsporozoite peptide-primed CD4+ T cell clones against Plasmodium yoelii liver stages. *J. Immunol.* **150:**1471–1478.
96. **Renia, L., M. S. Marussig, D. Grillot, S. Pied, G. Corradin, F. Miltgen, G. Del Giudice, and D. Mazier.** 1991. In vitro activity of CD4+ and CD8+ T lymphocytes from mice immunized with a synthetic malaria peptide. *Proc. Natl. Acad. Sci. USA* **88:**7963–7967.
97. **Renia, L., D. Mattei, J. Gonna, S. Pied, P. Dubois, F. Miltgen, A. Nussler, H. Matile, F. Menegaux, M. Gentilini, and D. Mazier.** 1990. A malaria heat-shock-like determining expressed on the infected hepatocyte surface is the target of antibody-dependent cell-mediated cytotoxic mechanisms by nonparenchymal liver cells. *Eur. J. Immunol.* **20:**1445–1449.
98. **Rieckmann, K. H., R. L. Beaudoin, J. S. Cassells, and D. W. Sell.** 1979. Use of attenuated sporozoites in the immunization of human volunteers against falciparum malaria. *Bull. W.H.O.* **57**(Suppl. 1):261–265.
99. **Rieckmann, K. H., P. E. Carson, R. L. Beaudoin, J. S. Cassells, and K. W. Sell.** 1974. Sporozoite induced immunity in man against an Ethiopian strain of Plasmodium falciparum. *Trans. R. Soc. Trop. Med. Hyg.* **68:**258–259.
100. **Robson, K. J., J. R. Hall, L. C. Davies, A. Crisanti, A. V. Hill, and T. E. Wellems.** 1990. Polymorphism of the TRAP gene of Plasmodium falciparum. *Proc. R. Soc. Lond. B Biol. Sci.* **242:**205–216.
101. **Robson, K. J., J. R. Hall, M. W. Jennings, T. J. Harris, K. Marsh, C. I. Newbold, V. E. Tate, and D. J. Weatherall.** 1988. A highly conserved amino-acid sequence in thrombospondin, properdin and in proteins from sporozoites and blood stages of a human malaria parasite. *Nature* (London) **335:**79–82.
102. **Rodrigues, M., R. S. Nussenzweig, P. Romero, and F. Zavala.** 1992. The in vivo cytotoxic activity of CD8+ T cell clones correlates with their levels of expression of adhesion molecules. *J. Exp. Med.* **175:**895–905.
103. **Rodrigues, M. M., A.-S. Cordey, G. Arreaza, G. Corradin, P. Romero, J. L. Maryanski, R. S. Nussenzweig, and F. Zavala.** 1991. CD8+ cytolytic T cell clones derived against the Plasmodium yoelii circumsporozoite protein protect against malaria. *Int. Immunol.* **3:**579–585.
104. **Rogers, W. O., A. Malik, S. Mellouk, K. Nakamura, M. D. Rogers, A. Szarfman, D. M. Gordon, A. K. Nussler, M. Aikawa, and S. L. Hoffman.** 1992. Characterization of Plasmodium falciparum sporozoite surface protein 2. *Proc. Natl. Acad. Sci. USA* **89:**9176–9180.
105. **Rogers, W. O., M. D. Rogers, R. C. Hedstrom, and S. L. Hoffman.** 1992. Characterization of the gene encoding sporozoite surface protein 2, a protective Plasmodium yoelii sporozoite antigen. *Mol. Biochem. Parasitol.* **53:**45–52.

106. **Romero, P., J. L. Maryanski, A. S. Cordey, G. Corradin, R. S. Nussenzweig, and F. Zavala.** 1990. Isolation and characterization of protective cytolytic T cells in a rodent malaria model system. *Immunol. Lett.* **25:**27–32.
107. **Romero, P., J. L. Maryanski, G. Corradin, R. S. Nussenzweig, V. Nussenzweig, and F. Zavala.** 1989. Cloned cytotoxic T cells recognize an epitope in the circumsporozoite protein and protect against malaria. *Nature* (London) **341:**323–325.
108. **Rosenberg, R., R. A. Wirtz, I. Schneider, and R. Burge.** 1990. An estimation of the number of malaria sporozoites ejected by a feeding mosquito. *Trans. R. Soc. Trop. Med. Hyg.* **84:**209–212.
109. **Sadoff, J. C., W. R. Ballou, L. S. Baron, W. R. Majarian, R. N. Brey, W. T. Hockmeyer, J. F. Young, S. J. Cryz, J. Ou, G. H. Lowell, and J. D. Chulay.** 1988. Oral Salmonella typhimurium vaccine expressing circumsporozoite protein protects against malaria. *Science* **240:**336–338.
110. **Sanchez, G. I., W. O. Rogers, S. Mellouk, and S. L. Hoffman.** 1994. Plasmodium falciparum: exported protein-1, a blood stage antigen, is expressed in liver stage parasites. *Exp. Parasitol.* **79:**59–62.
111. **Santoro, F., A. H. Cochrane, V. Nussenzweig, E. H. Nardin, R. S. Nussenzweig, R. W. Gwadz, and A. Ferreira.** 1983. Structural similarities among the protective antigens of sporozoites from different species of malaria parasites. *J. Biol. Chem.* **258:**3341–3345.
112. **Satchidanandam, V., F. Zavala, and B. Moss.** 1991. Studies using a recombinant vaccinia virus expressing the circumsporozoite protein of Plasmodium berghei. *Mol. Biochem. Parasitol.* **48:**89–100.
113. **Scheller, L. F., and A. F. Azad.** 1995. Maintenance of protective immunity against malaria by persistent hepatic parasites derived from irradiated sporozoites. *Proc. Natl. Acad. Sci. USA* **92:**4066–4068.
114. **Schofield, L., A. Ferreira, V. Nussenzweig, and R. S. Nussenzweig.** 1987. Antimalarial activity of alpha tumor necrosis factor and gamma-interferon. *Fed. Proc.* **46:**760.
115. **Schofield, L., J. Villaquiran, A. Ferreira, H. Schellekens, R. S. Nussenzweig, and V. Nussenzweig.** 1987. Gamma-interferon, CD8+ T cells and antibodies required for immunity to malaria sporozoites. *Nature* (London) **330:**664–666.
116. **Sedegah, M., R. L. Beaudoin, P. De la Vega, M. F. Leef, M. A. Ozcel, E. Jones, Y. Charoenvit, L. F. Yuan, M. Gross, W. R. Majarian, F. A. Robey, W. Weiss, and S. L. Hoffman.** 1988. Use of a vaccinia construct expressing the circumsporozoite protein in the analysis of protective immunity to Plasmodium yoelii, p. 295–309. *In* L. Lasky (ed.), *Technological Advances in Vaccine Development.* Alan R. Liss, Inc., New York.
117. **Sedegah, M., R. L. Beaudoin, W. R. Majarian, M. D. Cochran, C. H. Chiang, J. Sadoff, A. Aggarwal, Y. Charoenvit, and S. L. Hoffman.** 1990. Evaluation of vaccines designed to induce protective cellular immunity against the Plasmodium yoelii circumsporozoite protein: vaccinia, pseudorabies and salmonella transformed with circumsporozoite gene. *Bull. W.H.O.* **68**(Suppl.):109–114.
118. **Sedegah, M., C. H. Chiang, W. R. Weiss, S. Mellouk, M. D. Cochran, R. A. Houghton, R. L. Beaudoin, D. Smith, and S. L. Hoffman.** 1992. Recombinant pseudorabies virus carrying a plasmodium gene: herpesvirus as a new live viral vector for inducing T- and B-cell immunity. *Vaccine* **10:**578–584.
119. **Sedegah, M., F. Finkelman, and S. L. Hoffman.** 1994. Interleukin-12 induction of interferon gamma-dependent protection against malaria. *Proc. Natl. Acad. Sci. USA* **91:**10700–10702.
120. **Sedegah, M., R. Hedstrom, P. Hobart, and S. L. Hoffman.** 1994. Protection against

malaria by immunization with plasmid DNA encoding circumsporozoite protein. *Proc. Natl. Acad. Sci. USA* **91:**9866–9870.

121. **Sedegah, M., B. K. L. Sim, C. Mason, T. Nutman, A. Malik, C. Roberts, A. Johnson, J. Ochola, D. Koech, B. Were, and S. L. Hoffman.** 1992. Naturally acquired $CD8^+$ cytotoxic T lymphocytes against the Plasmodium falciparum circumsporozoite protein. *J. Immunol.* **149:**966–971.
122. **Seguin, M. C., W. R. Ballou, and C. A. Nacy.** 1989. Interactions of Plasmodium berghei sporozoites and murine Kupffer cells in vitro. *J. Immunol.* **143:**1716–1722.

122a. **Seguin, M. C., F. W. Klotz, I. Schneider, J. P. Weir, M. Goodbary, M. Slayter, J. J. Raney, J. U. Aniagolu, and S. J. Green.** 1994. Induction of nitric oxide synthase protects against malaria in mice exposed to irradiated Plasmodium berghei infected mosquitos: involvement of interferon γ and $CD8^+$ T cells. *J. Exp. Med.* **180:**353–358.

123. **Sigler, C. L., P. Leland, and M. R. Hollingdale.** 1984. In vitro infectivity of irradiated P. berghei sporozoites to cultured hepatoma cells. *Am. J. Trop. Med. Hyg.* **33:**544–547.
124. **Sina, B. J., V. E. Do Rosario, G. Woollett, K. Sakhuja, and M. R. Hollingdale.** 1993. Plasmodium falciparum sporozoite immunization protects against Plasmodium berghei sporozoite infection. *Exp. Parasitol.* **77:**129–135.
125. **Sina, B. J., C. Wright, C. T. Atkinson, R. Ballou, M. Aikawa, and M. Hollingdale.** 1995. Characterization of a sporozoite antigen common to Plasmodium falciparum and Plasmodium berghei. *Mol. Biochem. Parasitol.* 69:239–246.
126. **Smith, J. E., J. F. G. M. Meis, T. Ponnudurai, J. P. Verhave, and H. J. Moshage.** 1984. In vitro culture of exoerythrocytic forms of Plasmodium falciparum in adult human hepatocytes. *Lancet* **75:**757–758.
127. **Suhrbier, A., L. Winger, C. O'Dowd, K. Hodivala, and R. E. Sinden.** 1990. An antigen specific to the liver stage of rodent malaria recognized by a monoclonal antibody. *Parasite Immunol.* **12:**473–481.

127a. **Syin, C., and N. D. Goldman.** Submitted for publication.

128. **Szarfman, A., J. A. Lyon, D. Walliker, I. Quakyi, R. J. Howard, S. Sun, W. R. Ballou, K. Esser, W. T. London, R. A. Wirtz, and R. Carter.** 1988. Mature liver stages of cloned Plasmodium falciparum share epitopes with proteins from sporozoites and asexual blood stages. *Parasite Immunol.* **10:**339–351.
129. **Terzakis, J. A., J. P. Vanderberg, D. Foley, and S. Shustak.** 1979. Exoerythrocytic merozoites of Plasmodium berghei in rat hepatic Kupffer cells. *J. Protozool.* **26:**385–389.

129a. **Tine, J., et al.** Submitted for publication.

130. **Tsuji, M., P. Mombaerts, L. Lefrancois, R. S. Nussenzweig, F. Zavala, and S. Tonegawa.** 1994. Gamma delta T cells contribute to immunity against the liver stages of malaria in alpha beta T-cell-deficient mice. *Proc. Natl. Acad. Sci. USA* **91:**345–349.
131. **Tsuji, M., P. Romero, R. S. Nussenzweig, and F. Zavala.** 1990. CD4+ cytolytic T cell clone confers protection against murine malaria. *J. Exp. Med.* **172:**1353–1357.
132. **Uni, S., M. Aikawa, W. E. Collins, C. C. Campbell, and M. R. Hollingdale.** 1985. Electron microscopy of Plasmodium vivax exoerythrocytic stages grown in vitro in a hepatoma cell line. *Am. J. Trop. Med. Hyg.* **34:**1017–1021.
133. **Valmori, D., J. F. Romero, Y. Men, J. L. Maryanski, P. Romero, and G. Corradin.** 1994. Induction of a cytotoxic T cell response by co-injection of a T helper peptide and a cytotoxic T lymphocyte peptide in incomplete Freund's adjuvant (IFA): further enhancement by pre-injection of IFA alone. *Eur. J. Immunol.* **24:**1458–1462.
134. **van Pelt, J. F., J. Kleuskens, M. R. Hollingdale, J. P. Verhave, T. Ponnudurai, J. H. Meuwissen, and S. H. Yap.** 1991. Identification of plasma membrane proteins involved in the hepatocyte invasion of Plasmodium falciparum sporozoites. *Mol. Biochem. Parasitol.* **44:**225–232.

135. **Verhave, J. P., G. T. Strickland, H. A. Jaffe, and A. Ahmed.** 1978. Studies on the transfer of protective immunity with lymphoid cells from mice immune to malaria sporozoites. *J. Immunol.* **121:**1031–1033.
136. **Vreden, S. G. S.** 1994. The role of Kupffer cells in the clearance of malaria sporozoites from the circulation. *Parasitol. Today* **10:**304–308.
136a. **Walker-Jonah, A., et al.** Submitted for publication.
136b. **Wang, H., et al.** Submitted for publication.
136c. **Wang, R., et al.** Submitted for publication.
137. **Weiss, W. R., J. A. Berzovsky, R. A. Houghten, M. Sedegah, M. Hollingdale, and S. L. Hoffman.** 1992. A T cell clone directed at the circumsporozoite protein which protects mice against both P. yoelii and P. berghei. *J. Immunol.* **149:**2103–2109.
138. **Weiss, W. R., S. Mellouk, R. A. Houghten, M. Sedegah, S. Kumar, M. F. Good, J. A. Berzofsky, L. H. Miller, and S. L. Hoffman.** 1990. Cytotoxic T cells recognize a peptide from the circumsporozoite protein on malaria-infected hepatocytes. *J. Exp. Med.* **171:**763–773.
139. **Weiss, W. R., M. Sedegah, R. L. Beaudoin, L. H. Miller, and M. F. Good.** 1988. CD8+ T cells (cytotoxic/suppressors) are required for protection in mice immunized with malaria sporozoites. *Proc. Natl. Acad. Sci. USA* **85:**573–576.
140. **White, K., U. Krzych, D. M. Gordon, T. G. Porter, R. L. Richards, C. R. Alving, C. D. Deal, M. Hollingdale, C. Silverman, D. R. Sylvester, W. R. Ballou, and M. Gross.** 1993. Induction of cytolytic and antibody responses using Plasmodium falciparum repeatless circumsporozoite protein encapsulated in liposomes. *Vaccine* **11:**1341–1346.
141. **Wizel, B., R. Houghten, P. Church, J. A. Tine, D. E. Lanar, D. M. Gordon, W. R. Ballou, A. Sette, and S. L. Hoffman.** 1995. HLA-A2-restricted cytotoxic T lymphocyte responses to multiple Plasmodium falciparum sporozoite surface protein 2 epitopes in sporozoite-immunized volunteers. *J. Immunol.* **155:**766–775.
142. **Wizel, B., R. A. Houghten, K. Parker, J. E. Coligan, P. Church, D. M. Gordon, W. R. Ballou, and S. L. Hoffman.** 1995. Irradiated sporozoite vaccine induces HLA-B8-restricted cytotoxic T lymphocyte responses against two overlapping epitopes of the Plasmodium falciparum surface sporozoite protein 2. *J. Exp. Med.* **182:**1435–1445.
143. **Wizel, B., W. O. Rogers, R. A. Houghten, D. E. Lanar, J. A. Tine, and S. L. Hoffman.** 1994. Induction of murine cytotoxic T lymphocytes against Plasmodium falciparum sporozoite surface protein 2. *Eur. J. Immunol.* **24:**1487–1495.
144. **Yang, C., Y.-P. Shi, V. Udhayakumar, M. P. Alpers, M. M. Povoa, W. A. Hawley, W. E. Collins, and A. A. Lal.** 1995. Sequence variations in the non-repetitive regions of the liver stage-specific antigen-1 (LSA-1) of Plasmodium falciparum from field isolates. *Mol. Biochem. Parasitol.* **71:**291–294.
145. **Yang, Y. F., P. Tan Ariya, Y. D. Sharma, and A. Kilejian.** 1987. The primary structure of a Plasmodium falciparum polypeptide related to heat shock proteins. *Mol. Biochem. Parasitol.* **26:**61–67.
146. **Yoshida, N., R. S. Nussenzweig, P. Potocnjak, V. Nussenzweig, and M. Aikawa.** 1980. Hybridoma produces protective antibodies directed against the sporozoite stage of malaria parasite. *Science* **207:**71–73.
147. **Zhu, J., and M. R. Hollingdale.** 1991. Structure of Plasmodium falciparum liver stage antigen-1. *Mol. Biochem. Parasitol.* **48:**223–226.

Malaria Vaccine Development: A Multi-Immune Response Approach
Edited by Stephen L. Hoffman

Chapter 4

Preventing Merozoite Invasion of Erythrocytes

Anthony A. Holder

INTRODUCTION

The asexual multiplication of the malaria parasite in the vertebrate host takes place in either hepatocytes or erythrocytes. After differentiation, the parasite ruptures the host erythrocyte to release free extracellular merozoites; these then invade new erythrocytes. Multiplication within the bloodstream is a cyclic process of intracellular multiplication, release, and reinvasion. A merozoite is a specialized cell with an apical complex of organelles involved in cell recognition and penetration (reviewed in reference 5). These membrane-bound organelles comprise a pair of rhoptries and a number of smaller structures called micronemes.

The process of erythrocyte invasion is well described at the ultrastructural level but is still poorly understood in molecular terms (reviewed in reference 75). It involves several steps (3, 57). The parasite attaches to the surface of the erythrocyte and then reorients itself such that its apical end is in close apposition to the plasma membrane of the host cell. A close junction is formed between the two surfaces, and then the host cell membrane invaginates and the parasite moves into the resulting vacuole. The junction between parasite and host cell moves over the surface of the parasite as the organism enters the vacuole until the host cell membrane is finally resealed, leaving the parasite in the cell and surrounded by a parasitophorous vacuolar membrane. Dense bodies are another set of merozoite organelles involved in early events of the intracellular parasite life cycle immediately after invasion (43, 181).

The parasite molecules that interact with the host cell in invasion are thought to be on the merozoite surface and within the apical organelle complex. Some of these proteins function as ligands to bind to receptors on the erythrocyte surface. The biochemical events surrounding invasion are poorly understood, but there must be intracellular signaling to induce the changes that occur. The contents of the apical organelles appear to be discharged either onto the surface of the merozoite or into the host cell membrane and cytoskeleton, and these contents contribute protein and lipid to the formation of the parasitophorous vacuole (6, 170). These events must also disrupt the erythrocyte cytoskeleton to allow vacuole formation (54, 184).

Proteins in three locations have been identified as possible targets of responses that inhibit erythrocyte invasion. These proteins are merozoite surface proteins (MSPs), soluble proteins that may be loosely associated with the merozoite surface, and proteins in the apical organelles. Several proteins can be detected by surface iodination of merozoites (79); the two best-characterized MSPs are MSP-1 and MSP-2. For the purpose of this chapter, only membrane-bound proteins are considered to be surface proteins, but there are other components that are more or less strongly bound to the merozoite surface. These components include proteins in macromolecular complex with polypeptides derived from MSP-1 (167) and proteins that are secreted into the vacuole during the late stages of intracellular development. Specific components from the apical organelles can also be found on the surface, and these components are probably secreted from the apical organelles during specific events in the multistep process of invasion.

Interference with the cycle of merozoite release and invasion of new cells is probably largely mediated by antibody, although cellular mechanisms induced by merozoite antigens and effective against the free merozoite or the intracellular parasite cannot be excluded. For antibody to be effective, its target must be accessible. The merozoite is only briefly extracellular, and for antibody to be effective, a high antibody concentration is required. At least five antibody-mediated mechanisms can be envisaged: (i) agglutination of merozoites at or just before rupture of mature schizonts to prevent merozoite release and dispersal; (ii) binding to the surface of the merozoite to block the initial interaction with the surface of the erythrocyte; (iii) blockade of receptor-ligand interactions later in invasion, a process that may involve components of the apical organelle complex; (iv) inhibition of specific biochemical events during invasion; and (v) targeted cell-mediated killing or removal of merozoites.

The purpose of any vaccine developed to prevent erythrocyte invasion is the interruption of the asexual cycle in the bloodstream. The effect of such an interruption would be a reduction in the parasite load and therefore the clinical symptoms associated with this stage of the life cycle. Sterile immunity, if it could be achieved, not only would have a profound effect on malaria morbidity and mortality but also would bring about an interruption of transmission to the mosquito.

The efficacy of immune responses to merozoite antigens can be assessed both in vivo and in vitro. Although a number of experimental models for malaria can be used, one limitation is that for many *Plasmodium falciparum* merozoite antigens, no homologs have been identified. So far, homologs of only three merozoite antigens (MSP-1, AMA-1, and EBA 175; see below) have been positively identified in primate and rodent malaria species. In addition to direct immunization with native, recombinant, or synthetic antigens, passive transfer of immunity with specific antibody or adoptive transfer of cells can be used to study the roles of specific effector mechanisms. A number of in vitro tests such as antibody inhibition of parasite invasion, agglutination of merozoites, and antibody-dependent cell targeting have also been exploited.

PROTEINS ON THE MEROZOITE SURFACE

Because of their obvious accessibility, MSPs have long been considered good candidates for vaccine development.

MSP-1

MSP-1 has been found in all species of malaria parasites studied. It has been implicated as a target for protective immunity by a variety of criteria in studies of *P. falciparum* in nonhuman primates and in vitro, studies of laboratory parasite-host models, and seroepidemiological studies of naturally acquired immunity (53, 84).

MSP-1 is synthesized as a high-molecular-mass precursor (185 to 210 kDa) during the late stages of intracellular development within the erythrocyte and is deposited on the surfaces of the developing merozoite forms. At the time of erythrocyte rupture and release of merozoites into the circulation, the protein is processed by proteases into a number of fragments held together in a noncovalent complex on the merozoite surface (117). The complex is held on the surface of the merozoite by a membrane-bound 42-kDa fragment (MSP-1_{42}). When the parasite invades a new erythrocyte, a second proteolytic step (10, 15) cleaves MSP-1_{42} into a 33-kDa polypeptide (MSP-1_{33}), which is shed as part of the soluble complex with other products of MSP-1 processing (12), and a 19-kDa fragment (MSP-1_{19}), which remains membrane bound and is carried on the surface of the merozoite into the new erythrocyte (11). The function of MSP-1 is unknown (85), but the intact (unprocessed) molecule may bind to receptors containing sialic acid on the surfaces of erythrocytes (132, 172).

The *msp-1* genes from a number of *P. falciparum* lines and isolates have been sequenced. The gene is a single-copy gene, and one of two distinct primary sequences is present and expressed in a parasite clone. A comparison of these sequences makes it possible to divide them into 17 blocks in which the two sequences are either closely related or divergent (121, 173). Additional variation (for example, point mutations) is found in each type. Block 2 is exceptional in that it occurs in three major forms, two of which consist of distinct repetitive amino acid sequences that vary in length. Block 17 at the extreme C terminus of the protein (corresponding to MSP-1_{19}) is cysteine rich and comprises two epidermal growth factor (EGF)-like motifs (173) preceding a short stretch of hydrophobic amino acids that is replaced by glycosylphosphatidylinositol during biosynthesis (72, 121). Block 17 is highly conserved, with only four substitutions in different laboratory strains and limited variation in field isolates (99, 100, 179). The gene has also been sequenced for *Plasmodium vivax* (52, 73), *Plasmodium chabaudi* (49, 119), and *Plasmodium yoelii* (107). Sequences partially conserved between the species have been identified (36); they include the sequences in block 17.

What is the evidence that MSP-1 is a candidate vaccine molecule? In several studies, MSP-1 purified from *P. falciparum* partially (64) or completely (154) protected against challenge in primate models. Partial protection was also observed with peptides (131) or recombinant proteins (64, 80, 81, 87). Monoclonal antibodies (MAbs) to MSP-1 inhibit parasite development in vitro (11, 37, 138). The C terminus of MSP-1, from which

$MSP\text{-}1_{42}$ and $MSP\text{-}1_{19}$ are derived, has been studied in most detail as the target of a protective immune response (84).

Much of the interest in MSP-1 as a vaccine candidate is based on studies with *P. yoelii* in mice. Native *P. yoelii* MSP-1 protects mice against challenge infection (68, 86), and immunity is passively transferred with a MAb against an epitope in the EGF-like motifs (22,112). Recently, immunization with recombinant polypeptides consisting of the C terminus of *P. yoelii* MSP-1 (corresponding to $MSP\text{-}1_{19}$) resulted in very effective protection against this parasite; this protection was mediated by antibody and required both EGF-like motifs and the maintenance of conformational epitopes (44, 45, 108, 109). The protection was abolished by reduction and alkylation of the recombinant protein (108). Studies with *P. yoelii* native MSP-1 (68) and recombinant proteins (44, 107a) showed that other adjuvants in addition to Freund's complete adjuvant are effective.

The secondary and tertiary structures of the C terminus of MSP-1 are important for the protein's antigenicity, and efforts have focused on replicating this structure by having the protein expressed in heterologous cells. The 42-kDa protein has been expressed in insect cells (26, 124), and the 19-kDa protein has been made in yeast cells (101) and bacteria (21) and synthesized as a peptide (166). The integrity of disulfide-constrained epitopes, defined by antibody reactive with the native protein, is intact in the recombinant proteins. Reduction and alkylation abolish or severely reduce the recognition of *P. falciparum* $MSP\text{-}1_{19}$ by MAbs (21, 29) and polyclonal antibodies (61), and this treatment of native MSP-1 also abolishes the ability of the protein to bind these antibodies (110). Epitopes of inhibitory MAbs either are present on the individual EGF-like motifs or require both motifs in order to be present (29). Most of the antibody-binding sites in $MSP\text{-}1_{19}$ appear to be conformational; the two EGF-like motifs are required for the formation of dominant epitopes (61).

The immunogenicity of the native and recombinant proteins is strongly influenced by their structure and the adjuvant used. Immunization with native MSP-1 induces growth inhibitory antibodies but not if the protein has first been reduced (110). Similarly, immunization with an insect cell product (26) induces antibodies in rabbits that completely inhibit parasite growth in vitro, and the induction of functional activity depends on intact disulfide bonds in the immunogen. The antibodies induced are largely against the conserved block 17 sequences; they also cross-react with and inhibit the growth of different *P. falciparum* lines (92). The yeast-expressed $MSP\text{-}1_{19}$ could deplete these inhibitory antibodies, and in contrast, although its immunogenicity is dependent on the mouse strain and adjuvant used, the antibodies it induces do not inhibit growth in vitro (91). Protection in vivo or the induction of antibodies inhibitory in vitro is also dependent on the use of adjuvants. When native protein is used, some synthetic adjuvants and a monophosphoryl lipid A-liposome formulation is as good as or better than Freund's complete adjuvant at inducing growth inhibition (27, 93). Recently, the yeast MSP-1_{19} product was found to protect *Aotus nancymai* but not *Aotus vociferans* against challenge infection (106).

The available evidence suggests that antibody against MSP-1_{19} can inhibit parasite growth in vivo and in vitro. Other mechanisms may also be effective in vivo (106). However, the quality of the antibody (avidity, fine specificity, and possibly isotype) may be a crucial factor. Antibodies raised in mice in response to yeast-expressed MSP-1_{19} are not inhibitory in vitro (in contrast to those raised in response to the insect cell product) (91), and human antibodies selected on the first EGF motif are also not inhibitory despite competing in an enzyme-linked immunosorbent assay with a protective MAb (28). One explanation of why certain antibodies are effective is that they can inhibit the processing of MSP-1, whereas others are ineffective or even detrimental to the action of inhibitory antibodies. It has been suggested that the cleavage of MSP-1_{42} is a prerequisite for successful invasion, and an in vitro system to measure the conversion of MSP-1_{42} to MSP-1_{33} and MSP-1_{19} as an assay for the protease has been developed (10). In this assay, antibodies that inhibit parasite invasion inhibit the proteolytic cleavage (14), and this may be the mechanism by which they interfere with erythrocyte invasion. Some other antibodies are neutral, do not affect parasite invasion, and have no effect on the proteolytic cleavage. A third class of antibody do not interfere with parasite invasion but instead block the binding of inhibitory antibodies. The neutral antibodies bind different epitopes on MSP-1_{19} to the inhibitory antibodies, but the blocking antibodies bind to overlapping epitopes (190). Interestingly, antibodies to other parts of MSP-1 can also compete for binding to intact MSP-1 with the inhibitory antibodies.

Seroepidemiological studies of the human immune response to natural infection have also contributed to the evaluation of MSP-1 as a malaria vaccine candidate. Prospective longitudinal studies showed that both antibody levels and T-cell proliferative and cytokine responses to the C-terminal region of MSP-1 are associated with reduced susceptibility to clinical malaria (60, 145, 146). Cross-sectional studies showed an inverse association between antibodies to certain regions of the protein and the prevalence of malaria infection in some age groups (180). T-cell epitopes have been identified in both conserved (41, 160, 161) and variant (40, 183) regions of MSP-1. The human antibody response was short lived and strain specific (19, 70, 123), although the studies did not investigate the response to the conformational epitopes at the C terminus. A large percentage of adults living in areas of endemicity have antibodies to MSP-1_{19} (61, 183); these antibodies cross-react with different allelic products and are largely against conformational epitopes. The response to individual motifs is much less prevalent, but recently, a strong association between antibodies to the second EGF motif and resistance to both clinical malaria and high levels of parasitemia was reported (60).

MSP-2

A 43- to 56-kDa protein on the merozoite surface was identified; it is referred to here as MSP-2 (34, 63, 120, 141, 162, 169). The protein is recognized by antibodies that inhibit *P. falciparum* growth in vitro (34, 63). The protein is anchored on the surface by a glycosylphosphatidylinositol anchor but, unlike MSP-1_{19}, is not carried into the erythrocyte at invasion. Most studies of this protein have focused on its sequence and antigenic diversity.

The gene has been sequenced from a large number of lines and isolates (67, 113, 114, 163, 164, 175). The predicted size of the polypeptide is approximately 26 kDa, and the primary structure contains three distinct regions. After the putative signal peptide are about 20 residues that are highly conserved in different parasites. A central variable region comprises a block of tandem repeats that vary in number, length, and sequence and are flanked by variable nonrepetitive sequences. At the C terminus of the protein is a second highly conserved sequence of about 70 residues. Sequence comparisons show that all of the sequences fall into one of two families, although hybrid genes generated by intragenic recombination have also been detected (115). The two families are largely defined by the highly variable repetitive and nonrepetitive variable regions. One family is typified by the sequence in the FC27 line; the second family is typified by that in the IC1 and 3D7 lines. Studies of MSP-2 diversity in the field suggest that the two families are widely distributed; within one area, many different forms can be detected (66, 139, 165). Parallel studies of antibody reactivity defined two serotypes that correspond to the two sequence families (67, 164).

Studies of the importance of MSP-2 as a vaccine candidate have not been extensive. The epitopes recognized by some of the inhibitory MAbs have been mapped; these epitopes appear to be in the central variable region (67, 142). The immunogenicity of several peptides based on the MSP-2 sequence has also been studied with a variety of carriers and adjuvants, but few data regarding functional responses have been reported. Some peptides derived from the conserved sequence elicit antibody reactive with MSP-2 (97, 98), and T-cell epitopes recognized by mice (150) and humans (149) have been defined. No homologs of MSP-2 have been identified in other malaria species, although peptides from the conserved region of MSP-2 provide some protection when they are used to immunize mice challenged with *P. chabaudi* (153). In one cross-sectional survey (4), a high prevalence of antibodies to both forms of MSP-2 was detected. The presence of antibodies against nonrepetitive sequences was associated with fewer fever episodes and less anemia, but overall, antibody levels appeared to correlate with infection and the presence of parasites. A recently reported association between MSP-2 type and severity of disease (62) suggests that this protein is important in the acquisition of natural immunity.

To date, the evidence that MSP-2 is a vaccine candidate is largely based on its location and on MAb inhibition of parasite growth in vitro. The sequence and antigenic diversity also suggest that MSP-2 is under immunological pressure. However, no direct immunization studies have been reported. If MSP-2 is important in protection, then the possibility of developing a vaccine based on this molecule depends on using either a structure conserved between the families or two structures representative of the two forms.

SOLUBLE PROTEINS ASSOCIATED WITH THE MEROZOITE SURFACE

Several proteins are more or less loosely associated with the merozoite surface. These proteins are soluble secreted polypeptides exported to the parasitophorous vacuole during the later stages of intracellular parasite development. At least five such proteins have been identified. They are considered potential vaccine candidates because of their locations and their possible associations with the merozoite surface and because they have been implicated by additional immunological studies. For example, since these proteins are found in aggregates of merozoites formed when schizonts rupture in the presence of immune serum, antibodies to them may be involved in the agglutination of merozoites (111). The evidence suggesting that some of these proteins are important for vaccine development is limited.

SERA

Serine repeat antigen (SERA; also known as serine-rich protein and p113, p126, or p140) is a soluble protein synthesized by late-erythrocytic-stage parasites and secreted into the parasitophorous vacuole (51). At the end of schizogony, the protein is cleaved into 50- and 73-kDa fragments.

N-terminal sequencing showed that the 73-kDa fragment is composed of a 47-kDa polypeptide from the N-terminus disulfide linked to an 18-kDa fragment from the C terminus of SERA (48). Sequence similarities with cysteine proteases suggest that SERA itself may be a protease (82).

The *sera* gene encodes a polypeptide of approximately 985 amino acids that has a signal sequence but no hydrophobic membrane anchor (24, 104). Overall, the protein is very rich in serine, including a long stretch of contiguous serine residues; the number of serines in this stretch varies in different parasite lines. A second gene for a related protein without the polyserine region is also expressed (105).

The evidence that SERA may be a potential vaccine candidate is based largely on direct immunization data, although SERA-specific growth inhibitory MAbs have also been described (135), and the protein has been shown in antibody-merozoite aggregates (111). When parasite-derived protein (134) is used, *Saimiri* monkeys are partially protected, with a relatively low and delayed patent parasitemia, and recover without treatment. Similar results were obtained in a second study (50). Two recombinant antigens, SERA 1 (amino acids 24 through 285) and SERA N (residues 24 through 506), have been expressed in yeast cells for murine (8) and *Aotus* (96) immunization studies. Both proteins are recognized by a MAb that blocks parasite invasion. After immunization, some partial protection is observed. In additional studies, SERA 1 was partially protective on immunization if adjuvanted with Freund's complete or MF75.2 but not three other adjuvants (94), although MF75.2 adjuvant alone had some effect (95). Few studies of the human response to SERA have been carried out, although lymphoproliferative responses to recombinant and peptide fragments have been measured (148).

SPAM–MSP-3

The level of antibody-dependent cellular inhibition of parasite growth was used as an in vitro assay of collaboration between cytophilic antibodies and monocytes (17). Purified human immunoglobulin G (IgG) active in this assay was used to identify a nonrepetitive structure in a previously unknown protein, MSP-3 (127, 128). Subsequently, a protein with the same partial sequence was also reported as SPAM (secreted polymorphic antigen associated with merozoites), a soluble protein only loosely associated with the merozoite surface (118). The *spam* gene predicts a 43-kDa protein that is very hydrophilic and contains an unusual set of heptad repeats of the general structure Ala-X-X-Ala-X-X-X. This structure is known to form an amphipathic alpha-helix that gives bundles or coiled-coil structures, but its function in SPAM is unknown. The variation detected in different parasite lines is due to deletions and substitutions in nonrepetitive sequences.

GLURP

Although glutamate-rich protein (GLURP) has been examined quite extensively in field studies, there are no in vivo or in vitro studies to implicate it as a target of protective immunity. The apparent molecular mass of this soluble protein is 220 kDa, and it is expressed in liver and asexual blood stages. The gene encodes a 1,271-amino-acid polypeptide with a calculated molecular mass of 145 kDa, 22% glutamate residues, and two repetitive regions (16). The first repeat (R1) consists of six units with 15 or 16 residues each and two poorly conserved units of 50 residues each. The second repeat (R2) consists of 14 units of 19 to 20 residues each. The protein appears to be antigenically conserved in different isolates (58), but the number of repeat units may vary. Seroepidemiological data have been obtained in Liberia by using recombinant GLURP polypeptides (83). In the 5- to 9-year-old age group, high antibody levels correlated with low parasite densities. An age-dependent increase in IgM and IgG was detected in another study (59). The vaccine candidacy of GLURP is based on its location and these seroepidemiological studies that show a correlation between antibody response and low parasitemia.

Acidic Basic Repeat Antigen

Acidic basic repeat antigen, a soluble protein of the vacuole (32, 168), is also found in agglutinated clusters of merozoites (111). The *abra* gene codes for a 743-amino-acid hydrophilic polypeptide of 86 kDa (188). Two regions of repetitive sequence (eight hexapeptide repeats and sequences consisting mostly of Lys-Glu and Lys-Glu-Glu) occur close to the C terminus. This protein has been poorly studied, and the evidence to support its use as a vaccine candidate is weak.

S-Antigen

S-antigens are heat-stable, soluble proteins of the parasitophorous vacuole (42). Analysis of the gene indicates that it contains N- and C-terminal conserved domains and a large central repetitive amino acid block, whose sequence and antigenicity vary considerably in different parasites (9, 20, 38). Four families of S-antigen genes have been defined; the groupings are based on the N- and C-terminal sequences. The protein can be cross-linked to MSP-1 and therefore may be loosely associated with the merozoite surface (133). An S-antigen-specific MAb inhibits parasite growth in vitro (152), but there is no other evidence that these proteins are involved in protective immunity against the merozoite stage.

PROTEINS IN THE APICAL ORGANELLES

Proteins in the apical organelles are considered vaccine candidates because of experimental evidence that suggests that they are recognized by inhibitory antibodies. Some immunization data are also available. In addition, parasite ligands binding to erythrocyte receptors have been identified in these compartments, and it is possible that this interaction can be interfered with by an immune response. Two families of erythrocyte receptor-binding ligands have been identified. The first is a family of erythrocyte-binding proteins located in the micronemes (1, 2, 159). This family includes *P. falciparum* EBA 175 (175-kDa erythrocyte-binding antigen) and the Duffy-binding ligands of *Plasmodium knowlesi* and *P. vivax* (see below). The second family is less well characterized. Two *P. vivax* proteins with binding specificities for reticulocytes have been described (71). These proteins have an apical distribution, although their exact locations have not been established. A family of rhoptry proteins in *P. yoelii* that have weak but significant homology to one of these reticulocyte-binding proteins (102) bind to erythrocytes (128a). Antibodies to these rhoptry proteins confine the parasite to invasion of reticulocytes (69), which is consistent with the proteins being involved in the recognition of mature erythrocytes. Homologous proteins in *P. falciparum* have not been identified yet.

Microneme Proteins

Different species and clones of malaria parasites have different erythrocyte specificities. The parasite's ability to invade erythrocytes depends on the host species, the expression of particular molecules on the erythrocyte surface, and the age of the erythrocyte. Parasite ligand binding to specific receptors on the erythrocyte surface is an integral part of the invasion process. The best-characterized receptors are glycophorin A, which is a receptor for *P. falciparum*, and the Duffy blood group antigen, which is important in erythrocyte invasion by *P. vivax* and *P. knowlesi*. *P. falciparum* has a binding specificity for sialic acid residues on the surface glycoconjugate glycophorin. Cleavage of sialic acid from the erythrocyte surface prevents invasion by many *P. falciparum* strains, although some can invade sialic acid-deficient erythrocytes (122). The Duffy blood group antigen is a 35- to 46-kDa chemokine receptor (88). *P. falciparum* can invade Duffy-positive or -negative cells equally well, but human erythrocytes lacking the Duffy blood group are refractory to invasion by both *P. vivax* and *P. knowlesi.* Antibodies to antigenic determinants on the Duffy antigen (7) and the binding of chemokines (88) inhibit *P. knowlesi* and *P. vivax* invasion of human erythrocytes.

Soluble erythrocyte-binding proteins were identified in the supernatants of cultured parasites. A 175-kDa *P. falciparum* erythrocyte-binding antigen (EBA 175) (25) is a sialic acid-binding protein (103, 129) that preferentially binds to the Neu5Ac(α2,3)-Gal determinant on glycophorin A. Both modifications or deficiencies in glycophorin and trypsin treatment to remove this protein inhibit the invasion of human erythrocytes by *P. falciparum* and also abolish EBA 175 binding. Haynes et al. (78) identified a 135-kDa protein released from cultured *P. knowlesi* cells that bound to Duffy-positive but not Duffy-negative cells. Wertheimer and Barnwell (189) identified a similar 130- to 140-kDa *P. vivax* protein that bound only to cells susceptible to invasion. The effect of proteases on protein binding parallels the effect on invasion: chymotrypsin treatment cleaves the Duffy receptor, abolishes binding of the 130- to 140-kDa proteins, and inhibits invasion by both *P. vivax* and *P. knowlesi.* Immunolocalization studies indicate that these proteins are located in the micronemes (1, 159). No antibody reaction with rhoptries, dense granules, or the parasite surface was detected. Presumably, these proteins are released after contact between merozoite and erythrocyte, and this release may result in formation of the soluble forms detected in culture supernatants, which are smaller than the parasite-associated form, a finding consistent with shedding by proteolytic cleavage (130).

The gene for *P. falciparum* EBA 175 has been cloned and sequenced (2, 158). The gene for the *P. knowlesi* Duffy-binding protein was identified by expression screening with antibodies specific for the 135-kDa protein (1), and a homolog in *P. vivax* was cloned by cross-hybridization (65). Three homologous genes located on separate chromosomes are present in *P. knowlesi,* whereas *P. vivax* has a single gene. The products of two *P. knowlesi* genes are 135-kDa Duffy-binding proteins, and the third is a 138-kDa protein that binds only to rhesus erythrocytes. The overall gene structures and the amino acid identities between the *P. falciparum* 175-kDa protein and the Duffy blood group ligands of *P. vivax* and *P. knowlesi* suggest that the genes for all of these proteins belong to a single family. The open reading frame can be divided into seven blocks. Block I extends from the signal sequence and is a highly charged region. Block II is a cysteine-rich region in which the cysteines and many aromatic amino acids are conserved. Blocks III to V compose a central hydrophilic region containing some repetitive elements. Block VI is a second conserved cysteine-rich region, and block VII consists of a hydrophobic transmembrane sequence followed by a cytoplasmic domain. The *P. falciparum* protein contains two copies of the N-terminal cysteine-rich region in block II. Two EBA 175 alleles that differ in the sequence and location of a large insertion have been identified (185).

Block II has been shown by expression on the surface of COS cells to be the erythrocyte-binding domain (31, 157). When a cell-rosetting assay was used, this region bound directly to the surfaces of erythrocytes. No other region was active in this assay. The specificity of the binding was identical to that of the parasite proteins and mirrored the invasion properties of the parasites. Of the two copies of the cysteine-rich region in block II of EBA 175, only the second was functional. Binding of EBA 175 block II to glycophorin A was not dependent on sialic acid alone but required a contribution from the glycophorin A polypeptide backbone.

These results indicate that the specific interaction between parasite ligand (block II in the erythrocyte-binding proteins) and erythrocyte surface receptor is an essential step in invasion. The possibility that antibody against block II will interfere with this interaction needs to be investigated. Judging from the limited information available, EBA 175 block II is not variable, but variation has been detected in the corresponding region of the *P. knowlesi* protein and in field isolates of *P. vivax* (140, 182). Antibodies to a second region of EBA 175 also inhibit merozoite invasion (155, 156, 158). Some *P. falciparum* parasites can invade erythrocytes by a glycophorin A-independent pathway that does not involve EBA 175 (55); the parasite ligand involved has not been identified yet, but the results suggest a redundancy in the molecular interactions necessary for invasion and may have implications for a vaccine strategy designed to interfere with specific receptor-ligand interactions.

Rhoptry Proteins

The rhoptry organelles have a number of proteins, all with unknown functions but some implicated as targets of protective immunity. There is compartmentalization within the rhoptries, with different proteins not being uniformly distributed throughout. A 225-kDa protein derived from a 240-kDa precursor is located in the necks of the rhoptries (147), and a similar location has been described for apical membrane antigen 1 (AMA-1) (39). In the bodies of the rhoptries, two macromolecular complexes have been identified: a high-molecular-mass complex consisting of 150-, 130-, and 110-kDa proteins (56) and a low-molecular-mass complex consisting of rhoptry-associated protein 1 (RAP-1) and RAP-2 (80 and 40 kDa, respectively). The RAP-1–RAP-2 complex is associated with membranous structures released from merozoites (23). In addition, a 76-kDa serine protease (18), a 60-kDa protein (74), and a 52-kDa protein (171) may exist in these organelles.

AMA-1

AMA-1 is initially located in the necks of the rhoptry organelles (39) and then is at least partially transferred to the merozoite surface at about the time of schizont rupture. Both immunofluorescence studies and surface radiolabeling indicate that AMA-1 is located on the merozoite surface at the apical end of the parasite (174, 186). In early studies with *P. knowlesi*, MAbs that inhibit parasite invasion in vitro were identified, and these antibodies reacted with a 66-kDa protein (46), recognizing conformational, reduction-sensitive epitopes. The protein is synthesized late in schizogony and is processed to 44- and 42-kDa fragments at the time of schizont rupture. Fab fragments of the antibodies are more inhibitory than intact immunoglobulin (176). The antibodies do not block merozoite attachment to erythrocytes but appear to block apical reorientation. Partial protection against *P. knowlesi* infection in rhesus monkeys was observed in immunization experiments with affinity-purified antigen and saponin as adjuvant (47). *P. falciparum* AMA-1 is also synthesized late in the cycle. It has an apparent size of 83 kDa, and an N-terminal peptide is removed to produce a fragment at 62 kDa (39, 125). The two forms are initially located at the merozoite apex, but after merozoite release, the 66-kDa form is also detectable on the parasite surface. AMA-1 is taken into erythrocytes at invasion and can be detected in early ring forms (125).

The structure of *P. falciparum* AMA-1 predicted from the gene indicates that the protein is a type 1 integral membrane protein of 622 residues of 71,929 Da (136, 178). A 546-residue N-terminal "ectodomain" is followed by a 21-amino-acid transmembrane sequence and a 55-residue C-terminal "cytoplasmic" domain. The *ama-1* gene has also been cloned for other malaria species: *P. vivax* (30), *P. knowlesi* (186, 187), *P. chabaudi* (116), and *Plasmodium fragile* (137). The *P. falciparum* protein contains 53 amino acids close to the N terminus that the AMA-1 proteins of these other species lack. Otherwise, there is extensive homology, including 16 cysteines on the N-terminal side of the transmembrane domain, which suggests a highly disulfide-bonded structure. The cleavage event that generates the 62-kDa fragment probably occurs before the first conserved cysteine. Although the sequence is conserved within *P. falciparum* overall, in a comparison of five different alleles, 57 nucleotide differences resulting in 52 amino acid substitutions were found, suggesting that the protein is under selective pressure.

The *P. falciparum* protein has been expressed as a full-length protein in insect cells (126), purified by MAb affinity, and used in a seroepidemiological study (177). A very high prevalence of IgG (close to 100%) was detected in human populations in Guinea-Bissau and Senegal, but no correlation with parasite load or fever was observed. One immunization experiment with recombinant AMA-1 has been reported. *Saimiri* monkeys immunized with *P. fragile* AMA-1 expressed in baculovirus plus ISA 720, a Seppic adjuvant, largely recovered without needing treatment, whereas the controls had normal infections needing treatment (35). On rechallenge with *P. falciparum*, no parasites were seen in the immunized group, suggesting some heterologous protection.

RAP-1 and RAP-2

Several proteins have been identified in the rhoptries of *P. falciparum*. Two of these, RAP-1 (an 80-kDa protein) and RAP-2 (a 40-kDa protein), are tightly associated in a complex that has been localized by immunoelectron microscopy (39) and immunofluorescence with MAbs to the bodies of the rhoptry organelles of the schizont and merozoites. After invasion, the proteins are associated with the parasitophorous vacuole membrane-parasite plasma membrane (33). The RAP-1–RAP-2 complex contains a number of polypeptides derived from RAP-1 in addition to the distinct RAP-2 protein (90). The initial RAP-1 translation product of 84 kDa appears to be rapidly processed to an 80-kDa species (39). Further processing of part of the 80-kDa species to a 65-kDa polypeptide appears to occur late in schizogony, but the 65-kDa species is not present in free merozoites or ring stages. Additional 77-, 70-, and 40-kDa species may be derived from the 80-kDa species as artifacts during purification. The 80-, 77-, 70-, and 65-kDa polypeptides apparently are related, because they all react with a single MAb. The 80-kDa protein can be detected in ring stages. The complex is the target of inhibitory antibodies, which react with RAP-1 (77). The protein, purified by affinity chromatography and electroelution, partially protects *Saimiri* monkeys against a challenge infection (143).

The genes for both RAP-1 and RAP-2 have been cloned and sequenced. The *rap-1* gene codes for a 782-amino-acid polypeptide with a predicted size of 90 kDa (144). Close to the N terminus is a region very rich in serines. Eight cysteines are found in a region flanked on either side by sequences with the potential for amphipathic-helix formation. The *rap-2* gene encodes a protein of 398 amino acids (151). After cleavage of a signal peptide, the mature protein is 377 residues with a predicted mass of 44 kDa. It is a basic protein that lacks repetitive sequences. It contains some very hydrophobic regions, but none of these regions have the characteristics of a transmembrane domain.

The evidence that RAP-1 or RAP-2 is the target of an immune response that inhibits merozoite invasion is not strong. Several inhibitory

MAbs recognize a linear epitope adjacent to the protease cleavage site for conversion of the 80-kDa polypeptide to the 65-kDa fragment (77), but the significance of this epitope is unknown. RAP-1 has been expressed in recombinant form, and a high prevalence of human antibodies reacting with these polypeptides was detected in a cross-sectional survey (89). How MAbs to this protein are inhibitory in vitro or how protective immunity against the protein can be induced by immunization is not known. Perhaps the protein is exposed at some time during the invasion process.

CONCLUSIONS

Despite considerable effort to define protein antigens of the merozoite that could be the targets of a protective immune response, the evidence that many of these antigens are important is weak. MSP-1 has been studied in the most detail, and the evidence that a response to the C terminus of this molecule may prevent erythrocyte invasion is now considerable. Nevertheless, this idea has not been tested in human vaccine trials, and the importance of correct adjuvant and induction of an appropriate response still needs to be addressed. Understanding the biological functions of parasite proteins such as ligands for erythrocyte binding may facilitate a more rational approach to vaccine development.

REFERENCES

1. **Adams, J. H., D. E. Hudson, M. Torii, G. E. Ward, T. E. Wellems, M. Aikawa, and L. H. Miller.** 1990. The Duffy receptor family of *Plasmodium knowlesi* is located within the micronemes of invasive malaria merozoites. *Cell* **63:**141–153.
2. **Adams, J. H., B. K. L. Sim, S. A. Dolan, X. D. Fang, D. C. Kaslow, and L. H. Miller.** 1992. A family of erythrocyte binding proteins of malaria parasites. *Proc. Natl. Acad. Sci. USA* **89:**7085–7089.
3. **Aikawa, M., L. H. Miller, J. Johnson, and J. Rabbege.** 1978. Erythrocyte entry by malarial parasites: a moving junction between erythrocyte and parasite. *J. Cell Biol.* **77:**72–78.
4. **al-Yaman, F., B. Genton, R. F. Anders, M. Falk, T. Triglia, D. Lewis, J. Hii, H. P. Beck, and M. P. Alpers.** 1994. Relationship between humoral response to *Plasmodium falciparum* merozoite surface antigen-2 and malaria morbidity in a highly endemic area of Papua New Guinea. *Am. J. Trop. Med. Hyg.* **51:**593–602.
5. **Bannister, L. H., and A. R. Dluzewski.** 1990. The ultrastructure of red cell invasion in malaria infections—a review. *Blood Cells* **16:**257–292.
6. **Bannister, L. H., G. H. Mitchell, G. A. Butcher, and E. D. Dennis.** 1986. Lamellar membranes associated with rhoptries in erythrocytic merozoites of *Plasmodium knowlesi*: a clue to the mechanism of invasion. *Parasitology* **92:**291–303.
7. **Barnwell, J. W., M. E. Nichols, and P. Rubinstein.** 1989. *In vitro* evaluation of the role of the Duffy blood group in erythrocyte invasion by *Plasmodium vivax*. *J. Exp. Med.* **169:**1795–1802.
8. **Barr, P. J., J. Inselburg, K. M. Green, J. Kansopon, B. K. Hahm, H. L. Gibson, C. T.**

Leeng, D. J. Bzik, W. Li, and I. C. Bathurst. 1991. Immunogenicity of recombinant *Plasmodium falciparum* SERA proteins in rodents. *Mol. Biochem. Parasitol.* **45:**159–170.

9. **Bickle, Q., R. F. Anders, K. Day, and R. L. Coppel.** 1993. The S-antigen of *Plasmodium falciparum* repertoire and origin of diversity. *Mol. Biochem. Parasitol.* **61:**189–196.
10. **Blackman, M. J., J. A. Chappel, S. Shai, and A. A. Holder.** 1993. A conserved parasite serine protease processes the *Plasmodium falciparum* merozoite surface protein-1. *Mol. Biochem. Parasitol.* **62:**103–114.
11. **Blackman, M. J., H. G. Heidrich, S. Donachie, J. S. McBride, and A. A. Holder.** 1990. A single fragment of a malaria merozoite surface protein remains on the parasite during red cell invasion and is the target of invasion-inhibiting antibodies. *J. Exp. Med.* **172:**379–382.
12. **Blackman, M. J., and A. A. Holder.** 1992. Secondary processing of the *Plasmodium falciparum* merozoite surface protein-1 (MSP1) by a calcium-dependent membrane-bound serine protease—shedding of $MSP1_{33}$ as a noncovalently associated complex with other fragments of MSP1. *Mol. Biochem. Parasitol.* **50:**307–316.
13. **Blackman, M. J., I. T. Ling, S. C. Nicholls, and A. A. Holder.** 1991. Proteolytic processing of the *Plasmodium falciparum* merozoite surface protein-1 produces a membrane-bound fragment containing two epidermal growth factor-like domains. *Mol. Biochem. Parasitol.* **49:**29–34.
14. **Blackman, M. J., T. J. Scott-Finnigan, S. Shai, and A. A. Holder.** 1994. Antibodies inhibit the protease-mediated processing of a malaria merozoite surface protein. *J. Exp. Med.* **180:**389–393.
15. **Blackman, M. J., H. Whittle, and A. A. Holder.** 1991. Processing of the *Plasmodium falciparum* major merozoite surface protein-1—identification of a 33-kilodalton secondary processing product which is shed prior to erythrocyte invasion. *Mol. Biochem. Parasitol.* **49:**35–44.
16. **Borre, M. B., M. Dziegiel, B. Hogh, E. Petersen, K. Rieneck, E. Riley, J. F. Meis, M. Aikawa, K. Nakamura, M. Harada, A. Wind, P. H. Jakobsen, J. Cowland, S. Jepsen, N. H. Axelsen, and J. Vuust.** 1991. Primary structure and localization of a conserved immunogenic *Plasmodium falciparum* glutamate rich protein (GLURP) expressed in both the preerythrocytic and erythrocytic stages of the vertebrate life cycle. *Mol. Biochem. Parasitol.* **49:**119–132.
17. **Bouharountayoun, H., P. Attanath, A. Sabchareon, T. Chongsuphajaisiddhi, and P. Druilhe.** 1990. Antibodies that protect humans against *Plasmodium falciparum* blood stages do not on their own inhibit parasite growth and invasion *in vitro*, but act in cooperation with monocytes. *J. Exp. Med.* **172:**1633–1641.
18. **Braun-Breton, C., T. L. Rosenberry, and L. P. da Silva.** 1988. Induction of the proteolytic activity of a membrane protein in *Plasmodium falciparum* by phosphatidyl inositol-specific phospholipase C. *Nature* (London) **332:**457–459.
19. **Brown, A. E., H. K. Webster, J. A. Lyon, A. W. Thomas, B. Permpanich, and M. Gross.** 1991. Characterization of naturally acquired antibody responses to a recombinant fragment from the N-terminus of *Plasmodium falciparum* glycoprotein-195. *Am. J. Trop. Med. Hyg.* **45:**567–573.
20. **Brown, H., D. J. Kemp, N. Barzaga, G. V. Bown, R. F. Anders, and R. L. Coppel.** 1987. Sequence variation in S-antigen genes of *Plasmodium falciparum*. *Mol. Biol. Med.* **4:**365–376.
21. **Burghaus, P. A., and A. A. Holder.** 1994. Expression of the 19-kilodalton carboxy-terminal fragment of the *Plasmodium falciparum* merozoite surface protein-1 in *Escherichia coli* as a correctly folded protein. *Mol. Biochem. Parasitol.* **64:**165–169.
22. **Burns, J. M., W. R. Majarian, J. F. Young, T. M. Daly, and C. A. Long.** 1989. A protective monoclonal antibody recognizes an epitope in the carboxyl-terminal cysteine-

rich domain in the precursor of the major merozoite surface antigen of the rodent malarial parasite, *Plasmodium yoelii*. *J. Immunol.* **143:**2670–2676.

23. **Bushell, G. R., L. T. Ingram, C. A. Fardoulys, and J. A. Cooper.** 1988. An antigenic complex in the rhoptries of *Plasmodium falciparum*. *Mol. Biochem. Parasitol.* **28:**105–112.
24. **Bzik, D. J., W.-B. Li, T. Horii, and J. Inselburg.** 1988. Amino acid sequence of the serine-repeat antigen (SERA) of *Plasmodium falciparum* determined from cloned cDNA. *Mol. Biochem. Parasitol.* **30:**279–288.
25. **Camus, D., and T. J. Hadley.** 1985. A *Plasmodium falciparum* antigen that binds to host erythrocytes and merozoites. *Science* **230:**553–556.
26. **Chang, S. P., H. L. Gibson, C. T. Leeng, P. J. Barr, and G. S. N. Hui.** 1992. A carboxyl-terminal fragment of *Plasmodium falciparum* gp195 expressed by a recombinant baculovirus induces antibodies that completely inhibit parasite growth. *J. Immunol.* **149:**548–555.
27. **Chang, S. P., C. M. Nikaido, A. C. Hashimoto, C. Q. Hashior, B. T. Yokota, and G. S. N. Hui.** 1994. Regulation of antibody specificity to *Plasmodium falciparum* merozoite surface protein-1 by adjuvant and MHC haplotype. *J. Immunol.* **152:**3483–3490.
28. **Chappel, J. A., A. F. Egan, E. M. Riley, P. Druilhe, and A. A. Holder.** 1994. Naturally acquired human antibodies which recognize the first epidermal growth factor-like module in the *Plasmodium falciparum* merozoite surface protein-1 do not inhibit parasite growth in vitro. *Infect. Immun.* **62:**4488–4494.
29. **Chappel, J. A., and A. A. Holder.** 1993. Monoclonal antibodies that inhibit *Plasmodium falciparum* invasion *in vitro* recognise the first growth factor-like domain of merozoite surface protein-1. *Mol. Biochem. Parasitol.* **60:**303–312.
30. **Cheng, Q., and A. Saul.** 1994. Sequence analysis of the apical membrane antigen I (AMA-1) of *Plasmodium vivax*. *Mol. Biochem. Parasitol.* **65:**183–187.
31. **Chitnis, C. E., and L. H. Miller.** 1994. Identification of the erythrocyte binding domains of *Plasmodium vivax* and *Plasmodium knowlesi* proteins involved in erythrocyte invasion. *J. Exp. Med.* **180:**497–506.
32. **Chulay, J. D., J. A. Lyon, J. D. Haynes, A. I. Meierovics, C. T. Atkinson, and M. Aikawa.** 1987. Monoclonal antibody characterization of *Plasmodium falciparum* antigens in immune complexes formed when schizonts rupture in the presence of immune serum. *J. Immunol.* **139:**2768–2774.
33. **Clark, J. T., R. Anand, T. Akoglu, and J. S. McBride.** 1987. Identification and characterisation of protein associated with the rhoptry organelles of *Plasmodium falciparum* merozoites. *Parasitol. Res.* **73:**425–434.
34. **Clark, J. T., S. Donachie, R. Anand, C. F. Wilson, H.-G. Heidrich, and J. S. McBride.** 1988. 46-53 kD glycoprotein from the surface of *Plasmodium falciparum* merozoites. *Mol. Biochem. Parasitol.* **32:**15–24.
35. **Collins, W. E., D. Pye, P. E. Crewther, K. L. Vandenberg, G. G. Galland, A. J. Sulzer, D. J. Kemp, S. J. Edwards, R. L. Coppel, J. S. Sullivan, C. L. Morris, and R. E. Anders.** 1994. Protective immunity induced in squirrel monkeys with recombinant apical membrane antigen-1 of *Plasmodium fragile*. *Am. J. Trop. Med. Hyg.* **51:**711–719.
36. **Cooper, J. A.** 1993. Merozoite surface antigen-I of Plasmodium. *Parasitol. Today* **9:**50–54.
37. **Cooper, J. A., L. T. Cooper, and A. J. Saul.** 1992. Mapping of the region predominantly recognized by antibodies to the *Plasmodium falciparum* merozoite surface antigen MSA-1. *Mol. Biochem. Parasitol.* **51:**301–312.
38. **Cowman, A. F., R. B. Saint, R. L. Coppel, G. V. Brown, R. F. Anders, and D. J. Kemp.** 1985. Conserved sequences flank variable tandem repeats in two S-antigen genes of *Plasmodium falciparum*. *Cell* **40:**775–783.

39. **Crewther, P. E., J. G. Culvenor, A. Silva, J. A. Cooper, and R. F. Anders.** 1990. *Plasmodium falciparum*—2 antigens of similar size are located in different compartments of the rhoptry. *Exp. Parasitol.* **70:**193–206.
40. **Crisanti, A., K. Fruh, H. M. Müller, and H. Bujard.** 1990. The T-cell reactivity against the major merozoite protein of *Plasmodium falciparum. Immunol. Lett.* **25:**143–148.
41. **Crisanti, A., H.-M. Müller, C. Hilbich, F. Sinigaglia, H. Matile, M. McKay, J. Scaife, et al.** 1988. Epitopes recognized by human T cells map within the conserved part of the GP190 of *P. falciparum. Science* **240:**1324–1326.
42. **Culvenor, J. G., and P. E. Crewther.** 1990. S-antigen localization in the erythrocytic stages of *Plasmodium falciparum. J. Protozool.* **37:**59–65.
43. **Culvenor, J. G., K. P. Day, and R. F. Anders.** 1991. *Plasmodium falciparum* ring-infected erythrocyte surface antigen is released from merozoite dense granules after erythrocyte invasion. *Infect. Immun.* **59:**1183–1187.
44. **Daly, T. M., and C. A. Long.** 1993. A recombinant 15-kDa carboxyl-terminal fragment of *Plasmodium yoelii yoelii* 17XL merozoite surface protein-1 induces a protective immune response in mice. *Infect. Immun.* **61:**2462–2467.
45. **Daly, T. M., and C. A. Long.** 1995. Humoral response to a carboxyl-terminal region of the merozoite surface protein-1 plays a predominant role in controlling blood-stage infection in rodent malaria. *J. Immunol.* **155:**236–243.
46. **Deans, J. A., T. Alderson, A. W. Thomas, G. H. Mitchell, E. S. Lennox, and S. Cohen.** 1982. Rat monoclonal antibodies which inhibit the *in vitro* multiplication of *Plasmodium knowlesi. Clin. Exp. Immunol.* **49:**297–309.
47. **Deans, J. A., A. M. Knight, W. C. Jean, A. P. Waters, S. Cohen, and G. H. Mitchell.** 1988. Vaccination trials in rhesus monkeys with a minor, invariant *Plasmodium knowlesi* 66kDa antigen. *Parasite Immunol.* **10:**535–552.
48. **Debrabant, A., P. Maes, P. Delplace, J. F. Dubremetz, A. Tartar, and D. Camus.** 1992. Intramolecular mapping of *Plasmodium falciparum* p126 proteolytic fragments by N-terminal amino acid sequencing. *Mol. Biochem. Parasitol.* **53:**89–95.
49. **Deleersnijder, W., D. Hendrix, N. Bendahman, J. Hanegreefs, L. Brijs, C. Hamerscasterman, and R. Hamers.** 1990. Molecular cloning and sequence analysis of the gene encoding the major merozoite surface antigen of *Plasmodium chabaudi chabaudi* IP-PC1. *Mol. Biochem. Parasitol.* **43:**231–244.
50. **Delplace, P., A. Bhatia, M. Cagnard, D. Camus, G. Colombet, B. Debrabant, A. Haq, J. Weber, and A. Vernes.** 1988. Protein p126: a parasitophorous vacuole antigen associated with the release of *Plasmodium falciparum* merozoites. *Biol. Cell* **64:**215–221.
51. **Delplace, P., B. Fortier, G. Tronchin, J. F. Dubermetz, and A. Vernes.** 1987. Localization, biosynthesis, processing and isolation of a major 126 kDa antigen of the parasitophorous vacuole of *Plasmodium falciparum. Mol. Biochem. Parasitol.* **23:**193–201.
52. **Delportillo, H. A., S. Longacre, E. Khouri, and P. H. David.** 1991. Primary structure of the merozoite surface antigen-1 of *Plasmodium vivax* reveals sequences conserved between different Plasmodium species. *Proc. Natl. Acad. Sci. USA* **88:**4030–4034.
53. **Diggs, C. L., W. R. Ballou, and L. H. Miller.** 1993. The major merozoite surface protein as a malaria vaccine target. *Parasitol. Today* **9:**300–302.
54. **Dluzewski, A. R., G. H. Mitchell, P. R. Fryer, S. Griffiths, R. J. M. Wilson, and W. B. Gratzer.** 1992. Origins of the parasitophorous vacuole membrane of the malaria parasite, *Plasmodium falciparum,* in human red blood cells. *J. Cell Sci.* **102:**527–532.
55. **Dolan, S. A., J. L. Proctor, D. W. Alling, Y. Okubo, T. E. Wellems, and L. H. Miller.** 1994. Glycophorin B as an EBA-175 independent *Plasmodium falciparum* receptor of human erythrocytes. *Mol. Biochem. Parasitol.* **64:**55–63.
56. **Doury, J. C., S. Bonnefoy, N. Roger, J. F. Dubremetz, and O. Mercereau-Puijalon.**

1994. Analysis of the high molecular weight rhoptry complex of *Plasmodium falciparum* using monoclonal antibodies. *Parasitology* **108:**269–280.

57. **Dvorak, J. A., L. H. Miller, W. C. Whithouse, and T. Shiroishi.** 1975. Invasion of erythrocytes by malaria merozoites. *Science* **187:**748–750.
58. **Dziegiel, M., M. B. Borre, S. Jepsen, B. Hogh, E. Petersen, and J. Vuust.** 1991. Recombinant *Plasmodium falciparum* glutamate rich protein—purification and use in enzyme-linked immunosorbent assay. *Am. J. Trop. Med. Hyg.* **44:**306–313.
59. **Dziegiel, M., P. Rowe, S. Bennett, S. J. Allen, O. Olerup, A. Gottschau, M. Borre, and E. M. Riley.** 1993. Immunoglobulin-M and immunoglobulin-G antibody responses to *Plasmodium falciparum* glutamate-rich protein: correlation with clinical immunity in Gambian children. *Infect. Immun.* **61:**103–108.
60. **Egan, A., J. Morris, G. Barnish, S. Allen, B. Greenwood, D. Kaslow, A. Holder, and E. Riley.** Clinical immunity to *Plasmodium falciparum* malaria is associated with serum antibodies to the 19 kDa C-terminal fragment of the merozoite surface antigen, PfMSP-1. *J. Infect. Dis.*, in press.
61. **Egan, A. F., J. A. Chappel, P. A. Burghaus, J. S. Morris, J. S. McBride, A. A. Holder, D. C. Kaslow, and E. M. Riley.** 1995. Serum antibodies from malaria-exposed people recognize conserved epitopes formed by the two epidermal growth factor motifs of MSP1(19), the carboxy-terminal fragment of the major merozoite surface protein of *Plasmodium falciparum. Infect. Immun.* **63:**456–466.
62. **Engelbrecht, F., I. Felger, B. Genton, M. Alpers, and H. P. Beck.** 1995. *Plasmodium falciparum*: malaria morbidity is associated with specific merozoite surface antigen 2 genotypes. *Exp. Parasitol.* **81:**90–96.
63. **Epping, R. J., S. D. Goldstone, L. T. Ingram, J. A. Upcroft, R. Ramasamy, J. A. Cooper, G. R. Bushell, and H. M. Geysen.** 1988. An epitope recognised by inhibitory monoclonal antibodies that react with a 51 kilodalton merozoite surface antigen in *Plasmodium falciparum. Mol. Biochem. Parasitol.* **28:**1–10.
64. **Etlinger, H. M., P. Caspers, H. Matile, H. J. Schönfeld, D. Stüber, and B. Takacs.** 1991. Ability of recombinant or native proteins to protect monkeys against heterologous challenge with *Plasmodium falciparum. Infect. Immun.* **59:**3498–3503.
65. **Fang, X. D., D. C. Kaslow, J. H. Adams, and L. H. Miller.** 1991. Cloning of the *Plasmodium vivax* Duffy receptor. *Mol. Biochem. Parasitol.* **44:**125–132.
66. **Felger, I., L. Tavul, S. Kabintik, V. Marshall, B. Genton, M. Alpers, and H. P. Beck.** 1994. *Plasmodium falciparum*: extensive polymorphism in merozoite surface antigen 2 alleles in an area with endemic malaria in Papua New Guinea. *Exp. Parasitol.* **79:**106–116.
67. **Fenton, B., J. T. Clark, C. M. A. Khan, J. V. Robinson, D. Walliker, R. Ridley, J. G. Scaife, and J. S. McBride.** 1991. Structural and antigenic polymorphism of the 35- to 48-kDa merozoite surface antigen (MSA-2) of the malaria parasite *Plasmodium falciparum. Mol. Cell. Biol.* **11:**963–971.
68. **Freeman, R. R., and A. A. Holder.** 1983. Characteristics of the protective response of BALB/c mice immunized with a purified *Plasmodium yoelii* schizont antigen. *Clin. Exp. Immunol.* **54:**609–616.
69. **Freeman, R. R., A. J. Trejdosiewicz, and G. A. M. Cross.** 1980. Protective monoclonal antibodies recognising stage-specific merozoite antigens of a rodent malaria parasite. *Nature* (London) **284:**366–368.
70. **Fruh, K., O. Doumbo, H. M. Müller, O. Koita, J. McBride, A. Crisanti, Y. Toure, and H. Bujard.** 1991. Human antibody response to the major merozoite surface antigen of *Plasmodium falciparum* is strain specific and short-lived. *Infect. Immun.* **59:**1319–1324.

71. **Galinski, M. R., C. C. Medina, P. Ingravallo, and J. W. Barnwell.** 1992. A reticulocyte-binding protein complex of *Plasmodium vivax* merozoites. *Cell* **69:**1213–1226.
72. **Gerold, P., L. Schofield, M. J. Blackman, A. A. Holder, and R. T. Schwarz.** Structural analysis of the glycosyl-phosphatidylinositol membrane anchor of the merozoite surface proteins-1 and -2 of *Plasmodium falciparum. Mol. Biochem. Parasitol.* in press.
73. **Gibson, H. L., J. E. Tucker, D. C. Kaslow, A. U. Krettli, W. E. Collins, M. C. Kiefer, I. C. Bathurst, and P. J. Barr.** 1992. Structure and expression of the gene for Pv200, a major blood-stage surface antigen of *Plasmodium vivax. Mol. Biochem. Parasitol.* **50:**325–334.
74. **Grellier, P., E. Precigout, A. Valentin, B. Carcy, and J. Schrevel.** 1994. Characterization of a new 60 kDa apical protein of *Plasmodium falciparum* merozoite expressed in late schizogony. *Biol. Cell* **82:**129–138.
75. **Hadley, T. J., F. W. Klotz, and L. H. Miller.** 1986. Invasion of erythrocytes by malaria parasites: a cellular and molecular overview. *Annu. Rev. Microbiol.* **40:**451–477.
76. **Haldar, K., M. A. J. Ferguson, and G. A. M. Cross.** 1985. Acylation of a *Plasmodium falciparum* merozoite surface antigen via sn-1,2-diacyl glycerol. *J. Biol. Chem.* **260:**4969–4974.
77. **Harnyuttanakorn, P., J. S. McBride, S. Donachie, H. G. Heidrich, and R. G. Ridley.** 1992. Inhibitory monoclonal antibodies recognise epitopes adjacent to a proteolytic cleavage site on the RAP-1 protein of *Plasmodium falciparum. Mol. Biochem. Parasitol.* **55:**177–186.
78. **Haynes, J. D., J. P. Dalton, F. W. Klotz, M. H. McGinniss, T. J. Hadley, D. E. Hudson, and L. H. Miller.** 1988. Receptor-like specificity of a *Plasmodium knowlesi* malarial protein that binds to Duffy antigen ligands on erythrocytes. *J. Exp. Med.* **167:**1873–1881.
79. **Heidrich, H. G., W. Strych, and J. E. K. Mrema.** 1983. Identification of surface and internal antigens from spontaneously released *Plasmodium falciparum* merozoites by radio-iodination and metabolic labelling. *Z. Parasitenkd.* **69:**715–725.
80. **Herrera, M. A., F. Rosero, S. Herrera, P. Caspers, D. Rotmann, F. Sinigaglia, and U. Certa.** 1992. Protection against malaria in *Aotus* monkeys immunized with a recombinant blood-stage antigen fused to a universal T-cell epitope: correlation of serum gamma-interferon levels with protection. *Infect. Immun.* **60:**154–158.
81. **Herrera, S., M. A. Herrera, B. L. Perlaza, Y. Burki, P. Caspers, H. Dobeli, D. Rotmann, and U. Certa.** 1990. Immunization of *Aotus* monkeys with *Plasmodium falciparum* blood-stage recombinant proteins. *Proc. Natl. Acad. Sci. USA* **87:**4017–4021.
82. **Higgins, D. G., D. J. Mcconnell, and P. M. Sharp.** 1989. Malarial proteinase. *Nature* (London) **340:**604.
83. **Hogh, B., E. Petersen, M. Dziegiel, K. David, A. Hanson, M. Borre, A. Holm, J. Vuust, and S. Jepsen.** 1992. Antibodies to a recombinant glutamate-rich *Plasmodium falciparum* protein—evidence for protection of individuals living in a holoendemic area of Liberia. *Am. J. Trop. Med. Hyg.* **46:**307–313.
84. **Holder, A. A.** 1988. The precursor to major merozoite surface antigens: structure and role in immunity. *Prog. Allergy* **41:**72–97.
85. **Holder, A. A., and M. J. Blackman.** 1994. What is the function of MSP-1 on the malaria merozoite? *Parasitol. Today* **10:**182–184.
86. **Holder, A. A., and R. R. Freeman.** 1981. Immunization against blood-stage rodent malaria using purified antigens. *Nature* (London) **294:**361–366.
87. **Holder, A. A., R. R. Freeman, and S. C. Nicholls.** 1988. Immunization against *Plasmodium falciparum* with recombinant polypeptides produced in *Escherichia coli. Parasite Immunol.* **10:**607–617.

88. **Horuk, R., C. E. Chitnis, W. C. Darbonne, T. J. Colby, A. Rybicki, T. J. Hadley, and L. H. Miller.** 1993. A receptor for the malarial parasite *Plasmodium vivax*—the erythrocyte chemokine receptor. *Science* **261:**1182–1184.

89. **Howard, R. F., J. B. Jensen, and H. L. Franklin.** 1993. Reactivity profile of human anti-82-kDa rhoptry protein antibodies generated during natural infection with *Plasmodium falciparum*. *Infect. Immun.* **61:**2960–2965.

90. **Howard, R. F., and R. T. Reese.** 1990. *Plasmodium falciparum*—hetero-oligomeric complexes of rhoptry polypeptides. *Exp. Parasitol.* **71:**330–342.

91. **Hui, G. S. N., W. L. Gosnell, S. E. Case, C. Hashiro, C. Nikaido, A. Hashimoto, and D. C. Kaslow.** 1994. Immunogenicity of the C-terminal 19kDa fragment of the *Plasmodium falciparum* merozoite surface protein 1 (MSP1) YMSP119 expressed in *S. cerevisiae*. *J. Immunol.* **153:**2544–2553.

92. **Hui, G. S. N., C. Hashiro, C. Nikaido, S. E. Case, A. Hashimoto, H. Gibson, P. J. Barr, and S. P. Chang.** 1993. Immunological cross-reactivity of the C-terminal 42-kDa fragment of *Plasmodium falciparum* merozoite surface protein-1 expressed in baculovirus. *Infect. Immun.* **61:**3403–3411.

93. **Hui, G. S. N., L. Q. Tam, S. P. Chang, S. E. Case, C. Hashiro, W. A. Siddiqui, T. Shiba, S. Kusumoto, and S. Kotani.** 1991. Synthetic low-toxicity muramyl dipeptide and monophosphoryl lipid-A replace Freund's complete adjuvant in inducing growth-inhibitory antibodies to the *Plasmodium falciparum* major merozoite surface protein, gp195. *Infect. Immun.* **59:**1585–1591.

94. **Inselburg, J., I. C. Bathurst, J. Kansopon, G. L. Barchfeld, P. J. Barr, and R. N. Rossan.** 1993. Protective immunity induced in *Aotus* monkeys by a recombinant SERA protein of *Plasmodium falciparum*: adjuvant effects on induction of protective immunity. *Infect. Immun.* **61:**2041–2047.

95. **Inselburg, J., I. C. Bathurst, J. Kansopon, P. J. Barr, and R. Rossan.** 1993. Protective immunity induced in *Aotus* monkeys by a recombinant SERA protein of *Plasmodium falciparum*: further studies using SERA 1 and MF75.2 adjuvant. *Infect. Immun.* **61:**2048–2052.

96. **Inselburg, J., D. J. Bzik, W. B. Li, K. M. Green, J. Kansopon, B. K. Hahm, I. C. Bathurst, P. J. Barr, and R. N. Rossan.** 1991. Protective immunity induced in *Aotus* monkeys by recombinant SERA proteins of *Plasmodium falciparum*. *Infect. Immun.* **59:**1247–1250.

97. **Jones, G. L., H. M. Edmundson, R. Lord, L. Spencer, R. Mollard, and A. J. Saul.** 1991. Immunological fine structure of the variable and constant regions of a polymorphic malarial surface antigen from *Plasmodium falciparum*. *Mol. Biochem. Parasitol.* **48:**1–10.

98. **Jones, G. L., L. Spencer, R. Lord, R. Mollard, D. Pye, and A. Saul.** 1990. Peptide vaccines derived from a malarian surface antigen—effects of dose and adjuvants on immunogenicity. *Immunol. Lett.* **24:**253–260.

99. **Jongwutiwes, S., K. Tanabe, and H. Kanbara.** 1993. Sequence conservation in the C-terminal part of the precursor to the major merozoite surface proteins (MSP1) of *Plasmodium falciparum* from field isolates. *Mol. Biochem. Parasitol.* **59:**95–100.

100. **Kang, Y., and C. A. Long.** 1995. Sequence heterogeneity of the C-terminal, cys-rich region of the merozoite surface protein-1 (MSP-1) in field samples of *Plasmodium falciparum*. *Mol. Biochem. Parasitol.* **73:**103–110.

101. **Kaslow, D. C., G. S. N. Hui, and S. Kumar.** 1994. Expression and antigenicity of *Plasmodium falciparum* major merozoite surface protein (MSP1(19)) variants secreted from *Saccharomyces cerevisiae*. *Mol. Biochem. Parasitol.* **63:**283–289.

102. **Keen, J. K., K. A. Sinha, K. N. Brown, and A. A. Holder.** 1994. A gene coding for a

high-molecular mass rhoptry protein of *Plasmodium yoelii. Mol. Biochem. Parasitol.* **65:**171–177.

103. **Klotz, F. W., P. A. Orlandi, G. Reuter, S. J. Cohen, J. D. Haynes, R. Schauer, R. J. Howard, P. Palese, and L. H. Miller.** 1992. Binding of *Plasmodium falciparum* 175-kD erythrocyte binding antigen and invasion of murine erythrocytes requires N-acetylneuraminic acid but not its O-acetylated form. *Mol. Biochem. Parasitol.* **51:**49–54.
104. **Knapp, B., E. Hundt, U. Nau, and H. A. Kupper.** 1989. Molecular cloning, genomic structure and localization in a blood stage antigen of *Plasmodium falciparum* characterized by a serine stretch. *Mol. Biochem. Parasitol.* **32:**73–84.
105. **Knapp, B., U. Nau, E. Hundt, and H. A. Kupper.** 1991. A new blood stage antigen of *Plasmodium falciparum* highly homologous to the serine-stretch protein SERP. *Mol. Biochem. Parasitol.* **44:**1–13.
106. **Kumar, S., A. Yadava, D. B. Keister, J. H. Tian, M. Ohl, K. A. Perdue-Greenfield, L. H. Miller, and D. C. Kaslow.** 1995. Immunogenicity and *in vivo* efficacy of recombinant *Plasmodium falciparum* merozoite surface protein-1 in *Aotus* monkeys. *Mol. Med.* **1:**325–332.
107. **Lewis, A. P.** 1989. Cloning and analysis of the gene encoding the 230-kDa merozoite surface antigen of *Plasmodium yoelii. Mol. Biochem. Parasitol.* **36:**271–282.

107a. **Ling, I. T., et al.** Unpublished data.

108. **Ling, I. T., S. A. Ogun, and A. A. Holder.** 1994. Immunization against malaria with a recombinant protein. *Parasite Immunol.* **16:**63–67.
109. **Ling, I. T., S. A. Ogun, and A. A. Holder.** 1995. The combined epidermal growth factor-like modules of *Plasmodium yoelii* merozoite surface protein-1 are required for a protective immune response to the parasite. *Parasite Immunol.* **17:**425–433.
110. **Locher, C. P., and L. Q. Tam.** 1993. Reduction of disulfide bonds in *Plasmodium falciparum* gp195 abolishes the production of growth-inhibitory antibodies. *Vaccine* **11:**1119–1123.
111. **Lyon, J. A., J. D. Haynes, C. L. Diggs, J. D. Chulay, and J. M. Pratt-Rossiter.** 1986. *Plasmodium falciparum* antigens synthesized by schizonts and stabilized at the merozoite surface by antibodies when schizonts mature in the presence of growth inhibitory immune serum. *J. Immunol.* **136:**2252–2258.
112. **Majarian, W. R., T. M. Daly, W. P. Weidanz, and C. A. Long.** 1984. Passive immunization against murine malaria with IgG3 monoclonal antibody. *J. Immunol.* **132:**3131–3137.
113. **Marshall, V. M., R. L. Anthony, M. J. Bangs Purnomo, R. F. Anders, and R. L. Coppel.** 1994. Allelic variants of the *Plasmodium falciparum* merozoite surface antigen 2 (MSA-2) in a geographically restricted area of Irian Jaya. *Mol. Biochem. Parasitol.* **63:**13–21.
114. **Marshall, V. M., R. L. Coppel, R. F. Anders, and D. J. Kemp.** 1992. Two novel alleles within subfamilies of the merozoite surface antigen-2 (MSA-2) of *Plasmodium falciparum. Mol. Biochem. Parasitol.* **50:**181–184.
115. **Marshall, V. M., R. L. Coppel, R. K. Martin, A. M. J. Oduola, R. F. Anders, and D. J. Kemp.** 1991. A *Plasmodium falciparum* MSA-2 gene apparently generated by intragenic recombination between the two allelic families. *Mol. Biochem. Parasitol.* **45:**349–351.
116. **Marshall, V. M., M. G. Peterson, A. M. Lew, and D. J. Kemp.** 1989. Structure of the apical membrane antigen-I (AMA-1) of *Plasmodium chabaudi. Mol. Biochem. Parasitol.* **37:**281–283.
117. **McBride, J. S., and H.-G. Heidrich.** 1987. Fragments of the polymorphic Mr 185,000

glycoprotein from the surface of isolated *Plasmodium falciparum* merozoites form an antigenic complex. *Mol. Biochem. Parasitol.* **23:**71–84.

118. **McColl, D. J., A. Silva, M. Foley, J. F. J. Kun, J. M. Favaloro, J. K. Thompson, V. M. Marshall, R. L. Coppel, D. J. Kemp, and R. F. Anders.** 1994. Molecular variation in a novel polymorphic antigen associated with *Plasmodium falciparum* merozoites. *Mol. Biochem. Parasitol.* **68:**53–67.
119. **McKean, P. G., K. Odea, and K. N. Brown.** 1993. Nucleotide sequence analysis and epitope mapping of the merozoite surface protein 1 from *Plasmodium chabaudi chabaudi* AS. *Mol. Biochem. Parasitol.* **62:**199–209.
120. **Miettinen-Baumann, A., W. Strych, J. McBride, and H.-G. Heidrich.** 1988. A 46,000 dalton *Plasmodium falciparum* merozoite surface glycoprotein not related to the 185,000–195,000 dalton schizont precursor molecule: isolation and characterization. *Parasitol. Res.* **74:**317–323.
121. **Miller, L. H., T. Roberts, M. Shahabuddin, and T. F. McCutchan.** 1993. Analysis of sequence diversity in the *Plasmodium falciparum* merozoite surface protein-1 (MSP-1). *Mol. Biochem. Parasitol.* **59:**1–14.
122. **Mitchell, G. H., T. J. Hadley, M. H. McGinniss, F. W. Klotz, and L. H. Miller.** 1986. Invasion of erythrocytes by *Plasmodium falciparum* malaria parasites: evidence for receptor heterogeneity and two receptors. *Blood* **67:**1519–1521.
123. **Müller, H. M., K. Fruh, A. von Brunn, F. Esposito, S. Lombardi, A. Crisanti, and H. Bujard.** 1989. Development of the human immune response against the major surface protein (Gp190) of *Plasmodium falciparum. Infect. Immun.* **57:**3765–3769.
124. **Murphy, V. F., W. C. Rowan, M. J. Page, and A. A. Holder.** 1990. Expression of hybrid malaria antigens in insect cells and their engineering for correct folding and secretion. *Parasitology* **100:**177–183.
125. **Narum, D. L., and A. W. Thomas.** 1994. Differential localization of full-length and processed forms of PF83/AMA-1 an apical membrane antigen of *Plasmodium falciparum* merozoites. *Mol. Biochem. Parasitol.* **67:**59–68.
126. **Narum, D. L., G. W. Welling, and A. W. Thomas.** 1993. Ion-exchange immunoaffinity purification of a recombinant baculovirus *Plasmodium falciparum* apical membrane antigen, Pf83/AMA-1. *J. Chromatogr. A* **657:**357–363.
127. **Oeuvray, C., H. Bouharountayoun, M. C. Filgueira, H. Grasmasse, A. Tartar, and P. Druilhe.** 1993. Characterization of a *Plasmodium falciparum* merozoite surface antigen targetted by defense mechanisms developed in immune individuals. *C.R. Acad. Sci. Ser. III* **316:**395–399.
128. **Oeuvray, C., H. Bouharountayoun, H. Grasmasse, E. Bottius, T. Kaidoh, M. Aikawa, M. C. Filgueira, A. Tartar, and P. Druilhe.** 1994. Merozoite surface protein-3: a malaria protein inducing antibodies that promote *Plasmodium falciparum* killing by cooperation with blood monocytes. *Blood* **84:**1594–1602.

128a. **Ogun, S. A., and A. A. Holder.** A high-molecular-mass *Plasmodium yoelii* rhoptry protein binds to erythrocytes. *Mol. Biochem. Parasitol.*, in press.

129. **Olrandi, P. A., F. W. Klotz, and J. D. Haynes.** 1992. A malaria invasion receptor, the 175-kD erythrocyte binding antigen of *Plasmodium falciparum* recognizes the terminal Neu5Ac(alpha2-3)Gal- sequences of glycophorin-A. *J. Cell Biol.* **116:**901–909.
130. **Orlandi, P. A., B. K. L. Sim, J. D. Chulay, and J. D. Haynes.** 1990. Characterization of the 175-kD erythrocyte binding antigen of *Plasmodium falciparum. Mol. Biochem. Parasitol.* **40:**285–294.
131. **Patarroyo, M. E., P. Romero, M. L. Torres, P. Clavijo, A. Moreno, A. Martinez, R. Rodriguez, F. Guzman, and E. Caberas.** 1987. Induction of protective immunity

against experimental infection with malaria using synthetic peptides. *Nature* (London) **328:**629–632.

132. **Perkins, M., and L. J. Rocco.** 1988. Sialic acid-dependent binding of *Plasmodium falciparum* merozoite surface antigen, Pf200, to human erythrocytes. *J. Immunol.* **141:**3190–3196.
133. **Perkins, M. E., and L. J. Rocco.** 1990. Chemical cross linking of *Plasmodium falciparum* glycoprotein, Pf200 (190–205kDa), to the S-antigen at the merozoite surface. *Exp. Parasitol.* **70:**207–216.
134. **Perrin, L. H., B. Merkli, M. Loche, C. Chizzolini, J. Smart, and R. Richle.** 1984. Antimalarial immunity in *Saimiri* monkeys. Immunization with surface components of asexual blood stages. *J. Exp. Med.* **160:**441–451.
135. **Perrin, L. H., E. Ramirez, P. H. Lambert, and P. A. Miescher.** 1981. Inhibition of *P. falciparum* growth in human erythrocytes by monoclonal antibodies. *Nature* (London) **289:**301.
136. **Peterson, M. G., V. M. Marshall, J. A. Smythe, P. E. Crewther, A. Lew, A. Silva, R. F. Anders, and D. J. Kemp.** 1989. Integral membrane protein located in the apical complex of *Plasmodium falciparum. Mol. Cell. Biol.* **9:**3151–3154.
137. **Peterson, M. G., P. Nguyendinh, V. M. Marshall, J. F. Elliott, W. E. Collins, R. F. Anders, and D. J. Kemp.** 1990. Apical membrane antigen of *Plasmodium fragile. Mol. Biochem. Parasitol.* **39:**279–284.
138. **Pirson, P. J., and M. E. Perkins.** 1985. Characterization with monoclonal antibodies of a surface antigen of *Plasmodium falciparum* merozoites. *J. Immunol.* **134:**1946–1951.
139. **Prescott, N., A. W. Stowers, Q. Cheng, A. Bobogare, C. M. Rzepczyk, and A. Saul.** 1994. *Plasmodium falciparum* genetic diversity can be characterised using the polymorphic merozoite surface antigen 2 (MSA-2) gene as a single locus marker. *Mol. Biochem. Parasitol.* **63:**203–212.
140. **Prickett, M. D., T. R. Smarz, and J. H. Adams.** 1994. Dimorphism and intergenic recombination within the microneme protein (MP-1) gene family of *Plasmodium knowlesi. Mol. Biochem. Parasitol.* **63:**37–48.
141. **Ramasamy, R.** 1987. Studies on glycoproteins in the human malaria parasite *Plasmodium falciparum.* Identification of a myristilated 45kDa merozoite membrane glycoprotein. *Immunol. Cell Biol.* **65:**419–424.
142. **Ramasamy, R., G. Jones, and R. Lord.** 1990. Characterisation of an inhibitory monoclonal antibody defined epitope on a malaria vaccine candidate antigen. *Immunol. Lett.* **23:**305–310.
143. **Ridley, R. G., B. Takacs, H. Etlinger, and J. G. Scaife.** 1990. A rhoptry antigen of *Plasmodium falciparum* is protective in Saimiri monkeys. *Parasitology* **101:**187–192.
144. **Ridley, R. G., B. Takacs, H. W. Lahm, C. J. Delves, M. Goman, U. Certa, H. Matile, G. R. Woollett, and J. G. Scaife.** 1990. Characterisation and sequence of a protective rhoptry antigen from *Plasmodium falciparum. Mol. Biochem. Parasitol.* **41:**125–134.
145. **Riley, E. M., S. J. Allen, J. G. Wheeler, M. J. Blackman, S. Bennett, B. Takacs, H. J. Schonfeld, A. A. Holder, and B. M. Greenwood.** 1992. Naturally acquired cellular and humoral immune responses to the major merozoite surface antigen (PfMSP1) of *Plasmodium falciparum* are associated with reduced malaria morbidity. *Parasite Immunol.* **14:**321–337.
146. **Riley, E. M., S. Morris-Jones, M. J. Blackman, B. M. Greenwood, and A. A. Holder.** 1993. A longitudinal study of naturally acquired cellular and humoral immune responses to a merozoite surface protein (MSP1) of *Plasmodium falciparum* in an area of seasonal malaria transmission. *Parasite Immunol.* **15:**513–524.

147. **Roger, N., J.-F. Dubremetz, P. Delplace, B. Fortier, G. Tronchin, and A. Vernes.** 1988. Characterization of a 225 kilodalton rhoptry protein of *Plasmodium falciparum. Mol. Biochem. Parasitol.* **27:**135–142.
148. **Roussilhon, C., E. Hundt, M. Agrapart, W. Stuber, B. Knapp, P. Dubois, and J. J. Ballet.** 1990. Responses of T-cells from sensitized donors to recombinant and synthetic peptides corresponding to sequences of the *Plasmodium falciparum* SERP antigen. *Immunol. Lett.* **25:**149–154.
149. **Rzepczyk, C. M., P. A. Csurhes, E. P. Baxter, T. J. Doran, D. O. Irving, and N. Kere.** 1990. Amino acid sequences recognized by T-cells—studies on a merozoite surface antigen from the FCQ-27 PNG isolate of *Plasmodium falciparum. Immunol. Lett.* **25:**155–164.
150. **Rzepczyk, C. M., P. A. Csurhes, A. J. Saul, G. L. Jones, S. Dyer, D. Chee, N. Goss, and D. O. Irving.** 1992. Comparative study of the T-cell response to two allelic forms of a malarial vaccine candidate protein. *J. Immunol.* **148:**1197–1204.
151. **Saul, A., J. Cooper, D. Hauquitz, D. Irving, Q. Cheng, A. Stowers, and T. Limpaiboon.** 1992. The 42-kD rhoptry-associated protein of *Plasmodium falciparum. Mol. Biochem. Parasitol.* **50:**139–150.
152. **Saul, A., J. Cooper, L. Ingram, R. F. Anders, and G. V. Brown.** 1985. Invasion of erythrocytes *in vitro* by *Plasmodium falciparum* can be inhibited by monoclonal antibody directed against an S-antigen. *Parasite Immunol.* **7:**587–593.
153. **Saul, A., R. Lord, G. L. Jones, and L. Spencer.** 1992. Protective immunization with invariant peptides of the *Plasmodium falciparum* antigen MSA2. *J. Immunol.* **148:**208–211.
154. **Siddiqui, W. A., L. Q. Tam, K. J. Kramer, G. S. N. Hui, S. E. Case, K. M. Yamaga, S. P. Chang, E. B. T. Chan, and S.-C. Kan.** 1987. Merozoite surface coat precursor protein completely protects *Aotus* monkeys against *Plasmodium falciparum* malaria. *Proc. Natl. Acad. Sci USA* **84:**3014–3018.
155. **Sim, B. K. L.** 1995. EBA-175: an erythrocyte-binding ligand of *Plasmodium falciparum. Parasitol. Today* **11:**213–217.
156. **Sim, B. K. L., J. M. Carter, C. D. Deal, C. Holland, J. D. Haynes, and M. Gross.** 1994. *Plasmodium falciparum*: further characterization of a functionally active region of the merozoite invasion ligand EBA-175. *Exp. Parasitol.* **78:**259–268.
157. **Sim, B. K. L., C. E. Chitnis, K. Wasniowska, T. J. Hadley, and L. H. Miller.** 1994. Receptor and ligand domains for invasion of erythrocytes by *Plasmodium falciparum. Science* **264:**1941–1944.
158. **Sim, B. K. L., P. A. Orlandi, J. D. Haynes, F. W. Klotz, J. M. Carter, D. Camus, M. E. Zegans, and J. D. Chulay.** 1990. Primary structure of the 175k *Plasmodium falciparum* erythrocyte binding antigen and identification of a peptide which elicits antibodies that inhibit malaria merozoite invasion. *J. Cell Biol.* **111:**1877–1884.
159. **Sim, B. K. L., T. Toyoshima, J. D. Haynes, and M. Aikawa.** 1992. Localization of the 175-kD erythrocyte binding antigen in micronemes of *Plasmodium falciparum* merozoites. *Mol. Biochem. Parasitol.* **51:**157–160.
160. **Simitsek, P. D., E. Ramirez, and L. H. Perrin.** 1990. Structural diversity of *Plasmodium falciparum* Gp200 is detected by T-cells. *Eur. J. Immunol.* **20:**1755–1759.
161. **Sinigaglia, F., B. Takacs, H. Jacot, H. Matile, J. R. Pink, A. Crisanti, and H. Bujard.** 1988. Nonpolymorphic regions of p190, a protein of the *Plasmodium falciparum* erythrocytic stage, contains both T and B cell epitopes. *J. Immunol.* **140:**3568–3572.
162. **Smythe, J. A., R. L. Coppel, G. V. Brown, R. Ramasamy, D. J. Kemp, and R. F. Anders.** 1988. Identification of two integral membrane proteins of *Plasmodium falciparum. Proc. Natl. Acad. Sci. USA* **85:**5195–5199.

163. **Smythe, J. A., R. L. Coppel, K. P. Day, R. K. Martin, A. M. J. Oduola, D. J. Kemp, and R. F. Anders.** 1991. Structural diversity in the *Plasmodium falciparum* merozoite surface antigen-2. *Proc. Natl. Acad. Sci. USA* **88:**1751–1755.
164. **Smythe, J. A., M. G. Peterson, R. L. Coppel, A. J. Saul, D. J. Kemp, and R. F. Anders.** 1990. Structural diversity in the 45-kD merozoite surface antigen of *Plasmodium falciparum. Mol. Biochem. Parasitol.* **39:**227–234.
165. **Snewin, V. A., M. Herrera, G. Sanchez, A. Scherf, G. Langsley, and S. Herrera.** 1991. Polymorphism of the alleles of the merozoite surface antigens MSA1 and MSA2 in *Plasmodium falciparum* wild isolates from Colombia. *Mol. Biochem. Parasitol.* **49:**265–276.
166. **Spetzler, J. C., C. Rao, and J. P. Tam.** 1994. A novel strategy for the synthesis of the cysteine-rich protective antigen of the malaria merozoite surface protein (MSP-1)—knowledge-based strategy for disulfide formation. *Int. J. Pept. Protein Res.* **43:**351–358.
167. **Stafford, W. H. L., M. J. Blackman, A. Harris, S. Shai, M. Grainger, and A. A. Holder.** 1994. N-terminal amino acid sequence of the *Plasmodium falciparum* merozoite surface protein-1 polypeptides. *Mol. Biochem. Parasitol.* **66:**157–160.
168. **Stahl, H. D., A. E. Bianco, R. F. Crewther, R. F. Anders, A. P. Kyne, R. L. Coppel, G. F. Mitchell, D. J. Kemp, and G. V. Brown.** 1986. Sorting large numbers of clones expressing *Plasmodium falciparum* antigens in *Escherichia coli* by differential antibody screening. *Mol. Biol. Med.* **3:**351–368.
169. **Stanley, H. A., R. F. Howard, and R. T. Reese.** 1985. Recognition of a M_r 56K glycoprotein on the surface of *Plasmodium falciparum* merozoites by mouse monoclonal antibodies. *J. Immunol.* **134:**3439–3444.
170. **Stewart, M. J., S. Schulaman, and J. P. Vanderberg.** 1986. Rhoptry secretion of membranous whorls by *Plasmodium falciparum* merozoites. *Am. J. Trop. Med. Hyg.* **35:**37–44.
171. **Storey, E.** 1992. A polyclonal but not a monoclonal antibody to an Mr 52-kD protein responsible for a punctate fluorescence pattern in *Plasmodium falciparum* merozoites inhibits invasion *in vitro. Am. J. Trop. Med. Hyg.* **47:**663–674.
172. **Su, S. D., A. R. Sanadi, E. Ifon, and E. A. Davidson.** 1993. A monoclonal antibody capable of blocking the binding of PF200 (MSA-1) to human erythrocytes and inhibiting the invasion of *Plasmodium falciparum* merozoites into human erythrocytes. *J. Immunol.* **151:**2309–2317.
173. **Tanabe, K., M. Mackay, M. Goman, and J. G. Scaife.** 1987. Allelic dimorphism in a surface antigen gene of the malaria parasite *Plasmodium falciparum. J. Mol. Biol.* **195:**273–287.
174. **Thomas, A. W., L. H. Bannister, and A. P. Waters.** 1990. 66 kD-related antigens of *Plasmodium knowlesi* are merozoite surface antigens associated with the apical prominence. *Parasite Immunol.* **12:**105–113.
175. **Thomas, A. W., D. A. Carr, J. M. Carter, and J. A. Lyon.** 1990. Sequence comparison of allelic forms of the *Plasmodium falciparum* merozoite surface antigen MSA2. *Mol. Biochem. Parasitol.* **43:**211–220.
176. **Thomas, A. W., J. A. Deans, G. H. Mitchell, T. Alderson, and S. Cohen.** 1984. The Fab fragments of monoclonal IgG to a merozoite surface antigen inhibit *Plasmodium knowlesi* invasion of erythrocytes. *Mol. Biochem. Parasitol.* **13:**187–199.
177. **Thomas, A. W., J. F. Trape, C. Rogier, A. Goncalves, V. E. Rosario, and D. L. Narum.** 1994. High prevalence of natural antibodies against *Plasmodium falciparum* 83-kilodalton apical membrane antigen (PF83/AMA-1) as detected by capture-enzyme-linked immunosorbent assay using full-length baculovirus recombinant PF83/AMA-1. *Am. J. Trop. Med. Hyg.* **51:**730–740.
178. **Thomas, A. W., A. P. Waters, and D. Carr.** 1990. Analysis of variation in Pf83, an

erythrocytic merozoite vaccine candidate antigen of *Plasmodium falciparum*. *Mol. Biochem. Parasitol.* **42:**285–287.

179. **Tolle, R., H. Bujard, and J. A. Cooper.** 1995. *Plasmodium falciparum*: variations within the C-terminal region of merozoite surface antigen-1. *Exp. Parasitol.* **81:**47–54.
180. **Tolle, R., K. Fruh, O. Doumbo, O. Koita, M. Ndiaye, A. Fischer, K. Dietz, and H. Bujard.** 1993. A prospective study of the association between the human humoral immune response to *Plasmodium falciparum* blood stage antigen-gp190 and control of malarial infections. *Infect. Immun.* **61:**40–47.
181. **Torii, M., J. H. Adams, L. H. Miller, and M. Aikawa.** 1989. Release of merozoite dense granules during erythrocyte invasion by *Plasmodium knowlesi*. *Infect. Immun.* **57:**3230–3233.
182. **Tsuboi, T., S. H. I. Kappe, F. Al-Yaman, M. D. Prickett, M. Alpers, and J. H. Adams.** 1994. Natural variation within the principal adhesion domain of the *Plasmodium vivax* Duffy binding protein. *Infect. Immun.* **62:**5581–5586.
183. **Udhayakumar, V., D. Anyona, S. Kariuki, Y. P. Shi, P. B. Bloland, O. H. Branch, W. Weiss, B. L. Nahlen, D. C. Kaslow, and A. A. Lal.** 1995. Identification of T and B cell epitopes recognized by humans in the C-terminal 42-kDa domain of the *Plasmodium falciparum* merozoite surface protein (MSP)-1. *J. Immunol.* **154:**6022–6030.
184. **Ward, G. E., L. H. Miller, and J. A. Dvorak.** 1993. The origin of parasitophorous vacuole membrane lipids in malaria-infected erythrocytes. *J. Cell Sci.* **106:**237–248.
185. **Ware, L. A., K. C. Kain, B. K. L. Sim, J. D. Haynes, J. K. Baird, and D. E. Lanar.** 1993. Two alleles of the 175-kD *Plasmodium falciparum* erythrocyte binding antigen. *Mol. Biochem. Parasitol.* **60:**105–110.
186. **Waters, A. P., A. W. Thomas, J. A. Deans, G. H. Mitchell, D. E. Hudson, L. H. Miller, T. F. McCutchan, and S. Cohen.** 1990. A merozoite receptor protein from *Plasmodium knowlesi* is highly conserved and distributed throughout Plasmodium. *J. Biol. Chem.* **265:**17974–17979.
187. **Waters, A. P., A. W. Thomas, G. H. Mitchell, and T. F. McCutchan.** 1991. Intrageneric conservation and limited inter-strain variation in a protective minor surface antigen of *Plasmodium knowlesi* merozoites. *Mol. Biochem. Parasitol.* **44:**141–144.
188. **Weber, J. L., J. A. Lyon, R. H. Wolff, T. Hall, G. H. Lowell, and J. D. Chulay.** 1988. Primary structure of a *Plasmodium falciparum* malaria antigen located at the merozoite surface and within the parasitophorous vacuole. *J. Biol. Chem.* **263:**11421–11425.
189. **Wertheimer, S. P., and J. W. Barnwell.** 1989. *Plasmodium vivax* interaction with the human Duffy blood group glycoprotein—identification of a parasite receptor-like protein. *Exp. Parasitol.* **69:**340–350.
190. **Wilson, C. F., R. Anand, J. T. Clark, and J. S. McBride.** 1987. Topography of epitopes on a polymorphic schizont antigen of *Plasmodium falciparum* determined by the binding of monoclonal antibodies in a two-site radioimmunoassay. *Parasite Immunol.* **9:**737–746.

Malaria Vaccine Development: A Multi-Immune Response Approach
Edited by Stephen L. Hoffman

Chapter 5

Malaria Vaccines: Attacking Infected Erythrocytes

Klavs Berzins and Peter Perlmann

INTRODUCTION

The asexual blood stages of malaria parasites are responsible for the clinical symptoms of malaria. Therefore, vaccines against these stages are intended to prevent or reduce malaria-related morbidity and mortality by eliminating or reducing the parasite load. Although the antiparasite mechanisms involved in protective immune responses that arise during natural malarial infections are incompletely understood, several of a large number of blood-stage antigens have been identified as potential vaccine candidates. In this chapter, we discuss different immune mechanisms that may give rise to protection against asexual blood-stage infections both in experimental systems and in humans with *Plasmodium falciparum* malaria. We discuss efforts to develop vaccines that attack infected erythrocytes (RBCs) and focus on the potential vaccine antigens that are expressed on the surfaces of infected RBCs or the parasites. Immune mechanisms involved in preventing merozoite invasion of RBCs are discussed in chapter 4, and those preventing endothelial adherence of infected RBCs are discussed in chapter 6.

THE ASEXUAL ERYTHROCYTIC CYCLE

The erythrocytic cycle of the malaria parasites is initiated upon merozoite invasion of RBCs. Residing within the parasitophorous vacuole, the parasite undergoes development from the ring stage (immature trophozoites) to the trophozoite and then, after mitotic divisions, to the schizont stage, which contains up to 32 merozoites (*P. falciparum*). The erythrocytic part of parasite development is completed after 48 h (*P. falciparum*, *Plasmodium vivax*, and *Plasmodium ovale*) or 72 h (*Plasmodium malariae*).

During intraerythrocytic development, transport of parasite proteins across the parasite plasma membrane into the parasitophorous vacuole is intense. Selected proteins are transported further through the RBC cytoplasm either along membranous cisternae called Maurer's clefts or in vesicular compartments (129, 187). Some proteins either become associated with the RBC skeleton underneath the membrane or are inserted into the RBC membrane, while others are exported from the infected RBC into the surrounding medium. Recently, another membrane structure, called the parasitophorous duct and seemingly distinct from the Maurer's clefts, was identified as a ductlike continuity between the parasitophorous vacuole and the RBC membrane (188). Through this duct, macromolecules may diffuse between the external medium and the vacuolar space surrounding the parasite.

The intraerythrocytic parasite is surrounded by the parasitophorous vacuole membrane and the RBC plasma membrane and is thus shielded from direct attack by cells of the immune system. However, to be able to develop intracellularly, the parasite requires an extraerythrocytic supply of nutrition, thought to be imported through the RBC membrane by pathways formed by parasite proteins. Furthermore, some malaria parasites transport proteins to the RBC membrane to form knoblike structures that are important for the cytoadherence of infected RBCs to endothelial cells, a phenomenon crucial for the survival of the parasite in vivo (see chapter 6). Thus, the parasite exposes proteins on the RBC surface that constitute target antigens for different immune effector mechanisms. In addition, antibodies and other immune effector molecules may reach the parasite through the parasitophorous duct to exert their parasiticidal activities.

PROTECTIVE IMMUNE RESPONSES AGAINST ASEXUAL ERYTHROCYTIC STAGES IN RODENT MALARIA MODELS

In mouse malaria systems, the protective importance of antibodies relative to that of antibody-independent T-cell responses varies among different plasmodial species (54, 252). Thus, the humoral immune system appears to play a major role in protecting mice against *Plasmodium yoelii*, since infection with this parasite becomes lethal in mice deprived of their B cells but is nonlethal in normal mice. In contrast, mice depleted of B cells are able to control their infections with *Plasmodium chabaudi* or *Plasmodium vinckei*, although they cannot completely abolish parasitemia (33, 133). The involvement of both T and B cells in a *P. chabaudi* model was proven by reconstitution experiments in mice with severe combined immunodeficiency lacking both cell types. Here, transfer of immune or normal $CD4^+$ T cells protected the majority of mice against an otherwise lethal parasite challenge, but the mice were unable to clear their parasitemias (154). However, complete clearance of infection was obtained after simultaneous transfer of $CD4^+$ T cells from normal or immune donors and immune B cells. Clearance of parasitemia was correlated with the presence of antibodies.

Resolution of an acute infection with *P. chabaudi adami* was also obtained in athymic nude mice after transfer of a $CD4^+$ T-cell clone specific for an epitope of a soluble parasite antigen (28). It appears that in resistant mice, $CD4^+$ T cells of the gamma interferon (IFN-γ)-interleukin 2 (IL-2)-producing type (TH1) play a crucial role during the initial phase of protective immune response against *P. chabaudi chabaudi*, while IL-4–IL-5-producing $CD4^+$ T cells of the helper type (TH2), together with antibodies, are important during later phases of the response (138, 219, 231,

232). In contrast, in susceptible mouse strains, induction of a strong TH2 response early in infection is associated with a severe and lethal course of malaria (219). Moreover, the solid reinfection immunity to *P. vinckei* that develops in drug-cured mice appears to depend exclusively on TH1 cells (177). In any event, in all mouse malaria models studied for T-cell involvement, $CD4^+$ T cells are the most important cells in controlling blood-stage infection. Although involvement of $CD8^+$ T cells cannot be ruled out, their role seems to be a minor one (186, 223).

It is well established that the spleen has important functions in the host response to the asexual blood stages of malaria parasites (257). Architectural modifications of the spleen seen in infected mice are believed to be important for filtering parasitized RBCs from the blood for subsequent destruction in the spleen (253). This spleen alteration could be the explanation of recent findings in the *P. vinckei* model, in which reconstitution of immunity by transfer of immune $CD4^+$ T cells appeared to require the presence in the recipients of a spleen modified by preceding malaria infection, since no protection could be transferred to naive mice (133, 256). However, the presence of an intact spleen seems to be necessary in some rodent malaria models, though not in others, and efficient spleen-independent mechanisms for clearing blood-stage infections do exist (206).

REGULATION OF IMMUNITY TO *P. FALCIPARUM*

Antibodies

The immune response to asexual blood stages is polyspecific, involving a large number of antigens. Protective immunity against these stages is partly mediated by antibodies, as demonstrated for *P. falciparum* by the classic experiments of Cohen et al. (43) and McGregor et al. (152), in which passive transfer of immune immunoglobulin G (IgG) from adults had curative effects in Gambian children. Similarly, passive transfer of human IgG from immune individuals to *Aotus* monkeys conferred protection against *P. falciparum* challenge (59). The mechanism of protection and the specificity of the antibodies involved are unknown. An indication that antibodies reactive with a defined *P. falciparum* antigen, Pf155/RESA, may be active in parasite neutralization in vivo was recently obtained when *Aotus* monkeys receiving small amounts of human affinity-purified antibodies showed depressed parasitemias after *P. falciparum* challenge and three of four monkeys self-cured the infection (19). Experiments with *P. falciparum* in squirrel monkeys indicated that protection conferred by passive transfer of IgG from immune monkeys is obtained only with IgG fractions containing opsonic antibodies (83). Similar conclusions were drawn in recent passive transfer experiments in which a pool of IgG from immune Africans was given to Thai patients with *P. falciparum* malaria and drastic clinical and parasitological improvements were obtained (24). Although the effective IgG fraction did not inhibit merozoite invasion in *P. falciparum* cultures in vitro, it exerted an antibody-dependent cellular inhibition of parasite growth in cooperation with normal blood monocytes. Among the antiparasite antibodies, the cytophilic IgG isotypes (IgG1 and IgG3) predominated in the IgG pool as well as in sera from malaria-immune subjects (25). The target antigens in this context appear to be associated mainly with the surface of merozoites and not with the surface of RBCs (26). These and other observations from different animal models indicate that antibodies are important for clearance of parasite loads (53, 58, 252).

Cellular Regulation

Immunity to the asexual blood stages of the malaria parasite is controlled by the T-cell system (235). A simplified scheme of the cellular regulation of the immune response to *Plasmodium*-infected RBCs is shown in Fig. 1. The involvement of T cells in human blood-stage infection is usu-

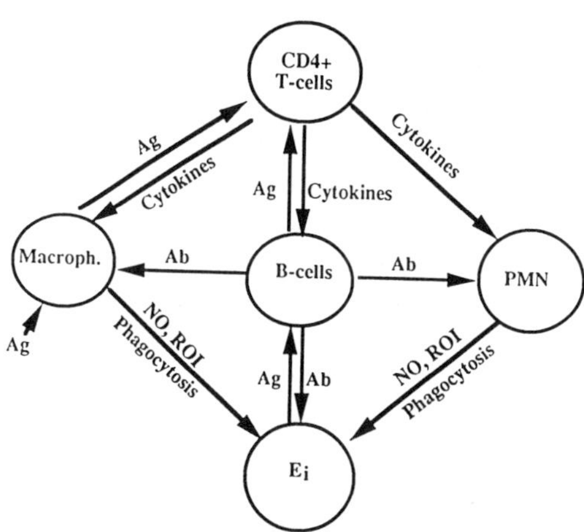

Figure 1. Cellular regulation (simplified) of the immune response to parasite-infected RBCs (E_i). Parasite antigen (Ag) is presented to $CD4^+$ T cells by B cells, macrophages, or other cells (not shown). Cytokines released from antigen-activated T cells in turn activate effector cells such as macrophages or neutrophils (polymorphonuclear leukocytes [PMN]). The activated effector cells can attack the infected RBCs and kill the parasites by a variety of mechanisms, including the release of TNF-α from the mononuclear cells and the induction of the formation of NO, reactive oxygen intermediates (ROI), and/or phagocytosis. Activated $CD4^+$ T cells also help B cells produce antiparasite antibodies (Ab), which mediate parasite clearance and amplify cell-mediated destruction of infected RBCs.

ally investigated by studying their responses to antigen in vitro. Thus, T lymphocytes from donors primed to *P. falciparum* by natural exposure (40, 235) but also T cells from naive donors (50) may be induced by parasite extracts to proliferate and/or release lymphokines in vitro in an antigen-specific manner. This responsiveness may be depressed in patients with acute infections (93) as well as in asymptomatic but parasitemic donors (233). Reduced specific T-cell responsiveness in vitro may have several causes, although it probably often reflects lymphocyte redistribution between blood and tissues rather than true immunosuppression (64, 108). As in mice, $CD4^+$ T cells have a major role in regulating the human immune response to *P. falciparum* blood-stage antigens. The T dependency of IgG antimalaria antibodies has been shown directly by assaying antibody secretion after T-cell stimulation with antigen in T-cell–B-cell cooperation systems in vitro (41, 121). However, in vitro stimulation of $CD4^+$ T cells with malarial antigens may also result in proliferation and/or IFN-γ secretion, neither of which is correlated with levels of serum antibodies against the corresponding antigens (71, 199, 239). In vitro stimulation may also induce IL-4 expression which in individual donors is correlated with neither proliferation nor IFN-γ release but correlates well with concentrations of the relevant serum antibodies (192, 238). These results suggest that the human response is controlled by distinct $CD4^+$ T-cell subsets that correspond to the TH1 and TH2 types found in *P. chabaudi*-infected mice (138). The role of $CD8^+$ T cells in human *P. falciparum* blood-stage infection is less well known, although such cells have been reported to inhibit parasite growth in vitro (70) and to be responsible for suppression and other regulatory deficiencies seen in malaria (162, 193). Major histocompatibility complex (MHC) class I-restricted $CD8^+$ T-cell clones with cytotoxic potential have also been obtained from acutely infected patients (213), but their role in blood-stage infection is unclear. It also remains to be established whether such antiparasite functions could be assigned to *P. falciparum*-specific γ/δ T cells, which are MHC dependent but not MHC restricted and which occur in increased numbers in the blood of acutely infected donors (65, 79, 196). A transient increase in circulating γ/δ cells associated with malarial paroxysm has also been observed in human *P. vivax* infections (176). Recently, the capacity of human γ/δ T cells to inhibit *P. falciparum* growth in vitro was demonstrated, but the target appeared to be extracellular merozoites rather than the infected RBCs (65). A role for γ/δ T cells in the resolution of blood-stage infections has also been indicated in different experimental malaria systems, where large increases in numbers of splenic γ/δ T cells were observed upon resolution of plasmodial infections (165, 243). However, mice lacking γ/δ T cells could control and reduce a primary infection of *P.*

chabaudi, although with a delay and with higher recrudescent parasitemias than intact control mice showed (139).

Genetic Regulation

Although it is well established that genetic factors of both parasites and hosts control the outcome of malaria infections, the nature of the host genes regulating antiparasite immune responses is poorly understood (1, 159, 251). In humans, certain HLA class I and II alleles have recently been implicated in resistance to severe *P. falciparum* malaria (91, 92) by mechanisms mediated by MHC-restricted cytotoxic T cells. Immunization of inbred mice with defined *P. falciparum* antigens such as Pf155/RESA has revealed the occurrence of a conventional MHC class II restriction (142), and a restriction also occurs with *P. falciparum*-specific human T-cell clones (47, 84). However, in outbred human populations, a correlation between HLA class II alleles and the immune response to this antigen has been difficult to demonstrate (158). In contrast, investigation of *P. falciparum*-primed mono- or dizygotic twins and their age-matched siblings from Liberia or Madagascar suggested that both their humoral and their cellular immune responses to defined malarial epitopes were subjected to genetic regulation (215) but that this regulation was due to genes that are probably located outside the MHC class II region. On the other hand, an association between HLA-DR4 and depressed antibody responses to the vaccine immunogen SPf66 in vaccinees (see chapter 9) has been reported (175). However, in general and not unexpectedly, in view of the great MHC polymorphism, finding MHC restrictions in outbred human populations sensitized to malaria by natural infections has been difficult (194, 237).

POSSIBLE MECHANISMS OF PARASITE NEUTRALIZATION

Neutralization of the erythrocytic stages of malaria parasites has been demonstrated to depend on antibodies acting alone or, more commonly, in conjunction with effector cells, but antibody-independent cell-mediated mechanisms have also been described (Fig. 1) (185, 236). Although antigens exposed on the surface of infected RBCs are the most obvious targets, antibodies may also reach the parasitophorous vacuolar space and the intracellular parasite either by being engulfed together with the merozoite at invasion or by diffusion through the parasitophorous duct (188). Antibodies in such intraerythrocytic locations could be responsible for inhibiting parasite growth, as seen in in vitro experiments, by blocking transport pathways or inhibiting enzymatic activities necessary for parasite release. However, the relevance of these mechanisms with regard to parasite neutralization in vivo remains to be demonstrated.

In vitro experiments with *P. yoelii*-infected RBCs have shown that surface-reactive antibodies may cause complement-mediated lysis (76). *P. falciparum*-infected human RBCs also activate the alternative complement pathway in the absence of antibodies (217). Activation resulted in the deposition of C3b on the surface of the RBCs and was augmented by the presence of antibodies but did not result in lysis. Thus, complement-mediated lysis of infected RBCs may not be an in vivo mechanism for parasite neutralization, but surface-deposited C3b most certainly facilitates opsonization and destruction of infected cells by monocytes or macrophages and polymorphonuclear leukocytes (201).

For identification of potentially protective malarial antigens, the effect of antibodies on parasite propagation in *P. falciparum* in vitro cultures is often studied. Inhibition of parasite growth in vitro may be caused by antibody interference with merozoite invasion (248), but antibodies may also affect intraerythrocytic parasite growth by reacting with parasite antigens on the RBC surface or in the parasitophorous vacuole (6). Furthermore, prior to merozoite release from the mature schizonts, antibodies may enter through the leaky RBC membrane and interfere with merozoite dispersal (82, 147, 148). This interference is accomplished by the formation of immune clusters of merozoites, and antibodies against a large number of parasite antigens, including several merozoite surface antigens as well as other schizont antigens, appear to be involved (148).

Experiments with *P. falciparum* in in vitro cultures demonstrate that several cell-mediated mechanisms with the capacity to neutralize the erythrocytic stages of the parasite may be involved in protective immunity. These mechanisms include antibody-dependent killing of the parasites by lymphocytes, monocytes, and neutrophils (24, 29, 34, 35, 126, 144). In these studies, sera from malaria-immune individuals were used as antibody source, and the antigens involved were not defined. Monocyte-derived macrophages (119, 168) and polymorphonuclear leukocytes (125, 166, 249) are able to kill late stages of the intraerythrocytic parasite in the absence of antibodies. This killing may to some extent be attributed to the expression of adhesin (e.g., CD36 and/or ICAM 1)-binding structures on late-stage-infected RBCs that may promote binding and phagocytosis by leukocytes (197). Human natural killer (NK) cells have also been reported to be cytotoxic against erythrocytic schizonts of *P. falciparum*, and this cytotoxicity was enhanced by the addition of IFN-α and/or IL-2 (171). As mentioned above, the major effector cells for parasite neutralization in blood-stage infection appear to be mononuclear phagocytes and polymorphonuclear leukocytes. There is no evidence for any direct antibody-independent cytotoxic effector role of T cells toward infected RBCs. The facts that human RBCs show a very low expression of MHC class I molecules (23) and lack both essential accessory molecules and antigen-processing machinery imply that malaria-infected RBCs do not constitute targets for cytotoxic T cells. Nevertheless, T cells are essential for parasite neutralization through the release of cytokines such as IFN-γ, which activate nonlymphocytic effector cells (56, 135) to control parasite growth through phagocytosis (218), generation of reactive oxygen intermediates (168), or L-arginine-derived NO (167, 220). Neither IFN-γ nor any other parasite inhibitory cytokine such as tumor necrosis factor alpha (TNF-α) (26, 80, 136, 228) controls parasite growth by directly attacking intraerythrocytic parasites (24, 42, 149). TNF-α production by mononuclear cells (and probably other leukocytes as well) is induced by phosphatidylinositol-containing moieties that constitute the anchor structures of some major plasmodial antigens (14). Although TNF-α has an important role in the defense against infection, overproduction of this factor has also been implicated in disease severity (229) and the pathogenesis of cerebral malaria (42, 81, 153).

POSSIBLE TARGET ANTIGENS FOR PROTECTIVE IMMUNE RESPONSES

Two main groups of parasite-derived antigens are regarded as potential target antigens for protective immune mechanisms because of their locations in the infected RBC: first, antigens exposed on the surface of infected RBCs, including both membrane antigens and secreted antigens, and second, antigens on the surface of the intracellular parasite or in the parasitophorous vacuole. Some of the major vaccine candidate antigens are listed in Table 1.

Antigens Exposed on the Surface of Infected RBCs

Parasite-derived proteins exposed on the surface of infected RBCs have been demonstrated by a great variety of methods (100), including immunofluorescence (98), immunogold labeling (99, 244), microagglutination, and inhibition of cytoadherence of infected RBCs to endothelial cells (244). In these studies, performed with sera from individuals living in regions where malaria is endemic, the antibodies usually detect a major, antigenically variant molecule on the surface of the infected RBC. This high-molecular-weight antigen, denoted PfEMP1 (*P. falciparum* erythrocyte membrane protein 1), is thought to be the major molecule mediating

Table 1. Antigens of asexual blood stages of *P. falciparum* considered as vaccine components[a]

Antigen	Alternative name	Mol wt (10^3)	Location	Antigenic diversity	Function	Repeats	Reference(s)
PfEMP1		250–300	RBC surface	Extensive	Cytoadherence	No	13, 222
Pf332		2,500	RBC surface	NK	NK	Yes	150
Rosettin		22	RBC surface	NK	Rosetting	NK	87, 247
		28	RBC surface	NK	Rosetting	NK	87, 247
MSP-1	PMMSA, gp195, MSA-1	185–220	Merozoite surface	Extensive	NK	Yes	95
MSP-2		45	Merozoite surface	Extensive	NK	Yes	216
Ag5.1	exp-1, CRA, QF116	23	Parasitophorous vacuole	Minor	Protein transport	No	101
HRP-2		65–85	Secreted	No	NK	Yes	103
SERP	SERA, p126, p140	113 or 126	Parasitophorous vacuole	Minor	Proteinase	Serine stretch	57, 130
41-3		41	RBC cytoplasm	NK	NK	NK	66
Pf155/RESA		155	Dense granules, RBC skeleton	Minor	NK	Yes	16, 72, 198
p96		96	Exoantigen	NK	NK	NK	61, 115
GLURP		220	Parasitophorous vacuole	Yes	NK	Yes	21

[a]NK, not known.

cytoadherence of infected RBCs to endothelial cells. A gene encoding PfEMP1 was recently cloned and identified as a member of a multigene family that forms the basis for the antigenic variation observed in *P. falciparum* (13, 22, 222) (for details, see chapter 6).

Another important antigen localized at the surface of *P. falciparum*-infected RBCs is Pf332, a giant protein with an M_r of approximately 2,500,000 (150) that comprises at least 1,400 amino acids arranged in highly degenerated, glutamic acid-rich, 11-amino-acid repeats (137, 150). Although the Pf332 gene displays a high degree of restriction fragment length polymorphism between different strains and clinical samples (150, 156), there are no apparent differences in antigenicity with regard to the dominant epitopes seen by polyclonal human sera or sera from immunized mice (149a). The surface exposure of Pf332 is proven by the human monoclonal antibody 33G2 (242), which reacts with the surface of late-stage-infected RBCs in immunofluorescence or agglutination (241). The optimal epitope recognized by monoclonal antibody 33G2 is the pentapeptide VTEEI (5), which together with related monoclonal-antibody-binding sequences occurs >40 times in Pf332 (137, 150). The antigen appears to be a major target for opsonizing antibodies present in sera from *P. falciparum*-hyperimmune *Saimiri* monkeys (85).

Two new parasite-derived polypeptides (M_r, 22,000 and 28,000) on the surface of *P. falciparum*-infected RBCs (30, 87) have been suggested as mediating binding to the infected cells of uninfected RBCs (rosette formation). In most parasite strains, these "rosettins" are not coexpressed (247). However, recent experiments indicate that the rosettins are members of a large family of size-polymorphic antigens associated with adhesion (246a). The involvement of these antigens as targets for parasite-neutralizing immune responses remains to be investigated.

Several *P. falciparum* antigens are associated with the cytoplasmic side of the RBC membrane by interaction with cytoskeletal components. In early blood-stage development, these antigens include Pf155/RESA (16, 72, 198) and a 110-kDa rhoptry protein (202). Of these antigens, Pf155/RESA has been studied the most. It is a polypeptide with an M_r of 155,000 that contains two immunodominant regions of tandemly repeated sequences (69), which include both B- and T-cell epitopes (122, 178, 239). Pf155/RESA is present in dense granules in the apical regions of merozoites and is released from the merozoites and translocated to the RBC membrane after invasion (8, 49). Although the immunodominant epitopes of Pf155/RESA are confined to the cytoplasmic side of the RBC membrane (16, 72, 198), a portion of the antigen appears to be accessible on the surface of infected RBCs (205). Evidence that immune responses to Pf155/RESA are protective includes efficient merozoite invasion inhibition in vitro by antibodies (18, 178), correlations between antibody levels and reduced parasitemias in seroepidemiological studies (10, 39, 55, 183, 192), and protection of *Aotus* monkeys by passive transfer of human antibodies (19).

Major antigens associated with the membrane of late-stage-infected RBCs are histidine-rich protein 1 (HRP-1) or KAHRP (230), PfEMP2 or MESA (46), and PfEMP3 (174). Although some have suggested that HRP-1 is localized to the submembranous part of the knob protrusions (230), it appears to be sensitive to trypsin digestion of intact RBCs (245). Furthermore, antibodies to HRP-1 are able to disrupt rosettes formed by the binding of noninfected RBCs to infected RBCs (30), indicating that a part of HRP-1 or a cross-reactive antigen is exposed on the RBC surface. PfEMP3, a recently described high-molecular-weight antigen (M_r, 315,000), shows the same localization in the infected RBC as HRP-1 (173, 174). Like HRP-1, PfEMP3 is thought to be involved in the formation of knobs or other perturbations of the RBC membrane, but it appears not to be essential for cytoadherence, and whether any part of the antigen is surface exposed is not clear.

Antigens Associated with the Surface of the Parasite

Merozoite surface protein 1 (MSP-1) of *P. falciparum* is the most-studied malarial asexual-blood-stage antigen. This glycoprotein (M_r, 185,000 to 220,000) is synthesized during schizogony. At the time of schizont rupture, it is proteolytically processed into fragments (95) that form noncovalently associated complexes on the surface of free merozoites (151). Secondary processing, when a 19-kDa fragment is formed, occurs prior to merozoite invasion. This fragment is the only part of MSP-1 carried into the RBC (20). It contains two cysteine-rich epidermal growth factor-like domains whose native conformation is crucial for reactivity with invasion-inhibitory antibodies (20, 36).

MSP-1 varies in size between different strains of *P. falciparum*, and its gene consists of strain-variable sequences separated by conserved and semiconserved sequences (95, 226). Genetic analyses of *P. falciparum* strains and isolates indicate that the antigen exists in two main allelic forms (226), but a third allelic form has also been described (184). Antibodies to MSP-1 raised in mice or rabbits identify conserved, allele- and strain-specific B-cell epitopes (95, 107, 170), but the natural antibody response in humans appears to be predominantly directed against nonconserved strain-specific regions (75, 163). Epitopes recognized by human T cells have also been mapped within nonpolymorphic regions of MSP-1 (48, 199, 210, 214).

MSP-2 is a polypeptide with an M_r of 45,000 whose N- and C-terminal regions are highly conserved but whose central part contains an extensive domain of repeats that vary in number, length, and sequence (216). The repeats are flanked by nonrepetitive variable regions that define two allelic families of the antigen. The central part of the antigen appears to be immunodominant with regard to humoral responses (118, 204), while T-cell epitopes have been identified in both the conserved and the variable regions of the antigen (199).

Recently, MSP-3, a 48-kDa merozoite surface antigen, was identified (169). The antigen appears to be a major target for the cytophilic antibodies in malaria-immune individuals that in cooperation with monocytes inhibit parasite growth in vitro (169).

Some parasite proteins are transported from the intracellular parasite through the parasitophorous vacuole membrane and RBC cytoplasm to the RBC surface. The *P. falciparum* antigen Ag5.1 (exp-1, CRA, QF116) was first identified by a monoclonal antibody that recognizes an epitope shared between erythrocytic stages and the sporozoite surface (101). The gene for Ag5.1 includes a region of sequence homology with the circumsporozoite protein (CS protein) repeat region that is probably the basis for this cross-reactivity (102). The antigen (M_r, 23,000) is acylated by myristic acid (124). When exported from the intraerythrocytic parasite, it becomes associated with the parasitophorous vacuole membrane and cytoplasmic vesicles of the RBC (123, 124, 211). It is believed to be involved in the transport of parasite proteins to the RBC surface (124, 211).

HRP-2 (M_r, 65,000 to 85,000) is rich in histidine (34%) and alanine (37%) arranged in numerous tandem repeats (103). HRP-2 is synthesized continuously from the early ring stage and is transported in a vesicular compartment to the RBC surface, from which it is secreted (105). HRP-2 is associated with the RBC membrane (103) and is found in the culture medium as well as in the plasma of *P. falciparum*-infected individuals (172).

The serine-stretch protein SERP (SERA, p126, p140) is characterized by a sequence of 37 consecutive serine residues (130). The antigen (M_r, 113,000 or 126,000) is associated with the membranes that surround the parasitophorous vacuole (57, 130). It is structurally similar to cysteine proteases and thus may function as a proteinase at the time of merozoite release from the infected RBC. Antibodies against SERP appear to inhibit merozoite dispersal from mature schizonts (148).

Exoantigens

Exoantigens are a heterogeneous group of soluble antigens that are released when *P. falciparum* schizonts rupture. They are found in the medium of in vitro cultures of the parasite (31, 117, 195, 254) and in the plasma of individuals infected with *P. falciparum* (255). In addition to some of the antigens already mentioned (HRP-2 and Pf155/RESA), this group also includes the heat-stable and serologically highly diverse S-antigens (104). The possible involvement of exoantigens in protective immunity is suggested by in vitro growth inhibition of *P. falciparum* by affinity-purified human antibodies and murine monoclonal antibodies (111, 116, 203). Furthermore, James et al. (113, 114) obtained depressed peak parasitemias and the appearance of crisis forms of the parasite in monkeys immunized with purified exoantigen mixtures, indicating that these antigens may induce parasite-neutralizing immune responses. Defined exoantigens presently considered as vaccine candidate antigens include a polypeptide with an M_r of 96,000 associated with protective immunity against *P. falciparum* in the squirrel monkey (61, 115, 120) and the glutamate-rich protein (GLURP; M_r, 22,000) found in the parasitophorous vacuole and on the surface of newly released merozoites (21).

Some exoantigens induce TNF and IL-6 production in human peripheral blood mononuclear cells in vitro, but this induction appears to be due mainly to phosphatidylinositol-containing moieties associated with the protein antigens (14, 15, 112). This phosphoglycolipid, which appears to be malaria specific, may become a major target in developing an antidisease vaccine against malaria (for details, see chapter 7).

PROGRESS TOWARD DEVELOPING VACCINES BASED ON SPECIFIC ANTIGENS

General

The antigenic complexity of the erythrocytic stages of *Plasmodium* spp. makes selection of vaccine candidate antigens difficult. Furthermore, since many of the relevant antigens (e.g., MSP-1 and MSP-2) display structural and antigenic polymorphisms, construction of a subunit vaccine may even make it necessary to select antigenic fragments that are both conserved and immunogenic. Using recombinant proteins or synthetic peptides as immunogens poses the additional problem that these components may not present the tertiary structure of the native antigen and thus may not give rise to antibody responses to assembled epitopes (170). However, a relatively high proportion of antibodies raised against recombinant malaria proteins or synthetic peptides react well with their corresponding native antigens, probably because their immunodominant structures often are linear epitopes located in regions of tandemly repeated amino acid sequences. The usefulness of epitopes located in the repeats has also been debated, primarily because of their frequent cross-reactivity, which is caused by the occurrence of similar but not identical repeat sequences in different antigens. This cross-reactivity has been claimed to prevent selection of high-affinity antibodies in the course of an infection, resulting in the accumulation of relatively low-affinity and therefore nonprotective antibodies (9). Be this as it may, this cross-reactivity would not reduce the usefulness of such epitopes for a subunit vaccine, since the epitopes can be incorporated into vaccine immunogens in a modified form that favors elicitation of non-cross-reacting specific antibodies of high affinity.

The first successful attempt to induce protection against asexual blood stages with a defined isolated antigen was made in the rodent *P. yoelii* system (96) with an affinity-purified protein (M_r, 230,000) analogous to MSP-1 in *P. falciparum* (94). However, in this as well as in most subsequent preclinical vaccination studies with defined asexual-blood-stage antigens, parasitemias developed in the vaccinated groups but were markedly depressed compared with those in the controls. Thus, successful vaccination apparently induced an immune response that depressed parasitemia enough to permit control of infection by the responses subsequently mounted. Protection in these experiments appeared to be antibody mediated, since similar results were obtained by passive transfer of monoclonal antibodies to the *P. yoelii* antigen (73). Similarly, passive transfer of a monoclonal antibody recognizing the MSP-1 analog of *P. chabaudi* protected mice against a *P. chabaudi* challenge (27, 140).

PfMSP-1

P. falciparum MSP-1 (PfMSP-1) has been investigated in several vaccination trials in monkeys. Siddiqui et al. (209) obtained complete protection against challenge with a homologous strain of *P. falciparum* in *Aotus* monkeys by using MSP-1 purified from cultured parasites and administered in Freund's complete adjuvant. Serum from the protected animals also inhibited growth of *P. falciparum* in vitro (106). In contrast, recombinant proteins or synthetic peptides corresponding to both conserved and variant MSP-1 sequences gave only partial protection in other monkey trials (38, 68, 86, 97). In one of these studies in which protection against a heterologous parasite strain was obtained, antibody levels and protection were correlated (68), suggesting that the conserved regions of MSP-1 may induce protective immune responses. While the complete MSP-1 molecule gives good antibody responses in experimental animals (37), shorter fragments and synthetic peptides containing a limited number of epitopes may give no or poor responses because of genetic (MHC) restrictions (132). Such restrictions are usually bypassed by presenting the peptide coupled to a carrier protein that provides a sufficient number of T-cell epitopes. In inbred mice, the restricted antibody response to a 20-amino-acid MSP-1 peptide from *P. falciparum* was overcome by coupling it to a tetanus toxoid-derived peptide representing a "promiscuous"-"universal" T-cell epitope recognized in the context of several different MHC class II molecules (132). Similar results had earlier been obtained by coupling the MSP-1 peptide to a 21-amino-acid peptide corresponding to the universal T-cell epitope CS.T3 of the *P. falciparum* CS protein (212). Furthermore, while immunization with a recombinant polypeptide (190L) representing an N-terminal fragment of MSP-1 gave only poor protection against *P. falciparum* challenge in *Aotus* monkeys, addition of the CS.T3 sequence to 190L by genetic engineering improved immunogenicity so much that three of four vaccinated monkeys became fully protected against challenge (88). There was no correlation between protection and either levels of antibodies to the immunizing antigens or the capacity of the sera to inhibit merozoite invasion in vitro. However, sera from the protected animals contained elevated levels of IFN-γ, indicating involvement of cell-mediated immune mechanisms in parasite neutralization. Much interest with regard to vaccine development is focused now on the C-terminal fragment MSP-1_{19}, because its capacity to induce protective immune responses is suggested both by experiments in the murine *P. yoelii* model (36, 51, 52, 143) and by a vaccination trial with the *P. falciparum* antigen in *Aotus* monkeys (134).

Other Potential Vaccine Candidates from *P. falciparum*

Only a few other defined asexual-blood stage antigens have been carried through to the construction of vaccine immunogens and testing. The first vaccine trial in *Aotus* monkeys with Pf155/RESA in Freund's complete adjuvant, using recombinant proteins expressing the immunodominant repeat regions of the antigen, gave partial protection against *P. falciparum* challenge (44). Subsequent vaccination trials in *Aotus* or *Saimiri* monkeys of various recombinant or synthetic immunogens based on this antigen failed to give protection (17, 45, 189), although in some studies, an inverse correlation between levels of parasitemia and serologic response to one of the repeat peptides was found (45, 189). No protection was obtained when monkeys were given injections with four different recombinant vaccinia virus preparations expressing Pf155/RESA, MSP-1, MSP-2, and the rhoptry antigen AMA-1 (189).

Using a band (M_r, 41,000) from separated *P. falciparum* proteins, Perrin et al. obtained partial protection in *Saimiri* monkeys (180). A recombinant protein corresponding to the major component in this band, an aldolase active antigen, was ineffective in inducing protection (89), and vaccination with other recombinant proteins indicated that a minor component in the band, designated 41-3, was the antigen responsible for the protection (66).

For MSP-2-based vaccines to *P. falciparum*, the murine *P. chabaudi* malaria offers a useful test system, as there is a high degree of cross-reactivity between the MSP-2 homologs of the two species (204). Synthetic peptides that correspond to invariant sequences of PfMSP-2 and are conjugated to diphtheria toxoid induce protection in mice against *P. chabaudi* challenge. There is an inverse correlation between the development of parasitemia and the antibody titer in the mice. Whether a similar degree of protection to *P. falciparum* can be obtained with these immunogens in monkeys or humans remains to be seen.

The *P. falciparum* antigen HRP-2 has been a vaccine component in several trials in *Aotus* monkeys, which usually were splenectomized before challenge with infected RBCs. Immunization with a recombinant fusion protein containing a 165-amino-acid sequence of HRP-2 in $Al(OH)_3$-lecithin-saponin adjuvant resulted in a delayed onset of parasitemia after challenge and a markedly depressed peak parasitemia (131). In subsequent vaccine trials, the immunogen consisted either of recombinant fusion proteins containing sequences of both HRP-2 and SERP or of an additional sequence from the conserved N-terminal region of MSP-1 (66, 127, 128). Monkeys receiving the hybrid molecules in polyalphaolefine as adjuvant developed high antibody responses against both HRP-2 and SERP. Although the control monkeys reached high peak parasitemias and most of them had to be treated with antimalarial agents, the parasitemias were strongly depressed in the vaccinated monkeys, which cleared their parasitemias without treatment.

The SERP antigen alone may also induce protective immunity, as demonstrated by immunization of *Saimiri* monkeys with the antigen (designated p140 or p126) purified from cultured parasites with either Freund's complete adjuvant (181) or the immunogen adsorbed onto aluminum hydroxide (57). A similar depression of parasitemias and self-cure was seen when *Aotus* monkeys were immunized with either a 262-amino-acid fragment (sequence 24 through 285) or a 483-amino-acid fragment (sequence 24 through 506), the latter occurring as part of a fusion protein with human IFN-γ (109, 110). The monkeys having the highest prechallenge levels of antipeptide antibodies did not develop countable parasitemias, which suggests a relationship between the humoral immune response to SERP and immune protection.

Ag5.1 also has the capacity to induce partial protection against a *P. falciparum* challenge in *Saimiri* monkeys when a recombinant antigen emulsified in Freund's adjuvant is used for immunization (32). The gene for Ag5.1 has further been used to make a construct coding for a hybrid molecule containing the sequences of Ag5.1 and 19 repeats of the *P. falciparum* CS protein (32). Immunization of mice, monkeys, and humans with this hybrid molecule gave rise to high titers of antibodies reactive with the CS protein and Ag5.1 but had no clear-cut protective effect against either sporozoite or blood-stage infections in humans (221).

Recently, partial protection was obtained in *Saimiri* monkeys that had experienced malaria and were then immunized with a mixture of recombinant fragments from both Pf332 and a 160-kDa *P. falciparum* protein (179). Protection correlated with the presence of opsonizing antibodies in the serum.

A synthetic peptide immunogen, SPf66, that contains sequences from three blood-stage antigens has been used in several monkey and human vaccine trials and is discussed in detail in chapter 9.

VACCINE-RELATED PROBLEMS

Trials in Monkeys

Although some of the vaccine trials referred to above have had promising results, the overall outcome with regard to protection is not entirely satisfactory. In essence, two major factors may account for this. The first factor involves problems inherent in the monkey model itself. The second factor concerns some general problems of subunit vaccine efficacy, many of which seem to be amenable to considerable improvement.

Trials in monkeys have contributed important information about which immunogens may be suitable for inclusion in a vaccine. However, these trials have serious limitations. Thus, species differences in the acquired immune response to a given vaccine may make comparisons of humans and monkeys difficult, particularly if protection requires induction of antibodies belonging to special immunoglobulin subclasses (25) or is elicited via particular cellular pathways. In addition, the number of animals that can be included in a trial is usually limited, necessitating the use of well-adapted and highly virulent parasites for challenge in order to make evaluation of the results possible at all. Moreover, the common use of infected RBCs rather than mosquito bites for parasite challenge is in itself an unnatural and unfavorable way of immunizing against blood-stage infection. For these and other reasons, it seems that vaccines that give only partial protection in monkeys might actually be more efficacious in humans subjected to natural transmission. While this fact does not mean that we should not test vaccines in monkeys before moving on to humans, the problems mentioned should be taken into account when trial results are evaluated.

Immunogenicity

Traditionally, in the most efficient vaccines, live attenuated pathogens serve as immunogens. Although use of attenuated malaria parasites may become possible in the future, it is presently not a realistic option for the rapid development of malaria vaccines. "Live" subunit vaccines against malaria have, however, been prepared by insertion of plasmodial genes into live vectors such as vaccinia virus or attenuated *Salmonella* bacteria. Attempts to immunize monkeys with vaccinia virus carrying the genes for several *P. falciparum* blood-stage antigens have thus far met with limited success (189). The potential of recombinant vaccinia virus vaccines in malaria is indicated by mouse and rabbit antibody responses to different *P. falciparum* antigens (67, 141, 234). Oral vaccination of mice with recombinant attenuated *Salmonella typhimurium* against the CS protein of *Plasmodium berghei* gives partial protection that is dependent on cytotoxic $CD8^+$ T cells (2, 200). Immunization of mice with *S. typhimurium* transformants expressing the *P. falciparum* merozoite antigens SERP and HRP-2 on their exterior surface gives rise to significant levels of antimalarial antibodies (207). However, ongoing vaccinations of monkeys with *S. typhimurium* expressing *P. falciparum* MSP-1 have thus far resulted only in delayed onset of parasitemia but no protection (132a).

A novel and potentially very useful development in vaccine strategies is based on immunization with naked DNA coding for the antigen of interest (60, 227). This approach was recently used in experimental malaria systems in which protection of mice against *P. yoelii* infection was obtained by immunization of plasmid DNA encoding the *P. yoelii* CS protein (208). Furthermore, vaccination of mice with DNA encoding the C terminus of *P. yoelii* MSP-1 induced partial protection against infection by asexual blood stages of this parasite (77).

Although insufficient immunogenicity is usually not a problem with live vaccines, it is a major concern with subunit vaccines administered in the conventional way without live vectors. In order to induce strong and long-lasting immune responses as well as a good memory function, the vaccine immunogen must have the capacity to efficiently stimulate the T-cell system. For vaccines to malaria parasite blood stages, efficient stimulation of $CD4^+$ T cells of both the TH1 and the TH2 phenotypes, which mediate antibody-independent delayed-type hypersensitivity reactions and provide antibody help (138), appears to be crucial. To provide the necessary T-helper epitopes, the malaria immunogen may be combined with a carrier protein either by recombinant DNA technology or by chemical coupling, as exemplified in the first two human vaccine trials against *P. falciparum* sporozoite infection (12, 90). More recently, good immune responses to *P. falciparum* sporozoites were obtained in humans vaccinated with CS protein epitopes conjugated to hepatitis B surface antigen (246) or to purified protein derivative from *Mycobacterium tuberculosis* in mice primed with *Mycobacterium bovis* BCG (146). The latter system is of particular interest because of the common natural priming of most humans to mycobacteria and the nature of the carrier protein, identified as a heat shock protein (145). To our knowledge, no application of these systems to vaccines against *P. falciparum* blood stages has been reported.

In order to induce an efficient parasite-specific memory that will ensure boosting following natural reinfection after vaccination, it is desirable to develop a vaccine in which both the carrier and the antigenic part of the immunogen are derived from the parasite. For this reason, efforts have been made to synthesize immunogens in which previously defined T- and B-cell epitopes from major malarial antigens are combined (78). To overcome the problem of MHC restriction, which is a major problem when the number of T-cell epitopes in a synthetic vaccine is limited, efforts have also been made to identify promiscuous or universal epitopes that can be presented to T cells by a large number of MHC alleles. As mentioned above, when linked to poorly immunogenic fragments of the *P. falciparum* antigen MSP-1, the T-cell epitope CS.T3 from the *P. falciparum* CS protein induces good anti-MSP-1 antibodies in both responder and nonresponder mice (132). A similar composite immunogen also induces protection against blood-stage infection in *Aotus* monkeys (88). As mentioned, protection in the latter model appears to be an antibody-independent expression of cellular immunity associated with elevated levels of IFN-γ in serum.

Instead of using linear peptides as immunogens, further improvements of peptide presentation have recently been achieved by using multiple antigenic peptide constructs in which short peptides containing T and B epitopes in a 1:1 ratio branch out from an oligolysine backbone (225). This system has been used in mice to induce strong immune responses to the CS proteins of *P. berghei*, *P. vivax*, and *P. falciparum* (157, 164, 182, 225, 250), and a few similar immunogens containing *P. falciparum* blood-stage epitopes have been produced (4, 7, 11). In rabbits, multiple antigenic peptides based on repeat sequences from Pf332 or Pf155/RESA induce antibodies with a high capacity to inhibit *P. falciparum* growth in vitro (7).

Adjuvants and Delivery Systems

Most of the immunogens described in the foregoing require incorporation into adjuvants in order to produce satisfactory vaccine effects. Because the most efficient experimental adjuvant, Freund's complete adjuvant, is not acceptable for human use and the commonly used aluminum hydroxide is insufficient, the search for better adjuvants for malaria vaccines has high priority. One promising new adjuvant system of low toxicity and pyrogenicity makes use of monophosphoryl lipid A incorporated into liposomes containing the antigen. In some cases, mycobacterial components such as muramyl dipeptide (MDP) are also included. Promising results have been obtained in monkeys and also in human volunteers vaccinated against *P. falciparum* sporozoites (74, 190, 191). Immunization of rabbits with *P. falciparum* MSP-1 in a similar adjuvant elicited immune responses fully comparable to those obtained with Freund's complete adjuvant (107).

Some of the most immunogenic plasmodial sequences are variant and therefore not suitable for inclusion in a vaccine. The choice of invariant and usually less immunogenic epitopes necessitates not only efficient adjuvants but also proper delivery systems that ensure sufficiently long retention of the immunogen in an active form and in the proper lymphoid organs. Live vectors used as immunogen sources are considered the most efficient delivery system in this respect. Another way of efficient delivery is incorporation of the immunogen into liposomes, as mentioned above. Similar delivery systems are the immunostimulating complexes (ISCOMS), which are stable lipid particles to which the antigen is attached by hydrophobic interactions. The particles that also contain adjuvant give rise to strong and long-lasting immune responses as well as to protection from infection (160, 161). ISCOM-based vaccines also appear to be efficient inducers of cytotoxic T cells of the $CD8^+$ phenotype (224, 240). A further well-defined delivery system consists of biodegradable microspheres composed of polylactide and polyglycolide in different proportions. These microspheres, which are acceptable for human use, can be programmed to release the immunogen over a very long time under controlled conditions (3, 62, 155) and provide excellent immune protection in experimental bacterial infections in mice (63). Experiments on using this system in transmission-blocking *P. falciparum* vaccines are ongoing (124a).

SUMMARY AND CONCLUSIONS

The primary purposes of malaria vaccines that attack infected RBCs are the prevention of mortality and the control of morbidity of the disease. In rodent malaria, the relative importance of antibody-dependent as opposed to antibody-independent cellular effector mechanisms that protect against blood-stage infection varies greatly, depending on both the plasmodial species and the mouse strain being infected. Although plasmodial disease in mice is very different from that in humans, many of the immune mechanisms that regulate malaria infection in rodents also apply to human infection.

In *P. falciparum* malaria, antibodies to antigens of the asexual blood stages are important for parasite neutralization, a process in which opsonizing IgG antibodies interacting with nonlymphoid effector cells appear to play a major role. Development of immunity to blood-stage infection in humans is regulated by the T-cell system. As in mice, malaria-specific CD4$^+$ T cells in humans appear to differ with regard to the lymphokines they produce (e.g., IL-4 versus IFN-γ), suggesting that these cells function either as helper cells for antibody production or as antibody-independent activators of macrophages or other effector cells. The possible roles of γ/δ-receptor-bearing T cells, which frequently occur in acute infection, and CD8$^+$ T cells in blood-stage infections remain to be elucidated. In any event, the central role of the T-cell system in the development of malaria immunity implies that immune responses to malarial epitopes are genetically restricted by the MHC system. Although this restriction is not easily seen in outbred human populations, it is a fact that has to be taken into account in vaccine development. Moreover, the intensity of human antibody- and T-cell responses to defined *P. falciparum* antigens is also genetically regulated by non-MHC genes. The nature of these genes or their products is unknown.

Because it is presently not possible to produce malaria vaccines based on attenuated or killed parasites, efforts in this area are focusing on the design of subunit vaccines. Several antigens have been identified as potential immunogens for a vaccine against the asexual blood stages of *P. falciparum*. With the exception of the synthetic peptides SPf66, which has already been tested in humans (chapter 9), all vaccine trials of the antigens described in this chapter have been in monkeys. Some of the trials suggested that protection can be achieved by immunization with single antigens such as MSP-1, but the efficacy of such vaccines in larger trials remains to be established. Promising results have also been obtained by using vaccines in which the immunogens were hybrid molecules including sequences from two or three antigens in the same construct. A number of reasons speak for the inclusion of several antigens into a vaccine, because including them may help in bypassing possible genetic restrictions and also will make it difficult for the parasite to evade the vaccine by deleting or mutating genes coding for vaccine immunogens.

Vaccine candidate antigens in addition to those described above may well exist. However, adding new antigens to the existing list is not the problem at this time. Rather, continued development of malaria vaccines in general has to focus on (i) improvement of the immunogenicity of already known vaccine candidates and (ii) optimization of adjuvant and delivery systems. The first point involves selection and possible modulation of proper B- and T-cell epitopes and their joining into efficient constructs such as the multiple-antigenic-peptide system. Several experiments have already documented the usefulness of including universal (promiscuous) T-cell epitopes into such synthetic vaccines to circumvent MHC restriction and to avoid carrier proteins of nonplasmodial origin. The second point concerns the choice of proper delivery systems, in the form either of live recombinant vectors or of liposomes or related vehicles. The present rapid advances in these areas make one hope that efficient vaccines attacking malaria-infected RBCs will be available for human trials within the not too distant future.

REFERENCES

1. **Abel, L., M. Cot, L. Mulder, P. Carnevale, and J. Feingold.** 1992. Segregation analysis detects a major gene controlling blood infection levels in human malaria. *Am. J. Hum. Genet.* **50:**1308–1317.
2. **Aggarwal, A., S. Kumar, R. Jaffe, D. Hone, M. Gross, and J. Sadoff.** 1990. Oral *Salmonella*: malaria circumsporozoite recombinants induce specific CD8+ cytotoxic T cells. *J. Exp. Med.* **172:**1083–1090.
3. **Aguado, M. T., and P.-H. Lambert.** 1992. Controlled-release vaccines—biodegradable polyactide/polyglycolide (PL/PG) microspheres as antigen vehicles. *Immunobiology* **184:**113–125.
4. **Ahlborg, N. 1995.** Synthesis of a diepitope multiple antigen peptide containing sequences from two malaria antigens using Fmoc chemistry. *J. Immunol. Methods* **179:**269–275.
5. **Ahlborg, N., K. Berzins, and P. Perlmann.** 1991. Definition of the epitope recognized by the *Plasmodium falciparum*-reactive human monoclonal antibody 33G2. *Mol. Biochem. Parasitol.* **46:**89–96.
6. **Ahlborg, N., J. Iqbal, L., Björk, S. Ståhl, P. Perlmann, and K. Berzins.** *Plasmodium falciparum*: differential parasite growth inhibition mediated by antibodies to the antigens Pf332 and Pf155/RESA. *Exp. Parasitol.*, in press.
7. **Ahlborg, N., J. Iqbal, M. Hansson, M. Uhlén, D. Mattei, P. Perlmann, S. Ståhl, and K. Berzins.** 1995. Immunogens containing sequences from antigen Pf332 induce *Plasmodium falciparum*-reactive antibodies which inhibit parasite growth but not cytoadherence. *Parasite Immunol.* **17:**341–352.
8. **Aikawa, M., M. Torii, A. Sjölander, K. Berzins, P. Perlmann, and L. H. Miller.** 1990. Pf155/RESA antigen is localized in dense granules of *Plasmodium falciparum* merozoites. *Exp. Parasitol.* **71:**326–329.
9. **Anders, R. F.** 1986. Multiple cross-reactivities amongst antigens of *Plasmodium falciparum* impair the development of protective immunity against malaria with special reference to oxidant stress. *Parasite Immunol.* **8:**529–539.
10. **Astagneau, P., C. Chougnet, J. P. Lepers, M. D. Andriamangatiana-Rason, and P. Deloron.** 1994. Antibodies to the 4-mer repeat of the ring-infected erythrocyte surface antigen (Pf155/RESA) protect against *Plasmodium falciparum* malaria. *Int. J. Epidemiol.* **23:**169–175.
11. **Baleux, F., and P. Dubois.** 1992. Novel version of multiple antigenic peptide allow-

ing incorporation on a cysteine functionalized lysine tree. *Int. J. Pept. Protein Res.* **40:**7–12.

12. **Ballou, W. R., S. L. Hoffman, J. A. Sherwood, M. R. Hollingdale, F. A. Neva, W. T. Hockmeyer, D. M. Gordon, I. Schneider, R. A. Wirtz, J. F. Young, G. F. Wasserman, P. Reeve, C. L. Diggs, and J. D. Chulay.** 1987. Safety and efficacy of a recombinant DNA *Plasmodium falciparum* sporozoite vaccine. *Lancet* **i:**1277.
13. **Baruch, D. I., B. L. Pasloske, H. B. Singh, X. H. Bi, X. C. Ma, M. Feldman, T. F. Taraschi, and R. J. Howard.** 1995. Cloning the *P. falciparum* gene encoding PfEMP1, a malarial variant antigen and adherence receptor on the surface of parasitized human erythrocytes. *Cell* **82:**77–87.
14. **Bate, C. A. W., and D. Kwiatkowski.** 1994. A monoclonal antibody that recognizes phosphatidylinositol inhibits induction of tumor necrosis factor alpha by different strains of *Plasmodium falciparum. Infect. Immun.* **62:**5261–5266.
15. **Bate, C. A. W., and D. P. Kwiatkowski.** 1994. Stimulators of tumour necrosis factor production released by damaged erythrocytes. *Immunology* **83:**256–261.
16. **Berzins, K.** 1991. Pf155/RESA is not a surface antigen of *Plasmodium falciparum*-infected erythrocytes. *Parasitol. Today* **7:**193–194.
17. **Berzins, K., S. Adams, J. R. Broderson, C. Chizzolini, M. Hansson, K. Lövgren, P. Millet, C. M. Morris, H. Perlmann, P. Perlmann, A. Sjölander, S. Ståhl, J. S. Sullivan, M. Troye-Blomberg, B. Wåhlin Flyg, and W. E. Collins.** 1995. Immunogenicity in *Aotus* monkeys of ISCOM formulated repeat sequences from the *Plasmodium falciparum* asexual blood stage antigen Pf155/RESA. *Vaccine Res.* **4:**121–133.
18. **Berzins, K., H. Perlmann, B. Wåhlin, J. Carlsson, M. Wahlgren, R. Udomsangpetch, A. Björkman, M. E. Patarroyo, and P. Perlmann.** 1986. Rabbit and human antibodies to a repeated amino acid sequence of a *Plasmodium falciparum* antigen, Pf155, react with the native protein and inhibit merozoite invasion. *Proc. Natl. Acad. Sci. USA* **83:**1065–1069.
19. **Berzins, K., H. Perlmann, B. Wåhlin, H.-P. Ekre, B. Högh, E. Petersen, B. Wellde, M. Schoenbechler, J. Williams, J. Chulay, and P. Perlmann.** 1991. Passive immunization of *Aotus* monkeys with human antibodies to the *Plasmodium falciparum* antigen Pf155/RESA. *Infect. Immun.* **59:**1500–1506.
20. **Blackman, M. J., H.-G. Heidrich, S. Donachie, J. S. McBride, and A. A. Holder.** 1990. A single fragment of a malaria merozoite surface protein remains on the parasite during red cell invasion and is the target of invasion-inhibiting antibodies. *J. Exp. Med.* **172:**379–382.
21. **Borre, M. B., M. Dziegiel, B. Högh, E. Petersen, K. Rieneck, E. Riley, J. F. Meis, M. Aikawa, K. Nakamura, M. Harada, A. Wind, P. H. Jakobsen, J. Cowland, S. Jepsen, N. H. Axelsen, and J. Vuust.** 1991. Primary structure and localization of a conserved immunogenic *Plasmodium falciparum* glutamate rich protein (GLURP) expressed in both the preerythrocytic and erythrocytic stages of the vertebrate life cycle. *Mol. Biochem. Parasitol.* **49:**119–132.
22. **Borst, P., W. Bitter, R. McCulloch, F. Van Leeuwen, and G. Rudenko.** 1995. Antigenic variation in malaria. *Cell* **82:**1–4.
23. **Botto, M., A. K.-L. So, C. M. Giles, P. D. Mason, and M. J. Walport.** 1990. HLA class I expression on erythrocytes and platelets from patients with systemic lupus erythematosus, rheumatoid arthritis and from normal subjects. *Br. J. Haematol.* **75:**106–111.
24. **Bouharoun-Tayoun, H., P. Attanath, A. Sabchareon, T. Chongsuphajaisiddhi, and P. Druilhe.** 1990. Antibodies that protect humans against *Plasmodium falciparum* blood stages do not on their own inhibit parasite growth and invasion in vitro, but act in cooperation with monocytes. *J. Exp. Med.* **172:**1633–1641.

25. **Bouharoun-Tayoun, H., and P. Druilhe.** 1992. *Plasmodium falciparum* malaria: evidence for an isotype imbalance which may be responsible for delayed acquisition of protective immunity. *Infect. Immun.* **60:**1473–1481.
26. **Bouharoun-Tayoun, H., C. Oeuvray, F. Lunel, and P. Druilhe.** 1995. Mechanisms underlying the monocyte-mediated antibody-dependent killing of *Plasmodium falciparum* asexual blood stages. *J. Exp. Med.* **182:**409–418.
27. **Boyle, D. B., C. I. Newbold, C. C. Smith, and K. N. Brown.** 1982. Monoclonal antibodies that protect in vivo against *Plasmodium chabaudi* recognize a 250,000-Dalton parasite polypeptide. *Infect. Immun.* **38:**94–102.
28. **Brake, D. A., C. A. Long, and W. P. Weidanz.** 1988. Adoptive protection against *Plasmodium chabaudi adami* malaria in athymic nude mice by a cloned T cell line. *J. Immunol.* **140:**1989–1993.
29. **Brown, J., and M. E. Smalley.** 1980. Specific antibody-dependent cellular cytotoxicity in human malaria. *Clin. Exp. Immunol.* **41:**423–429.
30. **Carlson, J., G. Holmquist, D. W. Taylor, P. Perlmann, and M. Wahlgren.** 1990. Antibodies to a histidine rich protein (PfHRP1) disrupt spontaneously formed *Plasmodium falciparum* erythrocyte rosettes. *Proc. Natl. Acad. Sci. USA* **87:**2511–2515.
31. **Carlsson, J., K. Berzins, H. Perlmann, and P. Perlmann.** 1991. Studies on Pf155/RESA and other soluble antigens from in vitro cultured *Plasmodium falciparum. Parasitol. Res.* **77:**27–32.
32. **Caspers, P., H. Etlinger, H. Matile, J. R. Pink, D. Stüber, and B. Takács.** 1991. A *Plasmodium falciparum* malaria vaccine candidate which contains epitopes from the circumsporozoite protein and a blood stage antigen, 5.1. *Mol. Biochem. Parasitol.* **47:**143–150.
33. **Cavacini, L., L. A. Parke, and W. P. Weidanz.** 1990. Resolution of acute malarial infections by T-cell-dependent non-antibody-mediated mechanisms of immunity. *Infect. Immun.* **58:**2946–2950.
34. **Celada, A., A. Cruchaud, and L. H. Perrin.** 1982. Opsonic activity of human immune serum on in vitro phagocytosis of *Plasmodium falciparum* infected red blood cells by monocytes. *Clin. Exp. Immunol.* **47:**635–644.
35. **Celada, A., A. Cruchaud, and L. H. Perrin.** 1983. Assessment of immune phagocytosis of *Plasmodium falciparum* infected red blood cells by human monocytes and polymorphonuclear leukocytes. A method for visualizing infected red blood cells ingested by phagocytes. *J. Immunol. Methods* **63:**263–271.
36. **Chang, S. P., H. L. Gibson, C. T. Leeng, P. J. Barr, and G. S. N. Hui.** 1992. A carboxyl-terminal fragment of *Plasmodium falciparum* gp195 expressed by a recombinant baculovirus induces antibodies that completely inhibit parasite growth. *J. Immunol.* **149:**548–555.
37. **Chang, S. P., G. S. N. Hui, A. Kato, and W. A. Siddiqui.** 1989. Generalized immunological recognition of the major merozoite surface antigen (gp195) of *Plasmodium falciparum. Proc. Natl. Acad. Sci. USA* **86:**6343–6347.
38. **Cheung, A., J. Leban, A. R. Shaw, B. Merkli, J. Stocker, C. Chizzolini, C. Sander, and L. H. Perrin.** 1986. Immunization with synthetic peptides of a *Plasmodium falciparum* surface antigen induces antimerozoite antibodies. *Proc. Natl. Acad. Sci. USA* **83:**8328–8332.
39. **Chizzolini, C., A. Dupont, J. P. Akue, M. H. Kaufmann, A. S. Verdini, A. Pessi, and G. Del Giudice.** 1988. Natural antibodies against three distinct and defined antigens of *Plasmodium falciparum* in residents of a mesoendemic area in Gabon. *Am. J. Trop. Med. Hyg.* **39:**150–156.
40. **Chizzolini, C., G. E. Grau, A. Geinoz, and D. Schrijvers.** 1990. T lymphocyte inter-

feron-gamma production induced by *Plasmodium falciparum* antigen is high in recently infected non-immune and low in immune subjects. *Clin. Exp. Immunol.* **79:**95–99.

41. **Chougnet, C., M. Troye-Blomberg, P. Deloron, L. Kabilan, J. P. Lepers, J. Savel, and P. Perlmann.** 1991. Human immune response in *Plasmodium falciparum* malaria—synthetic peptides corresponding to known epitopes of the Pf155/RESA antigen induce production of parasite-specific antibodies in vitro. *J. Immunol.* **147:**2295–2301.
42. **Clark, I. A., S. Ilschner, J. D. MacMicking, and W. B. Cowden.** 1990. TNF and *Plasmodium berghei* ANKA-induced cerebral malaria. *Immunol. Lett.* **25:**195–198.
43. **Cohen, S., I. A. McGregor, and S. Carrington.** 1961. Gamma-globulin and acquired immunity to human malaria. *Nature* (London) **192:**733–737.
44. **Collins, W. E., R. F. Anders, M. Pappaioanou, G. H. Campbell, G. V. Brown, D. J. Kemp, R. L. Coppel, J. C. Skinner, P. M. Andrysiak, J. M. Favaloro, L. M. Corcoran, J. R. Broderson, G. F. Mitchell, and C. C. Campbell.** 1986. Immunization of *Aotus* monkeys with recombinant proteins of an erythrocyte surface antigen of *Plasmodium falciparum. Nature* (London) **323:**259–262.
45. **Collins, W. E., R. F. Anders, T. K. Ruebush, D. J. Kemp, G. C. Woodrow, G. H. Campbell, G. V. Brown, D. O. Irving, N. Goss, V. K. Filipski, R. L. Coppel, J. R. Broderson, L. M. Thomas, D. Pye, J. C. Skinner, C. Wilson, P. S. Stanfill, and P. M. Procell.** 1991. Immunization of owl monkeys with the ring-infected erythrocyte surface antigen of *Plasmodium falciparum. Am. J. Trop. Med. Hyg.* **44:**34–41.
46. **Coppel, R. L., S. Lustigman, L. Murray, and R. F. Anders.** 1988. MESA is a *Plasmodium falciparum* phosphoprotein associated with the erythrocyte membrane skeleton. *Mol. Biochem. Parasitol.* **31:**223–232.
47. **Crisanti, A., K. Früh, H. M. Müller, and H. Bujard.** 1990. The T cell reactivity against the major merozoite protein of *Plasmodium falciparum. Immunol. Lett.* **25:**143–148.
48. **Crisanti, A., H.-M. Müller, C. Hilbich, F. Sinigaglia, H. Matile, M. McKay, J. Scaife, K. Beyreuther, and H. Bujard.** 1988. Epitopes recognized by human T-cells map within the conserved part of the GP190 of *P. falciparum. Science* **240:**1324–1326.
49. **Culvenor, J. G., K. P. Day, and R. F. Anders.** 1991. *Plasmodium falciparum* ring-infected erythrocyte surface antigen is released from merozoite dense granules after erythrocyte invasion. *Infect. Immun.* **59:**1183–1187.
50. **Currier, J., H. P. Beck, B. Currie, and M. F. Good.** 1995. Antigens released at schizont burst stimulate *Plasmodium falciparum*-specific CD4 T cells from non-exposed donors: potential for cross reactive memory T cells to cause disease. *Int. Immunol.* **7:**821–833.
51. **Daly, T. M., and C. A. Long.** 1993. A recombinant 15-kilodalton carboxyl-terminal fragment of *Plasmodium yoelii yoelii* 17XL merozoite surface protein 1 induces a protective immune response in mice. *Infect. Immun.* **61:**2462–2467.
52. **Daly, T. M., and C. A. Long.** 1995. Humoral response to a carboxyl-terminal region of the merozoite surface protein-1 plays a predominant role in controlling blood-stage infection in rodent malaria. *J. Immunol.* **155:**236–243.
53. **Deans, J. A., and S. Cohen.** 1983. Immunology of malaria. *Annu. Rev. Microbiol.* **37:**25–49.
54. **Del Giudice, G., G. E. Grau, and P.-H. Lambert.** 1988. Host responsiveness to malaria epitopes and immunopathology. *Prog. Allergy* **41:**288–333.
55. **Deloron, P., G. H. Campbell, D. Brandling-Bennett, J. M. Roberts, I. K. Schwartz, J. S. Odera, A. A. Lal, C. O. Osanga, V. de la Cruz, and T. M. McCutchan.** 1989. Antibodies to *Plasmodium falciparum* ring-infected erythrocyte surface antigen and *P. falciparum* and *P. malariae* circumsporozoite proteins: seasonal prevalence in Kenyan villages. *Am. J. Trop. Med. Hyg.* **41:**395–399.
56. **Deloron, P., C. Chougnet, J. P. Lepers, S. Tallet, and P. Coulanges.** 1991. Protective

value of elevated levels of gamma interferon in serum against exoerythrocytic stages of *Plasmodium falciparum*. *J. Clin. Microbiol.* **29:**1757–1760.

57. **Delplace, P., A. Bhatia, M., Cagnard, D. Camus, G. Colombet, A. Debrabant, J.-F. Dubremetz, N. Dubreuil, G. Prensier, B. Fortier, A. Haq, J. Weber, and A. Vernes.** 1988. Protein p126: a parasitophorous vacuole antigen associated with release of *Plasmodium falciparum* merozoites. *Biol. Cell* **64:**215–221.
58. **Diggs, C. L., F. Hines, and B. T. Wellde.** 1995. *Plasmodium falciparum*: passive immunization of *Aotus lemurinus griseimembra* with immune serum. *Exp. Parasitol.* **80:**291–296.
59. **Diggs, C. L., B. T. Wellde, J. S. Anderson, R. M. Weber, and E. Rodriguez.** 1972. The protective effect of African human immunoglobulin G in *Aotus trivirgatus* infected with Asian *Plasmodium falciparum*. *Proc. Helminthol. Soc. Wash.* **39:**449–456.
60. **Donnelly, J. J., J. B. Ulmer, and M. A. Liu.** 1994. Immunization with polynucloetides. A novel approach to vaccination. *Immunologist* **2:**20–26.
61. **Dubois, P., J.-P. Dedet, T. Fandeur, C. Roussilhon, M. Jendoubi, S. Pauillac, O. Mercereau-Puijalon, and L. Pereira da Silva.** 1984. Protective immunization of the squirrel monkey against asexual blood stages of *Plasmodium falciparum* by use of parasite protein fractions. *Proc. Natl. Acad. Sci. USA* **81:**229–232.
62. **Eldridge, J. H., J. K. Staas, J. A. Meulbroek, J. R. McGhee, T. R. Tice, and R. M. Gilley.** 1991. Biodegradable microspheres as a vaccine delivery system. *Mol. Immunol.* **28:**287–294.
63. **Eldridge, J. H., J. K. Staas, J. A. Meulbroek, T. R. Tice, and R. M. Gilley.** 1991. Biodegradable and biocompatible poly(DL-lactide-co-glycolide) microspheres as an adjuvant for staphylococcal enterotoxin B toxoid which enhances the level of toxin-neutralizing antibodies. *Infect. Immun.* **59:**2978–2986.
64. **Elhassan, I. M., L. Hviid, G. Satti, B. Akerstrom, P. H. Jakobsen, J. B. Jensen, and T. G. Theander.** 1994. Evidence of endothelial inflammation, T cell activation, and T cell reallocation in uncomplicated *Plasmodium falciparum* malaria. *Am. J. Trop. Med. Hyg.* **51:**372–379.
65. **Elloso, M. M., H. C. van der Heyde, J. A. vande Waa, D. D. Manning, and W. P. Weidanz.** 1994. Inhibition of *Plasmodium falciparum* in vitro by human $\gamma\delta$ T cells. *J. Immunol.* **153:**1187–1194.
66. **Enders, B., E. Hundt, and B. Knapp.** 1992. Strategies for the development of an antimalarial vaccine. *Vaccine* **10:**920–927.
67. **Etlinger, H. M., and W. Altenburger.** 1991. Overcoming inhibition of antibody responses to a malaria recombinant vaccinia virus caused by prior exposure to wild type virus. *Vaccine* **9:**470–472.
68. **Etlinger, H. M., P. Caspers, H. Matile, H. J. Schoenfeld, D. Stueber, and B. Takacs.** 1991. Ability of recombinant or native proteins to protect monkeys against heterologous challenge with *Plasmodium falciparum*. *Infect. Immun.* **59:**3498–3503.
69. **Favaloro, J. M., R. L. Coppel, L. M. Corcoran, S. J. Foote, G. V. Brown, R. F. Anders, and D. J. Kemp.** 1986. Structure of the RESA gene of *Plasmodium falciparum*. *Nucleic Acids Res.* **14:**8265–8277.
70. **Fell, A. H., J. Currier, and M. F. Good.** 1994. Inhibition of *Plasmodium falciparum* growth in vitro by CD4+ and CD8+ T cells from non-exposed donors. *Parasite Immunol.* **16:**579–586.
71. **Fievet, N., B. Maubert, M. Cot, C. Chougnet, B. Dubois, J. Bickii, F. Migot, J.-Y. Le Hesran, Y. Frobert, and P. Deloron.** 1995. Humoral and cellular immune responses to synthetic peptides from the *Plasmodium falciparum* blood-stage antigen, Pf155/RESA, in Cameroonian women. *Clin. Immunol. Immunopathol.* **76:**164–169.
72. **Foley, M., L. Tilley, W. H. Sawyer, and R. F. Anders.** 1991. The ring-infected

erythrocyte surface antigen of *Plasmodium falciparum* associates with spectrin in the erythrocyte membrane. *Mol. Biochem. Parasitol.* **46:**137–148.

73. **Freeman, R. R., A. J. Trejdosiewicz, and G. A. M. Cross.** 1980. Protective monoclonal antibodies recognising stage-specific merozoite antigens of a rodent malaria parasite. *Nature* (London) **284:**366–368.
74. **Fries, L. F., D. M. Gordon, R. L. Richards, J. E. Egan, M. R. Hollingdale, M. Gross, C. Silverman, and C. R. Alving.** 1992. Liposomal malaria vaccine in humans: a safe and potent adjuvant strategy. *Proc. Natl. Acad. Sci. USA* **89:**358–362.
75. **Früh, K., O. Doumbo, H.-M. Müller, O. Koita, J. McBride, A. Crisanti, Y. Touré, and H. Bujard.** 1991. Human antibody response to the major merozoite surface antigen of *Plasmodium falciparum* is strain specific and short-lived. *Infect. Immun.* **59:**1319–1324.
76. **Gabriel, J., and K. Berzins.** 1983. Specific lysis of *Plasmodium yoelii* infected mouse erythrocytes with antibody and complement. *Clin. Exp. Immunol.* **52:**129–134.
77. **Gardner, M. J., R. Wang, R. C. Hedstrom, C. Long, P. Hobart, and S. L. Hoffman.** 1995. Partial protection against *Plasmodium yoelii* by a DNA vaccine encoding the C-terminus of MSP-1, p. 56–57, abstr. PI4. *In 7th Malaria Meeting of the British Society for Parasitology*. British Society for Parasitology, London.
78. **Good, M. F., W. L. Maloy, M. N. Lunde, H. Margalit, J. L. Cornette, G. L. Smith, B. Moss, L. H. Miller, and J. A. Berzofsky.** 1987. Construction of synthetic immunogen: use of new T-helper epitope on malaria circumsporozoite protein. *Science* **235:**1059–1062.
79. **Goodier, M. R., C. Lundqvist, M.-L. Hammarstrom, M. Troye-Blomberg, and J. Langhorne.** 1995. Cytokine profiles for human Vg9+ T cells stimulated by *Plasmodium falciparum. Parasite Immunol.* **17:**413–423.
80. **Grau, G. E., G. Bieler, S. De Kossodo, F. Tacchini-Cotier, P. Vassalli, P. F. Piquet, and P.-H. Lambert.** 1990. Significance of cytokine production and adhesion molecules in malarial immunopathology. *Immunol. Lett.* **25:**189–194.
81. **Grau, G. E., T. E. Taylor, M. E. Molyneux, J. J. Virima, P. Vassalli, M. Hommel, and P.-H. Lambert.** 1989. Tumor necrosis factor and disease severity in children with *Plasmodium falciparum* malaria. *N. Engl. J. Med.* **320:**1586–1591.
82. **Green, T. J., M. Morhardt, R. G. Brackett, and R. L. Jacobs.** 1981. Serum inhibition of merozoite dispersal from *Plasmodium falciparum* schizonts: indicator of immune status. *Infect. Immun.* **31:**1203–1208.
83. **Groux, H., R. Perraut, O. Garraud, J. P. Poingt, and J. Gysin.** 1990. Functional characterization of the antibody-mediated protection against blood stages of *Plasmodium falciparum* in the monkey *Saimiri sciureus. Eur. J. Immunol.* **20:**2317–2323.
84. **Guttinger, M., P. Romagnoli, L. Vandel, R. Meloen, B. Takacs, J. R. L. Pink, and F. Sinigaglia.** 1991. HLA polymorphism and T cell recognition of a conserved region of p190, a malaria vaccine candidate. *Int. Immunol.* **3:**899–906.
85. **Gysin, J., S. Gavoille, D. Mattei, A. Scherf, S. Bonnefoy, O. Mercereau-Puijalon, T. Feldmann, J. Kun, B. Müller-Hill, and L. Pereira da Silva.** 1993. In vitro phagocytosis inhibition assay for the screening of potential candidate antigens for sub-unit vaccines against the asexual blood stage of *Plasmodium falciparum. J. Immunol. Methods* **159:**209–219.
86. **Hall, R., J. E. Hyde, M. Goman, D. L. Simmons, I. A. Hope, M. Mackay, J. Scaife, B. Merkli, R. Richle, and J. Stocker.** 1984. Major surface antigen gene of a human malaria parasite cloned and expressed in bacteria. *Nature* (London) **311:**379–382.
87. **Helmby, H., L. Cavelier, U. Pettersson, and M. Wahlgren.** 1993. Rosetting *Plasmodium falciparum*-infected erythrocytes express unique strain-specific antigens on their surface. *Infect. Immun.* **61:**284–288.

88. **Herrera, M. A., F. Rosero, S. Herrera, P. Caspers, D. Rotmann, F. Sinigaglia, and U. Certa.** 1992. Protection against malaria in *Aotus* monkeys immunized with a recombinant blood-stage antigen fused to a universal T-cell epitope: correlation of serum gamma interferon levels with protection. *Infect. Immun.* **60:**154–158.
89. **Herrera, S., M. A. Herrera, B. L. Perlaza, Y. Burki, P. Caspers, H. Döbeli, D. Rotmann, and U. Certa.** 1990. Immunization of *Aotus* monkeys with *Plasmodium falciparum* blood-stage recombinant proteins. *Proc. Natl. Acad. Sci. USA* **87:**4017–4021.
90. **Herrington, D. A., D. F. Clyde, G. Losonsky, M. Cortesia, J. R. Murphy, J. Davis, S. Baqar, A. M. Felix, E. P. Heimer, D. Gillessen, E. Nardin, R. S. Nussenzweig, V. Nussenzweig, M. R. Hollingdale, and M. M. Levine.** 1987. Safety and immunogenicity in man of a synthetic peptide malaria vaccine against *Plasmodium falciparum* sporozoites. *Nature* (London) **328:**257–259.
91. **Hill, A. V. S., C. E. M. Allsopp, D. Kwiatkowski, N. M. Anstey, P. Twumasi, P. A. Rowe, S. Bennett, D. Brewster, A. J. McMichael, and B. M. Greenwood.** 1991. Common West African HLA antigens are associated with protection from severe malaria. *Nature* (London) **352:**595–600.
92. **Hill, A. V. S., J. Elvin, A. C. Willis, M. Aidoo, C. E. M. Allsopp, F. M. Gotch, X. M. Gao, M. Takiguchi, B. M. Greenwood, A. R. M. Townsend, A. J. McMichael, and H. C. Whittle.** 1992. Molecular analysis of the association of HLA-B53 and resistance to severe malaria. *Nature* (London) **360:**434–439.
93. **Ho, M., and H. K. Webster.** 1989. Immunology of human malaria. A cellular perspective. *Parasite Immunol.* **11:**105–116.
94. **Holder, A., R. Freeman, and C. Newbold.** 1983. Serological cross-reaction between high molecular weight proteins synthesized in blood schizonts of *Plasmodium yoelii, Plasmodium chabaudi* and *Plasmodium falciparum. Mol. Biochem. Parasitol.* **9:**191–196.
95. **Holder, A. A.** 1988. The precursor to major merozoite surface antigens: structure and role in immunity. *Prog. Allergy* **41:**72–97.
96. **Holder, A. A., and R. R. Freeman.** 1981. Immunization against blood-stage rodent malaria using purified parasite antigens. *Nature* (London) **294:**361–364.
97. **Holder, A. A., R. R. Freeman, and S. C. Nicholls.** 1988. Immunization against *Plasmodium falciparum* with recombinant polypeptides produced in *Escherichia coli. Parasite Immunol.* **10:**607–617.
98. **Hommel, M., P. H. David, and L. D. Oligino.** 1983. Surface alterations of erythrocytes in *Plasmodium falciparum* malaria. Antigenic variation, antigenic diversity and the role of the spleen. *J. Exp. Med.* **157:**1137–1148.
99. **Hommel, M., M. Hughes, P. Bond, and J. M. Crampton.** 1991. Antibodies and DNA probes used to analyze variant populations of the Indochina-1 strain of *Plasmodium falciparum. Infect. Immun.* **59:**3975–3981.
100. **Hommel, M., and S. Semoff.** 1988. Expression and function of erythrocyte-associated surface antigens in malaria. *Biol. Cell* **64:**183–203.
101. **Hope, I. A., R. Hall, D. L. Simmons, J. E. Hyde, and J. G. Scaife.** 1984. Evidence for immunological cross-reaction between sporozoites and blood stages of a human malaria parasite. *Nature* (London) **308:**191–194.
102. **Hope, I. A., M. Mackay, J. E. Hyde, M. Goman, and J. Scaife.** 1985. The gene for an exported antigen of the malaria parasite *Plasmodium falciparum* cloned and expressed in *Escherichia coli. Nucleic Acids Res.* **13:**369–379.
103. **Howard, R. J.** 1988. Malaria proteins at the membrane of *Plasmodium falciparum*-infected erythrocytes and their involvement in cytoadherence to endothelial cells. *Prog. Allergy* **41:**98–147.
104. **Howard, R. J., L. J. Panton, K. Marsh, I. T. Ling, E. J. Winchell, and R. J. M. Wil-**

son. 1986. Antigenic diversity and size diversity of *Plasmodium falciparum* antigens in isolates from Gambian patients. I. S-antigens. *Parasite Immunol.* **8:**39–55.

105. **Howard, R. J., S. Uni, M. Aikawa, S. B. Aley, J. H. Leech, A. M. Lew, T. E. Wellems, J. Rener, and D. W. Taylor.** 1986. Secretion of a malarial histidine-rich protein (PfHRP II) from *Plasmodium falciparum* infected erythrocytes. *J. Cell Biol.* **103:**1269–1277.
106. **Hui, G. S. N., and W. A. Siddiqui.** 1987. Serum from Pf195 protected *Aotus* monkeys inhibit *Plasmodium falciparum* growth in vitro. *Exp. Parasitol.* **64:**519–522.
107. **Hui, G. S. N., L. Q. Tam, S. P. Chang, S. E. Case, C. Hashiro, W. A. Siddiqui, T. Shiba, S. Kusumoto, and S. Kotani.** 1991. Synthetic low-toxicity muramyl dipeptide and monophosphoryl lipid A replace Freund complete adjuvant in inducing growth-inhibitory antibodies to the *Plasmodium falciparum* major merozoite surface protein, gp195. *Infect. Immun.* **59:**1585–1591.
108. **Hviid, L., T. G. Theander, N. H. Abdulhadi, Y. A. Abu-Zeid, R. A. Bayoumi, and J. B. Jensen.** 1991. Transient depletion of T cells with high LFA-1 expression from peripheral circulation during acute *Plasmodium falciparum* malaria. *Eur. J. Immunol.* **21:**1249–1253.
109. **Inselburg, J., I. C. Bathurst, J. Kansopon, P. J. Barr, and R. Rossan.** 1993. Protective immunity induced in *Aotus* monkeys by a recombinant SERA protein of *Plasmodium falciparum*: further studies using SERA 1 and MF75.2 adjuvant. *Infect. Immun.* **61:**2048–2052.
110. **Inselburg, J., D. J. Bzik, W. B. Li, K. M. Green, J. Kansopon, B. K. Hahm, I. C. Bathurst, P. J. Barr, and R. N. Rossan.** 1991. Protective immunity induced in *Aotus* monkeys by recombinant SERA proteins of *Plasmodium falciparum. Infect. Immun.* **59:**1247–1250.
111. **Jakobsen, P. H., S. Jepsen, and R. Agger.** 1987. Inhibitory monoclonal antibodies to soluble *Plasmodium falciparum* antigens. *Parasitol. Res.* **73:**518–523.
112. **Jakobsen, P. H., R. Moon, R. G. Ridley, C. A. W. Bate, J. Taverne, M. B. Hansen, B. Takacs, J. H. L. Playfair, and J. S. McBride.** 1993. Tumour necrosis factor and interleukin-6 production induced by components associated with merozoite proteins of *Plasmodium falciparum. Parasite Immunol.* **15:**229–237.
113. **James, M. A., C. J. Fajfar-Whetstone, I. Kakoma, M. M. Buese, G. W. Clabaugh, R. Hansen, and M. Ristic.** 1991. Immunogenicity and protective efficacy of affinity-purified *Plasmodium falciparum* exoantigens in *Aotus nancymai* monkeys. *Trop. Med. Parasitol.* **42:**49–54.
114. **James, M. A., I. Kakoma, M. Ristic, and M. Cagnard.** 1985. Induction of protective immunity to *Plasmodium falciparum* in *Saimiri sciureus* monkeys with partially purified exoantigens. *Infect. Immun.* **49:**476–480.
115. **Jendoubi, M., P. Dubois, and L. Pereira da Silva.** 1985. Characterization of one polypeptide antigen potentially related to protective immunity against the blood infection by *Plasmodium falciparum* in the squirrel monkey. *J. Immunol.* **134:**1941–1945.
116. **Jepsen, S.** 1983. Inhibition of in vitro growth of *Plasmodium falciparum* by purified antimalarial human IgG antibodies. *Scand. J. Immunol.* **18:**567–571.
117. **Jepsen, S., and B. J. Andersen.** 1981. Immunoadsorbent isolation of antigens from the culture medium of in vitro cultivated *Plasmodium falciparum. Acta Pathol. Microbiol. Scand.* **89:**99–103.
118. **Jones, G. L., H. M. Edmundson, R. Lord, L. Spencer, R. Mollard, and A. J. Saul.** 1991. Immunological fine structure of the variable and constant regions of a polymorphic malarial surface antigen from *Plasmodium falciparum. Mol. Biochem. Parasitol.* **48:**1–10.
119. **Jones, K. R., B. J. Cottrell, G. A. Targett, and J. H. L. Playfair.** 1989. Killing of *Plas-*

modium falciparum by human monocyte-derived macrophages. *Parasite Immunol.* **11:**585–592.

120. **Jouin, H., P. Dubois, J. Gysin, T. Fandeur, O. Mercereau-Puijalon, and L. Pereira da Silva.** 1987. Characterization of a 96-kilodalton thermostable polypeptide antigen of *Plasmodium falciparum* related to protective immunity in the squirrel monkey. *Infect. Immun.* **55:**1387–1392.
121. **Kabilan, L., M. Troye-Blomberg, M. E. Patarroyo, A. Björkman, and P. Perlmann.** 1987. Regulation of the immune response in *Plasmodium falciparum* malaria. IV. T cell dependent production of immunoglobulin and anti-*P. falciparum* antibodies in vitro. *Clin. Exp. Immunol.* **68:**288–297.
122. **Kabilan, L., M. Troye-Blomberg, H. Perlmann, G. Andersson, B. Högh, E. Petersen, A. Björkman, and P. Perlmann.** 1988. T-cell epitopes in Pf155/RESA a major candidate for a *Plasmodium falciparum* malaria vaccine. *Proc. Natl. Acad. Sci. USA* **85:**5659–5663.
123. **Kara, U., D. Pye, R. Lord, C. Pam, H. Gould, M. Geysen, G. Jones, D. Stenzel, C. Kidson, and A. Saul.** 1989. Immune response to a synthetic peptide corresponding to an epitope of a parasitophorous vacuole membrane antigen from *Plasmodium falciparum. J. Immunol.* **143:**1334–1339.
124. **Kara, U. A. K., D. J. Stenzel, L. T. Ingram, G. R. Bushell, J. A. Lopez, and C. Kidson.** 1988. Inhibitory monoclonal antibody against a (myristylated) small-molecular-weight antigen from *Plasmodium falciparum* associated with the parasitophorous vacuole membrane. *Infect. Immun.* **56:**903–909.

124a. **Kaslow, D.** Personal communication.

125. **Kharazmi, A., and S. Jepsen.** 1984. Enhanced inhibition of in vitro multiplication of *Plasmodium falciparum* by stimulated human polymorphonuclear leucocytes. *Clin. Exp. Immunol.* **57:**287–292.
126. **Khusmith, S., and P. Druilhe.** 1983. Cooperation between antibodies and monocytes that inhibit in vitro proliferation of *Plasmodium falciparum. Infect. Immun.* **41:**219–223.
127. **Knapp, B., E. Hundt, B. Enders, and H. A. Küpper.** 1991. A recombinant hybrid protein as antigen for an anti-blood stage malaria vaccine. *Behring Inst. Mitt.* **88:**147–156.
128. **Knapp, B., E. Hundt, B. Enders, and H. A. Küpper.** 1992. Protection of *Aotus* monkeys from malaria infection by immunization with recombinant hybrid proteins. *Infect. Immun.* **60:**2397–2401.
129. **Knapp, B., E. Hundt, and K. R. Lingelbach.** 1991. Structure and possible function of *Plasmodium falciparum* proteins exported to the erythrocyte membrane. *Parasitol. Res.* **77:**277–282.
130. **Knapp, B., E. Hundt, U. Nau, and H. A. Küpper.** 1989. Molecular cloning, genomic structure and localization in a blood stage antigen of *Plasmodium falciparum* characterized by a serine stretch. *Mol. Biochem. Parasitol.* **32:**73–84.
131. **Knapp, B., A. Shaw, E. Hundt, B. Enders, and H. A. Küpper.** 1988. A histidine alanine rich recombinant antigen protects *Aotus* monkeys from *P. falciparum* infection. *Behring Inst. Mitt.* **82:**349–359.
132. **Kumar, A., R. Arora, P. Kaur, V. S. Chauhan, and P. Sharma.** 1992. "Universal" T helper cell determinants enhance immunogenicity of a *Plasmodium falciparum* merozoite surface antigen peptide. *J. Immunol.* **148:**1499–1505.

132a. **Kumar, S.** Personal communication.

133. **Kumar, S., M. F. Good, F. Dontfraid, J. M. Vinetz, and L. H. Miller.** 1989. Interdependence of CD4+ T cells and malarial spleen in immunity to *Plasmodium vinckei vinckei.* Relevance to vaccine development. *J. Immunol.* **143:**2017–2023.
134. **Kumar, S., A. Yadava, D. B. Keister, J. H. Tian, M. Ohl, K. A. Perdue-Greenfield,**

L. H. Miller, and D. C. Kaslow. 1995. Immunogenicity and in vivo efficacy of recombinant *Plasmodium falciparum* merozoite surface protein-1 in *Aotus* monkeys. *Mol. Med.* **1:**325–332.

135. **Kumaratilake, L. M., and A. Ferrante.** 1994. T-cell cytokines in malaria: their role in the regulation of neutrophil- and macrophage-mediated killing of *Plasmodium falciparum* asexual blood forms. *Res. Immunol.* **145:**423–429.
136. **Kumaratilake, L. M., D. A. Rathjen, P. Mack, F. Widmer, V. Prasertsiriroj, and A. Ferrante.** 1995. Synthetic tumor necrosis factor-α agonist peptide enhances human polymorphonuclear leukocyte-mediated killing of *Plasmodium falciparum* in vitro and suppresses *Plasmodium chabaudi* infection in mice. *J. Clin. Invest.* **95:**2315–2323.
137. **Kun, J., J. Hesselbach, M. Schreiber, A. Scherf, J. Gysin, D. Mattei, L. Pereira da Silva, and B. Müller-Hill.** 1991. Cloning and expression of genomic DNA sequences coding for putative erythrocyte membrane-associated antigens of *Plasmodium falciparum. Res. Immunol.* **142:**199–210.
138. **Langhorne, J.** 1989. The role of CD4+ T cells in the immune response to *Plasmodium chabaudi. Parasitol. Today* **5:**862–864.
139. **Langhorne, J., P. Mombaerts, and S. Tonegawa.** 1995. $\alpha\beta$ and $\gamma\delta$ T cells in the immune response to the erythrocytic stages of malaria in mice. *Int. Immunol.* **7:**1005–1011.
140. **Lew, A. M., C. J. Langford, R. F. Anders, D. J. Kemp, A. Saul, C. Fardoulys, M. Geysen, and M. Sheppard.** 1989. A protective monoclonal antibody recognizes a linear epitope in the precursor to the major merozoite antigens of *Plasmodium cabaudi adami. Proc. Natl. Acad. Sci. USA* **86:**3768–3772.
141. **Liew, A. M., C. J. Langford, D. Pye, S. Edwards, L. Corcoran, and R. F. Anders.** 1989. Class II restriction in mice to the malaria candidate vaccine ring infected erythrocyte surface antigen (RESA) as synthetic peptides or as expressed in recombinant vaccinia. *J. Immunol.* **142:**4012–4016.
142. **Liew, A. M., C. J. Langford, D. Pye, S. Edwards, L. Corcoran, and R. F. Anders.** 1989. Class II restriction in mice to the malaria candidate vaccine ring infected erythrocyte surface antigen (RESA) as synthetic peptides or as expressed in recombinant vaccinia. *J. Immunol.* **142:**4012–4016.
143. **Ling, I. T., S. A. Ogun, and A. A. Holder.** 1995. The combined epidermal growth factor-like modules of *Plasmodium yoelii* merozoite surface protein-1 are required for a protective immune response to the parasite. *Parasite Immunol.* **17:**425–433.
144. **Lunel, F., and P. Druilhe.** 1989. Effector cells involved in nonspecific and antibody-dependent mechanisms directed against *Plasmodium falciparum* blood stages in vitro. *Infect. Immun.* **57:**2043–2049.
145. **Lussow, A. R., C. Barrios, J. van Embden, R. Van der Zee, A. S. Verdini, A. Pessi, M. J. A. Louis, P.-H. Lambert, and G. Del Giudice.** 1991. Mycobacterial heat-shock proteins as carrier molecules. *Eur. J. Immunol.* **21:**2297–2302.
146. **Lussow, A. R., G. Del Giudice, L. Rénia, D. Mazier, J. P. Verhave, A. S. Verdini, A. Pessi, J. A. Louis, and P.-H. Lambert.** 1990. Use of a tuberculin purified protein derivative-Asn-Ala-Asn-Pro conjugate in bacillus Calmette-Guérin primed mice overcomes H-2 restriction of the antibody response and avoids the need for adjuvants. *Proc. Natl. Acad. Sci. USA* **87:**2960–2964.
147. **Lyon, J. A., J. D. Haynes, C. L. Diggs, J. D. Chulay, and J. M. Pratt-Rossiter.** 1986. *Plasmodium falciparum* antigens synthesized by schizonts and stabilized at the merozoite surface by antibodies when schizonts mature in the presence of growth inhibitory immune serum. *J. Immunol.* **136:**2252–2258.
148. **Lyon, J. A., A. W. Thomas, T. Hall, and J. D. Chulay.** 1989. Specificities of antibod-

ies that inhibit merozoite dispersal from malaria-infected erythrocytes. *Mol. Biochem. Parasitol.* **36:**77–86.

149. **Maheshwari, R. K., C. W. Czarniecki, G. P. Dutta, S. K. Puri, B. N. Dhawan, and R. M. Friedman.** 1986. Recombinant human gamma interferon inhibits simian malaria. *Infect. Immun.* **53:**628–630.

149a. **Mattei, D.** Personal communication.

150. **Mattei, D., and A. Scherf.** 1992. The Pf332 gene of *Plasmodium falciparum* codes for a giant protein that is translocated from the parasite to the membrane of infected erythrocytes. *Gene* **110:**71–79.

151. **McBride, J. S., and H.-G. Heidrich.** 1987. Fragments of the polymorphic M_r 185,000 glycoprotein from the surface of isolated *Plasmodium falciparum* merozoites form an antigenic complex. *Mol. Biochem. Parasitol.* **23:**71–84.

152. **McGregor, I. A., S. C. Carrington, and S. Cohen.** 1963. Treatment of East African *Plasmodium falciparum* malaria with West African human gammaglobulin. *Trans. R. Soc. Trop. Med. Hyg.* **57:**170–175.

153. **McGuire, W., A. V. S. Hill, C. E. M. Allsopp, B. M. Greenwood, and D. Kwiatkowski.** 1994. Variation in the TNF-α promoter region associated with susceptibility to cerebral malaria. *Nature* (London) **371:**508–511.

154. **Meding, S. J., and J. Langhorne.** 1991. CD4+ T cells and B cells are necessary for the transfer of protective immunity to *Plasmodium chabaudi chabaudi. Eur. J. Immunol.* **21:**1433–1438.

155. **Men, Y., C. Thomasin, H. P. Merkle, B. Gander, and G. Corradin.** 1995. A single administration of tetanus toxoid in biodegradable microspheres elicits T cell and antibody responses similar or superior to those obtained with aluminium hydroxide. *Vaccine* **13:**683–689.

156. **Mercereau-Puijalon, O., C. Jacquemot, and J. L. Sarthou.** 1991. A study of the genomic diversity of *Plasmodium falciparum* in Senegal. 1. Typing by Southern blot analysis. *Acta Trop.* **49:**281–292.

157. **Migliorini, P., B. Betschart, and G. Corradin.** 1993. Malaria vaccine: immunization of mice with a synthetic T cell helper epitope alone leads to protective immunity. *Eur. J. Immunol.* **23:**582–585.

158. **Migot, F., C. Chougnet, B. Perichon, P.-M. Danze, J.-P. Lepers, R. Krishnamoorthy, and P. Deloron.** 1995. Lack of correlation between HLA class II alleles and immune responses to Pf155/ring-infected erythrocyte surface antigen (RESA) from *Plasmodium falciparum* in Madagascar. *Am. J. Trop. Med. Hyg.* **52:**252–257.

159. **Miller, L. H.** 1994. Impact of malaria on genetic polymorphism and genetic diseases in Africans and African Americans. *Proc. Natl. Acad. Sci. USA* **91:**2415–2419.

160. **Morein, B., K. Lövgren, S. Höglund, and B. Sundquist.** 1987. The ISCOM: an immunostimulating complex. *Immunol. Today* **8:**333–338.

161. **Morein, B., K. Lövgren, B. Rönnberg, A. Sjölander, and M. Villacrés-Eriksson.** 1995. Immunostimulating complexes. Clinical potential in vaccine development. *Clin. Immunother.* **3:**461–475.

162. **Mshana, R. N., S. McLean, and J. Boulandi.** 1990. In vitro cell-mediated immune responses to *Plasmodium falciparum* schizont antigens in adults from a malaria endemic area—CD8+ lymphocytes inhibit the response of low responder individuals. *Int. Immunol.* **2:**1121–1133.

163. **Müller, H.-M., K. Früh, A. von Brunn, F. Esposito, S. Lombardi, A. Crisanti, and H. Bujard.** 1989. Development of the human immune response against the major surface protein (gp190) of *Plasmodium falciparum. Infect. Immun.* **57:**3765–3769.

164. **Munesinghe, D. Y., P. Clavijo, M. C. Calle, R. S. Nussenzweig, and E. Nardin.** 1991.

Immunogenicity of multiple antigen peptides (MAP) containing T and B cell epitopes of the repeat region of the *P. falciparum* circumsporozoite protein. *Eur. J. Immunol.* **21:**3015–3020.

165. **Nakazawa, S., A. E. Brown, Y. Maeno, C. D. Smith, and M. Aikawa.** 1994. Malaria-induced increase of splenic $\gamma\delta$ T cells in humans, monkeys, and mice. *Exp. Parasitol.* **79:**391–398.
166. **Nnalue, N. A., and M. J. Friedman.** 1988. Evidence for a neutrophil-mediated protective response in malaria. *Parasite Immunol.* **10:**47–58.
167. **Nussler, A., J. C. Drapier, L. Renia, S. Pied, F. Miltgen, M. Gentilini, and D. Mazier.** 1991. L-Arginine-dependent destruction of intrahepatic malaria parasites in response to tumor necrosis factor and/or interleukin-6 stimulation. *Eur. J. Immunol.* **21:**227–230.
168. **Ockenhouse, C. F., S. Schulman, and H. L. Shear.** 1984. Induction of crisis forms in the human malaria parasite, *Plasmodium falciparum* by γ-interferon-activated, monocyte-derived macrophages. *J. Immunol.* **133:**1601–1608.
169. **Oeuvray, C., H. Bouharoun-Tayoun, H. Gras-Masse, E. Bottius, T. Kaidoh, M. Aikawa, M.-C. Filgueira, A. Tartar, and P. Druilhe.** 1994. Merozoite surface protein-3: a malaria protein inducing antibodies that promote *Plasmodium falciparum* killing by cooperation with blood monocytes. *Blood* **84:**1594–1602.
170. **Olafsson, P., H. Matile, and U. Certa.** 1992. *Plasmodium falciparum*: the repetitive MSA-1 surface protein of the RO-71 isolate is recognized by mouse antibody against the nonrepetitive repeat block of RO-33. *Exp. Parasitol.* **74:**381–389.
171. **Orago, A. S. S., and C. A. Facer.** 1991. Cytotoxicity of human natural killer (NK) cell subsets for *Plasmodium falciparum* erythrocytic schizonts: stimulation by cytokines and inhibition by neomycin. *Clin. Exp. Immunol.* **86:**22–29.
172. **Parra, M. E., C. B. Evans, and D. W. Taylor.** 1991. Identification of *Plasmodium falciparum* histidine-rich protein 2 in the plasma of humans with malaria. *J. Clin. Microbiol.* **29:**1629–1634.
173. **Pasloske, B. L., D. I. Baruch, C. Ma, T. F. Taraschi, J. A. Gormley, and R. J. Howard.** 1994. PfEMP3 and HRP1: co-expressed genes localized to chromosome 2 of *Plasmodium falciparum*. *Gene* **144:**131–136.
174. **Pasloske, B. L., D. I. Baruch, M. R. van Schravendijk, S. M. Handunnetti, M. Aikawa, H. Fujioka, T. F. Taraschi, J. A. Gormley, and R. J. Howard.** 1993. Cloning and characterization of a *Plasmodium falciparum* gene encoding a novel high-molecular weight host membrane-associated protein, PfEMP3. *Mol. Biochem. Parasitol.* **59:**59–72.
175. **Patarroyo, M. E., J. Vinasco, R. Amador, F. Espejo, Y. Silva, A. Moreno, M. Rojas, A. L. Mora, M. Salcedo, V. Valero, A. K. Goldberg, and J. Kalil.** 1991. Genetic control of the immune response to a synthetic vaccine against *Plasmodium falciparum*. *Parasite. Immunol.* **13:**509–516.
176. **Perera, M. K., R. Carter, R. Goonewardene, and K. M. Mendis.** 1994. Transient increase in circulating γ/δ T cells during *Plasmodium vivax* malarial paroxysms. *J. Exp. Med.* **179:**311–315.
177. **Perlmann, H., S. Kumar, J. M. Vinetz, M. Kullberg, L. H. Miller, and P. Perlmann.** 1995. Cellular mechanisms in the immune response to malaria in *Plasmodium vinckei*-infected mice. *Infect. Immun.* **63:**3987–3993.
178. **Perlmann, H., P. Perlmann, K. Berzins, B. Wåhlin, M. Troye-Blomberg, M. Hagstedt, I. Andersson, B. Högh, E. Petersen, and A. Björkman.** 1989. Dissection of the human antibody response to the malaria antigen Pf155/RESA into epitope specific components. *Immunol. Rev.* **112:**115–132.
179. **Perraut, R., O. Mercereau-Puijalon, D. Mattei, E. Bourreau, O. Garraud, B. Bonnemains, L. Pereira da Silva, and J.-C. Michel.** 1995. Induction of opsonizing anti-

bodies after injection of recombinant *Plasmodium falciparum* vaccine candidate antigens in preimmune *Saimiri sciureus* monkeys. *Infect. Immun.* **63:**554–562.

180. **Perrin, L. H., B. Merkli, M. S. Gabra, J. W. Stocker, C. Chizzolini, and R. Richle.** 1985. Immunization with a *Plasmodium falciparum* merozoite surface antigen induces a partial immunity in monkeys. *J. Clin. Invest.* **75:**1718–1721.
181. **Perrin, L. H., B. Berkli, M. Loche, C. Chizzolini, J. Smart, and R. Richle.** 1984. Antimalarial immunity in *Saimiri* monkeys. Immunization with surface components of asexual blood stages. *J. Exp. Med.* **160:**441–451.
182. **Pessi, A., D. Valmori, P. Migliorini, C. Tougne, E. Bianchi, P. H. Lambert, G. Corradin, and G. Del Giudice.** 1991. Lack of H-2 restriction of the *Plasmodium falciparum* (NANP) sequence as multiple antigen peptide. *Eur. J. Immunol.* **21:**2273–2276.
183. **Petersen, E., B. Högh, N. T. Marbiah, H. Perlmann, M. Willcox, E. Dolopaie, A. P. Hansson, A. Björkman, and P. Perlmann.** 1990. A longitudinal study of antibodies to the *Plasmodium falciparum* antigen Pf155/RESA and immunity to malaria infection in adult Liberians. *Trans. R. Soc. Trop. Med. Hyg.* **84:**339–345.
184. **Peterson, G., R. L. Coppel, M. B. Moloney, and D. J. Kemp.** 1988. Third form of the precursor to the major merozoite surface antigens of *Plasmodium falciparum*. *Mol. Cell. Biol.* **8:**2664–2667.
185. **Phillips, S.** 1994. Effector mechanisms against asexual erythrocytic stages of *Plasmodium*. *Immunol. Lett.* **41:**109–114.
186. **Podoba, J. E., and M. M. Stevenson.** 1991. $CD4^+$ and $CD8^+$ T lymphocytes both contribute to acquired immunity to blood-stage *Plasmodium chabaudi* AS. *Infect. Immun.* **59:**51–58.
187. **Pouvelle, B., J. A. Gormley, and T. F. Taraschi.** 1994. Characterization of trafficking pathways and membrane genesis in malaria-infected erythrocytes. *Mol. Biochem. Parasitol.* **66:**83–96.
188. **Pouvelle, B., R. Spiegel, L. Hsiao, R. J. Howard, R. L. Morris, A. P. Thomas, and T. F. Taraschi.** 1991. Direct access to serum macromolecules by intraerythrocytic malaria parasites. *Nature* (London) **353:**73–75.
189. **Pye, D., S. J. Edwards, R. F. Anders, C. M. O'Brien, P. Franchina, L. N. Corcoran, C. Monger, M. G. Peterson, K. L. Vandenberg, J. A. Smythe, S. R. Westley, R. L. Coppel, T. L. Webster, D. J. Kemp, A. W. Hampson, and C. J. Langford.** 1991. Failure of recombinant vaccinia viruses expressing *Plasmodium falciparum* antigens to protect *Saimiri* monkeys against malaria. *Infect. Immun.* **59:**2403–2411.
190. **Richards, R. L., J. Swartz G. M., C. Schulz, M. D. Haye, G. S. Ward, W. R. Ballou, J. D. Chulay, W. T. Hockmeyer, S. L. Berman, and C. R. Alving.** 1989. Immunogenicity of liposomal malaria sporozoite antigen in monkeys: adjuvant effects of aluminium hydroxide and non-pyrogenic liposomal lipid A. *Vaccine* **7:**506.
191. **Rickman, L. S., D. M. Gordon, R. Wistar, U. Krzych, M. Gross, M. R. Hollingdale, J. E. Egan, J. D. Chulay, and S. L. Hoffman.** 1991. Use of adjuvant containing mycobacterial cell-wall skeleton, monophosphoryl lipid A, and squalane in malaria circumsporozoite protein vaccine. *Lancet* **337:**998–1001.
192. **Riley, E. M., S. J. Allen, M. Troye-Blomberg, S. Bennett, H. Perlmann, G. Andersson, L. Smedman, P. Perlmann, and B. M. Greenwood.** 1991. Association between immune recognition of the malaria vaccine candidate antigen Pf155/RESA and resistance to clinical disease: a prospective study in a malaria-endemic region of West Africa. *Trans. R. Soc. Trop. Med. Hyg.* **85:**436–443.
193. **Riley, E. M., O. Jobe, and H. C. Whittle.** 1989. $CD8^+$ T cells inhibit *Plasmodium falciparum*-induced lymphocyte proliferation and gamma interferon production in cell preparations from some malaria-immune individuals. *Infect. Immun.* **57:**1281–1284.

194. **Riley, E. M., O. Olerup, S. Bennett, P. Rowe, S. J. Allen, M. J. Blackman, M. Troye-Blomberg, A. A. Holder, and B. M. Greenwood.** 1992. MHC and malaria: the relationship between HLA class II alleles and immune responses to *Plasmodium falciparum*. *Int. Immunol.* **4:**1055–1063.
195. **Rodriguez Da Silva, L., M. Loche, R. Dayal, and L. H. Perrin.** 1983. *Plasmodium falciparum* polypeptides released during in vitro cultivation. *Bull. W.H.O.* **61:**105–112.
196. **Roussilhon, C., M. Agrapart, P. Guglielmi, A. Bensussan, P. Brasseur, and J. J. Ballet.** 1994. Human TcR$\gamma\delta^+$ lymphocyte response on primary exposure to *Plasmodium falciparum*. *Clin. Exp. Immunol.* **95:**91–97.
197. **Ruangjirachuporn, W., B. A. Afzelius, H. Helmby, A. V. S. Hill, B. M. Greenwood, J. Carlson, K. Berzins, P. Perlmann, and M. Wahlgren.** 1992. Ultrastructural analysis of fresh *Plasmodium falciparum* infected erythrocytes and their cytoadherence to human leukocytes. *Am. J. Trop. Med. Hyg.* **46:**511–519.
198. **Ruangjirachuporn, W., R. Udomsangpetch, J. Carlsson, D. Drenckhahn, P. Perlmann, and K. Berzins.** 1991. *Plasmodium falciparum*: analysis of the interaction of antigen Pf155/RESA with the erythrocyte membrane. *Exp. Parasitol.* **73:**62–72.
199. **Rzepczyk, C. M., R. Ramasamy, D. A. Mutch, P. C.-L. Ho, D. Battistutta, K. L. Anderson, D. Parkinson, T. J. Doran, and M. Honeyman.** 1989. Analysis of human T cell response to *Plasmodium falciparum* merozoite surface antigens. *Eur. J. Immunol.* **19:**1797–1802.
200. **Sadoff, J. C., W. R. Ballou, L. S. Baron, W. R. Majarian, R. N. Brey, W. T. Hockmeyer, J. F. Young, S. J. Cryz, J. Ou, G. H. Lowell, and J. D. Chulay.** 1988. Oral *Salmonella typhimurium* vaccine expressing circumsporozoite protein protects against malaria. *Science* **240:**336–338.
201. **Salmon, D., J. L. Vilde, B. Andrieu, R. Simonovic, and J. Lebras.** 1986. Role of immune serum and complement in stimulation of the metabolic burst of human neutrophils by *Plasmodium falciparum*. *Infect. Immun.* **51:**801–806.
202. **Sam-Yellowe, T. Y., H. Shio, and M. E. Perkins.** 1988. Secretion of *Plasmodium falciparum* rhoptry protein into the plasma membrane of host erythrocytes. *J. Cell Biol.* **106:**1507–1513.
203. **Saul, A., J. Cooper, L. Ingram, R. F. Anders, and G. V. Brown.** 1985. Invasion of erythrocytes in vitro by *Plasmodium falciparum* can be inhibited by monoclonal antibody directed against an S antigen. *Parasite Immunol.* **7:**587–593.
204. **Saul, A., R. Lord, G. L. Jones, and L. Spencer.** 1992. Protective immunization with invariant peptides of the *Plasmodium falciparum* antigen MSA2. *J. Immunol.* **148:**208–211.
205. **Saul, A., W. L. Maloy, E. P. Rock, and R. J. Howard.** 1988. A portion of the Pf155/RESA antigen of *Plasmodium falciparum* is accessible on the surface of infected erythrocytes. *Immunol. Cell Biol.* **66:**269–276.
206. **Sayles, P. C., A. J. Cooley, and D. L. Wassom.** 1991. A spleen is not necessary to resolve infections with *Plasmodium yoelii*. *Am. J. Trop. Med. Hyg.* **44:**42–48.
207. **Schorr, J., B. Knapp, E. Hundt, H. A. Küpper, and E. Amann.** 1991. Surface expression of malarial antigens in *Salmonella typhimurium*: induction of serum antibody response upon oral vaccination of mice. *Vaccine* **9:**675–681.
208. **Sedegah, M., R. Hedstrom, P. Hobart, and S. L. Hoffman.** 1994. Protection against malaria by immunization with plasmid DNA encoding circumsporozoite protein. *Proc. Natl. Acad. Sci. USA* **91:**9866–9870.
209. **Siddiqui, W. A., L. Q. Tam, K. J. Kramer, G. S. N. Hui, S. E. Case, K. M. Yamaga, S. P. Chang, E. B. T. Chan, and S.-C. Kan.** 1987. Merozoite surface coat precursor protein completely protects *Aotus* monkeys against *Plasmodium falciparum* malaria. *Proc. Natl. Acad. Sci. USA* **84:**3014–3018.

210. **Simitsek, P. D., E. Ramirez, and L. H. Perrin.** 1990. Structural diversity of *Plasmodium falciparum* gp200 is detected by T cells. *Eur. J. Immunol.* **20:**1755–1759.
211. **Simmons, D., G. Woollett, M. Bergin-Cartwright, D. Kay, and J. Scaife.** 1987. A malaria protein exported into a new compartment within the host erythrocyte. *EMBO J.* **6:**485–491.
212. **Sinigaglia, F., M. Guttinger, J. Kilgus, D. M. Doran, H. Matile, H. Etlinger, A. Trzeciak, D. Gillessen, and J. R. L. Pink.** 1988. A malaria T-cell epitope recognized in association with most mouse and human MHC class II molecules. *Nature* (London) **336:**778–780.
213. **Sinigaglia, F., H. Matile, and J. R. L. Pink.** 1987. *Plasmodium falciparum*-specific human T cell clones: evidence for helper and cytotoxic activities. *Eur. J. Immunol.* **17:**187–192.
214. **Sinigaglia, F., B. Takacs, H. Jacot, H. Matile, J. R. L. Pink, A. Crisanti, and H. Bujard.** 1988. Nonpolymorphic regions of p190, a protein of the *Plasmodium falciparum* erythrocytic stage, contain both T and B cell epitopes. *J. Immunol.* **140:**3568–3572.
215. **Sjöberg, K., J. P. Lepers, L. Raharimalala, Å. Larsson, O. Olerup, N. T. Marbiah, M. Troye-Blomberg, and P. Perlmann.** 1992. Genetic regulation of human anti-malarial antibodies in twins. *Proc. Natl. Acad. Sci. USA* **89:**2101–2104.
216. **Smythe, J. A., R. L. Coppel, K. P. Day, R. K. Martin, A. M. J. Oduola, D. J. Kemp, and R. F. Anders.** 1991. Structural diversity in the *Plasmodium falciparum* merozoite surface antigen-2. *Proc. Natl. Acad. Sci. USA* **88:**1751–1755.
217. **Stanley, H. A., J. T. Mayes, N. R. Cooper, and R. T. Reese.** 1984. Complement activation by the surface of *Plasmodium falciparum* infected erythrocytes. *Mol. Immunol.* **21:**145–150.
218. **Stevenson, M. M., E. Ghadirian, N. C. Phillips, D. Rea, and J. E. Podoba.** 1989. Role of mononuclear phagocytosis in elimination of *Plasmodium chabaudi* AS infection. *Parasite Immunol.* **11:**529–544.
219. **Stevenson, M. M., and M. F. Tam.** 1993. Differential induction of helper T cell subsets during blood-stage *Plasmodium chabaudi* AS infection in resistant and susceptible mice. *Clin. Exp. Immunol.* **92:**77–83.
220. **Stevenson, M. M., M. F. Tam, S. F. Wolf, and A. Sher.** 1995. IL-12-induced protection against blood-stage *Plasmodium chabaudi* AS requires IFN-γ and TNF-α and occurs via a nitric oxide-dependent mechanism. *J. Immunol.* **155:**2545-2556.
221. **Stürchler, D., M. Just, R. Berger, R. Reber-Liske, H. Matile, H. Etlinger, B. Takacs, C. Rudin, and M. Fernex.** 1992. Evaluation of 5.1-$[NANP]_{19}$, a recombinant *Plasmodium falciparum* vaccine candidate, in adults. *Trop. Geogr. Med.* **44:**9–14.
222. **Su, X. Z., V. M. Heatwole, S. P. Wertheimer, F. Guinet, J. A. Herrfeldt, D. S. Peterson, J. A. Ravetch, and T. E. Wellems.** 1995. The large diverse gene family *var* encodes proteins involved in cytoadherence and antigenic variation of *Plasmodium falciparum*-infected erythrocytes. *Cell* **82:**89–100.
223. **Süss, G., K. Eichmann, E. Kury, A. Linke, and J. Langhorne.** 1988. Roles of CD4- and CD8-bearing T lymphocytes in the immune response to the erythrocytic stages of *Plasmodium chabaudi. Infect. Immun.* **56:**3081–3088.
224. **Takahashi, H., T. Takeshita, B. Morein, S. Putney, R. N. Germain, and J. A. Berzofsky.** 1990. Induction of CD8+ cytotoxic T cells by immunization with purified HIV-1 envelope protein in ISCOMs. *Nature* (London) **334:**873–875.
225. **Tam, J. P., P. Clavijo, Y. Lu, V. Nussenzweig, R. Nussenzweig, and F. Zavala.** 1990. Incorporation of T and B epitopes of the circumsporozoite protein in a chemically defined synthetic vaccine against malaria. *J. Exp. Med.* **171:**299–306.
226. **Tanabe, K., M. Mackay, M. Goman, and J. G. Scaife.** 1987. Allelic dimorphism in a

surface antigen gene of the malaria parasite *Plasmodium falciparum*. *J. Mol. Biol.* **195:**273–287.

227. **Tang, D., M. DeVit, and S. A. Johnston.** 1992. Genetic immunization is a simple method for eliciting an immune response. *Nature* (London) **356:**152–154.
228. **Taverne, J., C. A. W. Bate, D. A. Sarker, D. A. Meager, G. A. W. Rook, and J. H. L. Playfair.** 1990. Human and murine macrophages produce TNF in response to soluble antigens of *Plasmodium falciparum*. *Parasite Immunol.* **12:**33–43.
229. **Taverne, J., N. Sheikh, J. B. de Souza, J. H. L. Playfair, L. Probert, and J. Kollias.** 1994. Anaemia and resistance to malaria in transgenic mice expressing human tumour necrosis factor. *Immunology* **82:**397–403.
230. **Taylor, D. W., M. Parra, G. B. Chapman, M. E. Stearna, J. Rener, M. Aikawa, S. Uni, S. B. Aley, L. J. Panton, and R. J. Howard.** 1987. Localisation of *Plasmodium falciparum* histidine-rich protein 1 in the erythrocyte skeleton under knobs. *Mol. Biochem. Parasitol.* **25:**165–174.
231. **Taylor-Robinson, A. W., and R. S. Phillips.** 1993. Protective $CD4^+$ T cell lines raised against *Plasmodium chabaudi* show characteristics of either Th1 or Th2 cells. *Parasite Immunol.* **15:**301–310.
232. **Taylor-Robinson, A. W., and R. S. Phillips.** 1994. B cells are required for the switch from Th1- to Th2-regulated immune responses to *Plasmodium chabaudi chabaudi* infection. *Infect. Immun.* **62:**2490–2498.
233. **Theander, T. G., I. C. Bygbjerg, B. J. Andersen, S. Jepsen, A. Kharazmi, and N. Ödum.** 1986. Suppression of parasite-specific response in *Plasmodium falciparum* malaria. A longitudinal study of blood mononuclear cell proliferation and subset composition. *Scand. J. Immunol.* **24:**73–81.
234. **Tine, J. A., V. Conseil, P. Delplace, C. De Taisne, D. Camus, and E. Paoletti.** 1993. Immunogenicity of the *Plasmodium falciparum* serine repeat antigen (p126) expressed by vaccinia virus. *Infect. Immun.* **61:**3933–3941.
235. **Troye-Blomberg, M., K. Berzins, and P. Perlmann.** 1994. T-cell control of immunity to the asexual blood stages of the malaria parasite. *Crit. Rev. Immunol.* **14:**131–155.
236. **Troye-Blomberg, M., and P. Perlmann.** 1991. Malaria immunity: an overview with emphasis on T cell function, p. 1–46. *In* M. F. Good and A. J. Saul (ed.), *Molecular Immunological Considerations in Malaria Vaccine Devleopment*. CRC Press, Inc., Boca Raton, Fla.
237. **Troye-Blomberg, M., O. Olerup, Å. Larsson, K. Sjöberg, H. Perlmann, E. Riley, J.-P. Lepers, and P. Perlmann.** 1991. Failure to detect MHC class II associations of the human immune response induced by repeated malaria infections to the *Plasmodium falciparum* antigen Pf155/RESA. *Int. Immunol.* *3:*1043–1051.
238. **Troye-Blomberg, M., E. M. Riley, L. Kabilan, M. Holmberg, H. Perlmann, U. Andersson, C. H. Heusser, and P. Perlmann.** 1990. Production by activated human T cells of interleukin 4 but not interferon-γ is associated with elevated levels of serum antibodies to activating malaria antigens. *Proc. Natl. Acad. Sci. USA* **87:**5484–5488.
239. **Troye-Blomberg, M., E. M. Riley, H. Perlmann, G. Andersson, Å. Larsson, R. W. Snow, S. J. Allen, R. A. Houghten, O. Olerup, B. M. Greenwood, and P. Perlmann.** 1989. T and B cell responses of *Plasmodium falciparum* malaria immune individuals to sequences in different regions of the *P. falciparum* antigen Pf155/RESA. *J. Immunol.* **143:**3043–3048.
240. **Trudel, M., F. Nadon, C. Séguin, S. Brault, Y. Lusignan, and S. Lemieux.** 1992. Initiation of cytotoxic T-cell response and protection of Balb/c mice by vaccination with an experimental ISCOMs respiratory syncytial virus subunit vaccine. *Vaccine* **10:**107–112.

241. **Udomsangpetch, R., M. Aikawa, K. Berzins, M. Wahlgren, and P. Perlmann.** 1989. Cytoadherence of knobless *Plasmodium falciparum* infected erythrocytes and its inhibition by a human monoclonal antibody. *Nature* (London) **338:**763–765.
242. **Udomsangpetch, R., J. Carlsson, B. Wåhlin, G. Holmquist, L. S. Ozaki, A. Scherf, D. Mattei, O. Mercereau-Puijalon, S. Uni, M. Aikawa, K. Berzins, and P. Perlmann.** 1989. Reactivity of the human monoclonal antibody 33G2 with repeated sequences of three distinct *Plasmodium falciparum* antigens. *J. Immunol.* **142:**3620–3626.
243. **van der Heyde, H. C., M. M. Elloso, D. C. Roopenian, D. D. Manning, and W. P. Weidanz.** 1993. Expansion of the $CD4^-$, $CD8^-$ $\gamma\delta$ T cell subsets in the spleens of mice during non-lethal blood stage malaria. *Eur. J. Immunol.* **23:**1846–1850.
244. **van Schravendijk, M. R., E. P. Rock, K. Marsh, Y. Ito, M. Aikawa, J. Neequaye, D. Ofori-Adjei, R. Rodriguez, M. E. Patarroyo, and R. J. Howard.** 1991. Characterization and localization of *Plasmodium falciparum* surface antigens on infected erythrocytes from West African patients. *Blood* **78:**226–236.
245. **Vernot-Hernandez, J. P., and H. G. Heidrich.** 1985. The relationship to knobs of the 92,000 D protein specific for knobby strains of *Plasmodium falciparum. Z. Parasitenkd.* **71:**41–51.
246. **Vreden, S. G. S., J. P. Verhave, T. Oettinger, R. W. Sauerwein, and J. H. E. T. Meuwissen.** 1991. Phase I clinical trial of a recombinant malaria vaccine consisting of the circumsporozoite repeat region of *Plasmodium falciparum* coupled to hepatitis B surface antigen. *Am. J. Trop. Med. Hyg.* **1991:**533–538.
246a. **Wahlgren, M., and V. Fernandez.** Personal communication.
247. **Wahlgren, M., V. Fernandez, C. Scholander, and J. Carlson.** 1994. Rosetting. *Parasitol. Today* **10:**73–79.
248. **Wåhlin, B., M. Wahlgren, H. Perlmann, K. Berzins, A. Björkman, M. E. Patarroyo, and P. Perlmann.** 1984. Human antibodies to a M_r 155,000 *Plasmodium falciparum* antigen efficiently inhibit merozoite invasion. *Proc. Natl. Acad. Sci. USA* **81:**7912–7916.
249. **Waki, S.** 1994. Antibody-dependent neutrophil-mediated parasite killing in non-lethal rodent malaria. *Parasite Immunol.* **16:**587–591.
250. **Wang, R. B., Y. Charoenvit, G. Corradin, R. Porrozzi, R. L. Hunter, G. Glenn, C. R. Alving, P. Church, and S. L. Hoffman.** 1995. Induction of protective polyclonal antibodies by immunization with a *Plasmodium yoelii* circumsporozoite protein multiple antigen peptide vaccine. *J. Immunol.* **154:**2784–2793.
251. **Wassom, D. L., and E. A. B. Kelly.** 1990. The role of the major histocompatibility complex in resistance to parasite infections. *Crit. Rev. Immunol.* **10:**31–52.
252. **Weidanz, W. P., and C. A. Long.** 1988. The role of T cells in immunity to malaria. *Prog. Allergy* **41:**215–252.
253. **Weiss, L.** 1990. The spleen in malaria: the role of barrier cells. *Immunol. Lett.* **25:**165–172.
254. **Wilson, R. J. M.** 1974. The production of antigens by *Plasmodium falciparum* in vitro. *Int. J. Parasitol.* **4:**537–547.
255. **Wilson, R. J. M., I. A. McGregor, and M. E. Wilson.** 1973. The stability and fractionation of malarial antigens from the blood of Africans infected with *Plasmodium falciparum. Int. J. Parasitol.* **3:**511–520.
256. **Winkel, K. D., and M. F. Good.** 1991. Inability of *Plasmodium vinckei* immune spleen cells to transfer protection to recipient mice exposed to vaccine vectors or heterologous species of *Plasmodium. Parasite Immunol.* **13:**517–530.
257. **Yap, G. S., and M. M. Stevenson.** 1994. Differential requirements for an intact spleen in induction and expression of B-cell-dependent immunity to *Plasmodium chabaudi* AS. *Infect. Immun.* **62:**4219–4225.

Malaria Vaccine Development: A Multi-Immune Response Approach
Edited by Stephen L. Hoffman

Chapter 6

Preventing Cytoadherence of Infected Erythrocytes to Endothelial Cells and Noninfected Erythrocytes

Graham V. Brown and Stephen J. Rogerson

CYTOADHERENCE IN *PLASMODIUM FALCIPARUM* MALARIA

Adherence of erythrocytes infected with mature parasites to the endothelium of postcapillary venules is characteristic of *P. falciparum* infection. Mature parasites are virtually never seen in the peripheral circulation because they are sequestered in deep tissues as part of the usual cycle of replication in symptomatic and asymptomatic individuals. Sequestration is believed to prevent passage of these infected cells through the spleen, where the altered erythrocytes infected with *P. falciparum* would be removed, but this phenomenon does not appear to be a normal part of the life cycle of the other *Plasmodium* species that infect humans. Severe *P. falciparum* malaria is associated with marked sequestration of parasitized erythrocytes in brain, placenta, lung, or elsewhere, and patients dying from *P. falciparum* malaria have a disproportionate amount of sequestration in vital organs (2, 49, 61). Although there is debate about the relative contributions of vascular obstruction and cytotoxin release by sequestered parasites to the pathology of cerebral malaria, it is clear that there may be synergy between these two mechanisms (reviewed by Clark and Rockett [27] and Berendt and colleagues [9]).

Ultrastructural studies have shown knoblike protrusions at the surfaces of *P. falciparum*-infected cells (48, 81), but other species sharing these characteristics (e.g., *Plasmodium malariae*) do not undergo deep vascular schizogony, and knobs are not necessary for cytoadherence in vitro (11, 86). The knobs are parasite dependent but may be partly derived from the host. A major parasite product, the *P. falciparum* knob-associated histidine-rich protein, is present in the knob but is not exposed at the external surface (32). These structural alterations may facilitate adherence by (i) altering the flow characteristics of cells by decreasing cell deformability or (ii) modifying the cell membrane to allow the integration of parasite ligands.

Udeinya and colleagues showed that erythrocytes infected with mature stages of *P. falciparum* bind to some cultured human umbilical vein endothelial cells (HUVEC), and knobs appear to be the site of attachment (85). Further studies indicated that parasitized cells could bind to amelanotic melanoma cells and that the binding could be inhibited in a strain-specific manner by immune monkey serum (84).

HOST RECEPTORS FOR CYTOADHERENCE

Many lines of evidence suggest that CD36 is an important receptor for *P. falciparum*-infected cells. Expression of CD36 on target cells correlates with the adhesion of some isolates (54, 59). Infected erythrocytes (IRBC) adhere to COS cells transfected with CD36 cDNA (58), and cytoadherence can be blocked by treatment of cells expressing CD36 with monoclonal antibody OKM5 directed against CD36 (6). Purified human CD36 binds to many human IRBC isolates (55), and a large proportion of field isolates contain parasites with the ability to adhere to CD36 immobilized on plastic (62). CD36 has been identified on platelets, monocytes, melanoma cells, human brain capillary endothelial cells (74), and human dermal microvascular endothelial cells but not large-vessel endothelial cells (79). Two sequences of CD36 responsible for its cytoadherence of IRBC have been localized to a single putative disulfide loop (5).

Several other potential receptors for IRBC have been identified. These include intercellular adhesion molecule 1 (ICAM-1) (8), vascular cell adhesion molecule 1 (VCAM-1), E-selectin (57), thrombospondin (64, 73), and chondroitin sulfate A (CSA) (66). CD36, ICAM-1, CSA, and thrombospondin are detectable on resting human cerebral endothelium (3). ICAM-1 expression is upregulated and expression of VCAM-1 and E-selectin is induced by cytokines such as tumor necrosis factor alpha and gamma interferon (8, 45) and directly by parasitized erythrocytes (83) or by glycosylphosphatidylinositol purified from parasitized erythrocytes (71). E-selectin and VCAM-1 were detected on brain microvascular endothelium from 10 of 13 patients with cerebral malaria but not in patients dying of other causes (57). In Thai adults dying of cerebral malaria, sequestration was most marked in the cerebral circulation (82). Expression of ICAM-1 and E-selectin was increased on cerebral vessels, whereas staining for CD36 and thrombospondin was sparse. Expression of ICAM-1, CD36, and E-selectin colocalized with sequestration, and expression of ICAM-1, E-selectin, and VCAM-1 was also increased in lung, kidney, and muscle vasculature.

Attempts to find a correlation between clinical outcome and in vitro adherence to particular cells or to these receptors have been largely unsuccessful, although Ho et al. reported an association between severe malaria and adherence of patient isolates to C32 melanoma cells (39). No adherence pattern (apart from rosetting, described below) has been associated with cerebral malaria.

In addition to adhering to endothelial surfaces, *P. falciparum*-infected erythrocytes are clearly able to adhere to many other human cell types. Ruangjirachuporn and colleagues (69) showed that primary clinical isolates can adhere to monocytes, lymphocytes, neutrophils, and plasma cells and that neutrophils and monocytes can phagocytose intact IRBC. Parasites in long-term culture appeared more likely to bind if they were also able to bind to melanoma cells. Interestingly, rosetting strains (see below) bound only at a low level, possibly because binding was blocked by uninfected erythrocytes. Anti-CD36 monoclonal antibody had a minimal effect on the degree of binding, which suggested that other receptors were involved, and anti-ICAM-1 monoclonal antibodies inhibited cellular interaction by approximately 40%. The importance of the state of cell activation and differentiation is highlighted by the observation that the relative levels of CD36 and ICAM-1 on the surfaces of monocytes change as adherent monocytes differentiate in vitro (88). The relevance of these findings to immunity in vivo is uncertain, but in *Plasmodium knowlesi* malaria, some evidence indicates that protection correlates with levels of opsonizing antibody (18–20).

The static assays in routine use may well be poorly representative of binding in vivo. Using a sealed chamber under physiological flow conditions, Cooke et al. (28) showed that some parasitized cells could be trapped and immobilized from flow over an endothelial cell layer or a receptor-coated surface, after which two types of adhesion were observed. Some cells became anchored tightly, whereas others formed bonds that allowed cells to roll across the surface. "Rolling" adhesion to HUVEC could be blocked by monoclonal antibody to ICAM-1, whereas adhesion to CD36 was static and could not be inhibited by anti-ICAM-1 antibody. A major finding was the heterogeneity in the binding capacities of isolates from different individuals, but no correlation with severity of disease was found.

No discussion of this topic would be complete without acknowledging that host receptors differ in different organs and can easily be upregulated in response to cytokines. For example, specific receptors on the syncytiotrophoblasts of the placenta may be critical for sequestration in this organ during pregnancy.

ANTIGENS ON THE SURFACES OF IRBC

Antigens expressed on the surfaces of *P. knowlesi*-infected cells exhibit great diversity (17), and Hommel et al. (40) used immunofluorescence to demonstrate the diversity of parasite antigens expressed on the surfaces of *Saimiri* cells infected with *P. falciparum* and recognized by the sera of immune animals. This diversity has also been demonstrated in human *P. falciparum* infections (34, 50, 76). Apart from the demonstrated diversity of antigens at the surfaces of infected cells, Biggs and colleagues showed that individual *P. falciparum* parasites can alter the antigens expressed on the IRBC surface (12). This process, known as antigenic variation, provides a mechanism for eluding the immune response to surface antigens that occurs with other parasites (13). In addition, there appears to be a link between antigenicity and phenotype that confers on different antigenic types the ability to bind to different receptors (10, 65). In vitro selection of parasites that bind well to melanoma cells concurrently selects for an antigenically variant population under experimental conditions that may mimic the many cycles of sequestration and multiplication that occur during the course of infection in vivo (10). We hypothesize that only a subset of parasites may have the ability to bind to particular receptors.

IMMUNE RESPONSE TO SURFACE ANTIGENS

Acute malaria infection is associated with development of a specific immune response to molecules on the surfaces of infected cells (50), but evidence for a protective role for these molecules is scanty; however, there are no documented examples to date of individuals having strain-specific agglutinating antibodies at the time of infection by those strains. A study in The Gambia of various measurements of immunity to blood-stage antigens (51) found that the only predictive index for protection against clinical malaria is the titer of agglutinating activity in serum before the onset of the transmission season. Assays were performed with a single wild isolate, and even better correlation might have been obtained if sera could have been tested against several isolates. The finding that the presence of antibodies had a strong negative relationship with clinical malaria but not with parasitemia was of interest, as it is consistent with a role for antibodies in modifying pathological outcome.

Some evidence suggesting a role for opsonizing antibody in protection was in vitro data demonstrating that antibody known to be therapeutic in the treatment of children with malaria facilitates opsonization (14).

POSSIBLE MODE OF ACTION OF IMMUNE RESPONSE TO SURFACE ANTIGENS

Antibodies to surface molecules could be important for several reasons. They may interfere with or reverse cytoadherence (84), thus leading to increased parasite vulnerability and death in the spleen, but they may also cause antibody-dependent cytotoxicity (15), opsonization (25), or cell lysis. The Gambian studies mentioned above (51) provided no evidence that the presence of antibodies correlated with protection against clinical malaria, although the authors highlighted inherent difficulties that may have caused false-negative results in their study. Acknowledging that tissue sequestration may be the end result of interaction between IRBC expressing particular antigenic types and host cells upregulated by particular cytokines, a combined strategy of neutralizing both the cytokine (to reduce receptor expression) and the parasite neoantigen may be particularly beneficial. On the other hand, induction of an antibody response that prevents interaction with cells of the host immune system (e.g., monocytes, neutrophils) could be deleterious.

ROSETTING

Some but not all erythrocytes infected with *P. falciparum* have the ability to adhere to uninfected cells in a process known as rosetting. The importance of this phenomenon in pathogenesis was highlighted in a study from West Africa of parasites from children with cerebral and nonsevere malaria. This study suggested a correlation between rosette formation and cerebral malaria (22). The same study further supported a role for rosetting in cerebral malaria by reporting that individuals with nonsevere malaria were far more likely to have antibody capable of disrupting rosettes than were children with cerebral malaria. A smaller study from Madagascar (63) supported that association, but in a larger study from Papua New Guinea, neither rosette formation nor lack of antirosetting antibodies correlated with disease severity (4). In a study of 154 fresh isolates from Kenyan children, rosette formation was associated with severe malaria but not specifically with cerebral malaria (68). In this study, only 3 of 150 plasma samples gave >50% disruption of rosettes of the R29 rosetting clone.

CD36 present on uninfected cells may act as a host receptor for rosetting of some strains and field isolates (36), since monoclonal antibody to CD36 is able to reverse rosetting, but this phenomenon is not universal. Rosettes formed by many isolates are not disrupted by anti-CD36 monoclonal antibody (85), and many isolates that adhere to CD36 apparently do not form rosettes. Antibodies of different specificities can disrupt other isolates, e.g., anti-HRP-1 monoclonal antibody MAb89 reversed rosettes of rosetting strain PA1 (23). Direct estimates of the forces of interaction between IRBC and the uninfected cells in a rosette suggest that these forces are approximately five times stronger than those that bind parasitized cells to vascular endothelium (53).

It remains to be seen whether inhibition or reversal of rosette formation is important for protection of individuals in areas of different endemicity, but even if natural infection does not induce this protective response, immunization leading to disruption or prevention of rosette formation could be an important component of prevention of disease.

ANTIGENS ON THE SURFACES OF INFECTED CELLS THAT ARE POTENTIALLY IMPORTANT FOR INDUCTION OF PROTECTIVE IMMUNITY

PfEMP1

In contrast to the surface antigens of sporozoites and merozoites, which are exposed only briefly, parasite-dependent antigens on the surfaces of IRBC are obvious targets for an immune response for half of the asexual cycle.

The best-characterized surface molecule of IRBC and the prime candidate for a parasite-derived ligand involved in host-parasite adhesion is the molecule described by Leech et al. (47) and now referred to as PfEMP1 (*P. falciparum* erythrocyte membrane protein 1). This high-molecular-weight protein is accessible to surface labeling by lactoperoxidase-catalyzed radioiodination of trophozoite-infected cells (under conditions allowing minimal internal labeling) and varies in molecular weight among different isolates. Its sensitivity to low concentrations of trypsin (<10 μg/ml) correlates with loss of cytoadherence (41). PfEMP1 appears to be insoluble in Triton X-100 but soluble in sodium dodecyl sulfate, and its parasite origin is demonstrated by metabolic labeling with tritiated isoleucine. Immunoprecipitation of this molecule with variant-specific immune serum correlates with the ability of the serum to inhibit cytoadherence (12).

A family of 50 to 150 *var* genes that encode PfEMP1 was recently identified (7, 78). *var* genes occur on multiple parasite chromosomes, often in clusters. The gene comprises two exons, which correspond to the extra- and intracellular domains of a transmembrane protein. In the extracellular domain are two to four cysteine-rich domains with homology to the *P. falciparum* EBA-175 erythrocyte-binding protein, termed Duffy binding-like (DBL) domains. DBL1 and the adjacent cysteine-rich interdomain region form a relatively conserved head region; DBL2, DBL3, and DBL4 are less conserved and may undergo shuffling and deletion. Sequences are highly diverse but lack significant repeats; cloned lines may demonstrate rearrangements in *var* sequences (78), and transcription of distinct *var* genes corresponds to the expression of distinct variant antigens at the erythrocyte surface (75).

Baruch et al. used a rabbit serum that immunoprecipitated PfEMP1 to identify a genomic DNA clone expressing a *P. falciparum* antigen (7). Serum from animals immunized with this protein blocked immunoprecipitation of PfEMP1 of the original strain but not of other strains and gave strain-specific agglutination of IRBC. The serum also reacted by immunofluorescence and immunoelectron microscopy with the IRBC surface, with reactivity being localized to the region of the knobs. Finally, adherence of IRBC to CD36 was blocked by sera raised to a recombinant *var*-derived protein. Whether *var* genes can also support binding to other receptors, and if they can, whether separate *var* genes encode proteins binding to different receptors or whether different domains bind to different receptors is unknown. These and other questions such as those about coexpression of more than one *var* gene and the links between antigenicity and binding phenotype are now accessible to molecular and genetic investigation.

Many studies of laboratory and field isolates have demonstrated extreme variability in parasite-induced surface antigens on the IRBC (40, 50), and there is considerable debate as to whether immune individuals are able to develop agglutinating antibody reactive to all isolates. After animal studies with *P. knowlesi*, Brown and Brown (17) speculated on two types of immunity, one variant specific and the other transcending variation. The evidence for a degree of heterologous protection in humans is considerable (for example, see references 14 and 44), but most studies are unable to provide any data on the site of action of either form of immunity.

A conserved adhesion domain common to all molecules of this very diverse group is often assumed to be the final common pathway for binding, but there is very little evidence that naturally exposed individuals develop antibodies capable of reacting with all isolates, the so-called "panagglutinating antibodies," as opposed to antibodies against a diverse range of antigenic types. Studies of immune populations in various countries (for example, see references 34 and 76) have not confirmed the presence of panagglutinating-antibody responses reported in adult Gambians (50) or in African sera described by Aguiar et al. (1). The number of episodes of clinical disease experienced by an immune individual is significantly smaller than the repertoire of antigenic types circulating, so we conclude that immune adults who are not at risk of clinical malaria must be protected from pathogenic variants either because of priming for an enhanced and rapid response to a new variant or because they are protected from clinical malaria by other, more efficient mechanisms (76), such as immunity to other antigens or parasite toxins (60).

Lack of detection of a panagglutinating immune response resulting from naturally acquired infection does not exclude an approach to vaccination that uses this molecule. The constant domains may be weakly immunogenic after short-term exposure or may be exposed only after initial weak attachment to a receptor such as CD36 or thrombospondin. Delivery of the adhesion domain with an appropriate adjuvant could be more immunogenic than clinical infection. It is quite possible that *P. falciparum* (like the yeast *Candida albicans*) expresses an integrin analog to facilitate adhesion (35). This molecular mimicry by the parasite may mean that cross-reactive autoantibodies are harmful or that the binding domain is so similar to the host that it is not recognized as foreign by the immune system.

Antigenic diversity would be a major problem for vaccines based on the variable regions of this molecule unless it could be shown that pathogenic strains constitute only a small proportion of all possible variants. Recently, we demonstrated an important link between antigenicity and adhesive phenotype by showing that selection for a particular phenotype in vitro is associated with alteration in surface antigenicity (10). These results suggest that in vivo, only a subset of parasite antigenic types with the ability to bind to relevant receptors might need to be neutralized to prevent complications secondary to adhesion. If this hypothesis is true, perhaps only a limited number of antigenic variants will need to be included in a vaccine for prevention of pathological adhesion.

Sequestrin

There have been many attempts to identify the parasite molecule that interacts with CD36. Ockenhouse et al. identified a protein of approximately 270 kDa on the surface of IRBC that reacted with monospecific polyclonal anti-idiotype antibodies induced against OKM5 that mimicked the binding region of CD36 (56). This molecule, known as sequestrin, was detected in Triton extracts of infected cells. Sequestrin appears to contain a conserved functional binding domain found in wild-type and laboratory-adapted parasite strains but absent from nonadherent IRBC. Publication of the sequence of sequestrin will allow us to determine whether it is related to *var* gene family proteins or whether it provides a different target for vaccine development.

Modified Host Components

Malaria parasites may modify components of the host erythrocyte to produce "neoantigens" that are not expressed in uninfected cells. Band 3 is one such major erythrocyte membrane protein that becomes concentrated and apparently redistributed in the knob structures referred to previously (72).

Naturally occurring autoantibodies directed against modified band 3 appear to bind to knobby cells (90) and immunoprecipitate a trypsin-sensitive radioiodinatable surface protein with an M_r of >240,000 (91). A truncated form of band 3 (M_r 85,000) with the same solubility characteristics as PfEMP1 can be immunoprecipitated from knobby cells. A monoclonal antibody that blocks cytoadherence also recognizes modified band 3, and the monoclonal antibody reacts with all the isolates of *P. falciparum* that have been tested (31, 91).

Further evidence that modified human band 3 may be an important ligand for interaction of infected cells with the host was provided in a series of experiments in which peptides were used to block or reverse cytoadherence (29). Synthetic peptides corresponding to predicted external loop regions of band 3 were tested for their abilities to inhibit the adherence of IRBC to C32 melanoma cells. Several peptides inhibited cytoadherence, and two peptides reversed adhesion in vitro by up to 60% (29).

Infusion of selected band 3 peptides into *P. falciparum*-infected monkeys is associated with an increase in the numbers of mature forms in peripheral blood 24 h later. Surprisingly, apart from the effect of one peptide at high doses, administration of peptides caused mature forms to appear in the peripheral blood some time later, in contrast to the early effects of infusion of immune serum referred to previously (33). It is possible that the doses of peptides infused were insufficient to release sequestered parasites but that the peptide was present in amounts sufficient to occupy receptor sites long enough to prevent adherence of the next cycle of parasitized cells. One unexplained observation was the autoagglutination of infected cells that were circulating 24 h after peptide infusion.

Recently, Crandall et al. (30) reported that a peptide derived from amino acid residues 546 through 553 of band 3 and exposed on the surfaces of IRBC specifically binds to CD36 and inhibits cytoadherence to CD36-transfected CHO cells (37). This peptide, termed Pfalhesin, does not affect adherence to ICAM-1 (30).

Rosettins

The surface molecules of IRBC involved in the interaction with uninfected cells are known as rosettins. Using antisera that could disrupt rosettes, Carlson and colleagues identified a 28,000-M_r parasite-derived molecule by immunoprecipitation and a weakly stained 90,000 -M_r band on immunoblot (23). Further studies suggested that rosetting is mediated by more than one lectinlike interaction, since different parasites display different rosetting characteristics when grown in human erythrocytes of different blood groups. Parasites produce small rosettes in blood group O erythrocytes but larger rosettes in the erythrocytes of group A or B. Blood group A-preferring parasites could be completely disrupted by heparin when they were grown in O or B cells but not when they were grown in group A cells. The same result was found with group B-preferring parasites (24). Further evidence for multiple receptors was the finding that trisaccharides mimicking the terminal regions of A or B blood group carbohydrate antigens can inhibit rosette formation of the preferred group erythrocytes. This provides strong evidence that the blood group antigens are specific, but not sole, receptors for rosette formation (21).

Another candidate rosettin, with an M_r of 22,000, was precipitated from sodium dodecyl sulfate extracts of surface-radioiodinated IRBC. This molecule was identified only in erythrocytes infected with rosetting parasite strains that were susceptible to rosette disruption by antibody with that specificity. Purified immunoglobulin G against the molecule immunoprecipitated three parasite-derived polypeptides (M_rs of 22,000, 45,000 [doublet], and 50,000) from biosynthetically labeled parasites (38).

The 28,000- and 22,000-M_r rosettins apparently define two separate mechanisms for rosetting. The 28,000-M_r molecule is absent from nonrosetting strains and undetectable in IRBC of rosettes insensitive to anti-28,000-M_r-rosettin antibodies. The 22,000-M_r molecule appears to define a separate family of rosetting strains that can be disrupted by specific antibody. These molecules may be targets of agglutination or "variant-restricted antisera," but there is as yet no evidence that they are involved in endothelial cell binding.

Recent descriptions of a range of sulfated glycoconjugates that inhibit rosette formation (67) may provide new tools for identification and affinity purification of surface molecules involved in this interaction.

Multiple Ligands

P. falciparum-infected cells are clearly able to interact with several different receptors, and recently, we showed that individual cells are able to bind to more than one receptor (26). Whether more than one parasite molecule is required to mediate all these functions or whether multiple domains of a single molecule could explain different adhesion-receptor interactions has not been determined. Serial or simultaneous expression of several ligands may facilitate removal of cells from flow, with initial weak binding to endothelial cells being followed by strong adhesion in processes analogous to those of leukocyte rolling and sticking. The finding of coexpression of multiple ligands by one infected cell suggests that individual cells may use several mechanisms of adhesion and that an immune response against more than one mechanism may be essential for blocking cytoadherence.

APPROACHES TO VACCINATION THAT USE SURFACE-ACCESSIBLE MOLECULES

Active Immunization

As stated in the introduction to this chapter, cytoadherence to endothelium leading to parasite sequestration is a normal part of the life cycle of *P. falciparum*, but pathological sequestration in vital organs leads to disease. Vaccination to prevent pathological sequestration could prevent disease without preventing infection, thereby inducing an immune status similar to that of immune adults in an area of endemicity who have frequent infections in the presence of stable clinical immunity (so-called concomitant immunity).

Prevention or reversal of cytoadherence

The in vitro studies described above suggest that strain-specific antibody can prevent or reverse the adherence of IRBC to melanoma cells (84). Nonspecifically, this could also occur if agglutination prevented IRBC from reaching host target cells or led to parasite clearance. In vivo, prevention of adherence would render parasites susceptible to clearance by the spleen and to opsonization. Ideally, vaccination based on this strategy would involve the use of a conserved region of the adherence molecule to induce either a panagglutinating antibody or an antibody that blocked the region of the parasite responsible for adhesion. Such a response was seen when rabbits were immunized with recombinant protein from the cysteine-rich interdomain region of PfEMP1 from the Malayan Camp line: the rabbits produced antibodies that blocked adhesion of that line to CD36 (7). It remains to be seen whether single or combined antibodies could block adherence of a wide range of patient isolates (especially if pathogenic "binding" isolates were identified) and whether established cytoadherence could be reversed.

Prevention or reversal of rosette formation

If disruption of flow is the explanation for an association between rosetting and cerebral malaria, then possibly rosettins or their component peptides can be used as immunogens for prevention of those episodes of cerebral malaria dependent on these ligands. It is far from clear that the full range of rosettins has been identified. Other researchers have proposed that apart from restoring blood flow, rosette-disrupting antibody may release IRBC, so that the IRBC become susceptible to phagocytosis (69).

Passive Immunization

Udeinya et al. (85) described serum from an immune *Aotus* monkey that abolished the attachment of IRBC to cultured endothelial cells, and David and colleagues (33) described experiments with *Aotus* monkeys in which an infusion of immune serum when parasites were sequestered led to release of parasites into the peripheral circulation. Experiments using passive transfer of immune serum to children with recrudescent malaria in Thailand (70) caused a decrease in asexual parasitemia and a relief of symptoms, but at no stage were pigmented or segmented parasites detected in the peripheral circulation. Treatment of Malawian children with immunoglobulin as part of the management of cerebral malaria (80) did not provide evidence that sequestered parasites could be dislodged in this way. Immunoglobulin did not hasten parasite clearance, despite the fact that the preparation could reverse and inhibit binding to melanoma cells (but not to U937 cells or human monocytes, whose receptors may more closely resemble those of the human cerebral vascular endothelium). Possibly, the major receptor-ligand interaction responsible for sequestration in these patients was not reversible by the preparation used. The possibility of using specific immunotherapy for reversal of cytoadherence cannot be discounted by these negative results.

An alternative approach to passive immunization was tested in vitro by Staunton and colleagues (77), who constructed an immunoadhesion molecule containing part of ICAM-1 fused to two domains of human immunoglobulin G1 heavy chain. This molecule inhibited the adherence of IRBC to ICAM-1 and promoted phagocytosis of IRBC by human monocytes. Lymphocyte binding to ICAM-1 was not affected.

CANDIDATE VACCINE MOLECULES

PfEMP1

PfEMP1 is obviously a prime candidate for an antiadhesion vaccine. Its sequence, however, is highly variable, and this variability accounts for the molecule's variability in antigenicity, but the head region (the first DBL domain and the adjacent cysteine-rich interdomain region) is relatively conserved, as is the putative cytoplasmic domain (78). These domains may be candidates for vaccine components, especially if they are shown to include binding domains for other receptors such as CSA or ICAM-1 or for those identified in the future as correlating with sequestration in vital organs in addition to binding domains for CD36, as discussed above. A multicomponent vaccine could theoretically induce an immune response to a range of adhesion domains if the number of relevant variants could be identified. Such vaccination might also prime individuals for rapid, enhanced response to new variants. Little is known of the role of T-cell-mediated immunity in this type of response or how such immunity could be achieved by immunization.

Pf332

The human monoclonal antibody 33G2 inhibits cytoadherence of IRBC from a range of parasites to melanoma cells (86). The monoclonal antibody reacts with several sequences, but the optimal target antigen appears to be Pf332 (87), a giant protein of 2.5 MDa that contains tens of peptapeptide sequences with the potential to bind this monoclonal antibody (52). Although monoclonal antibody 33G2 inhibits a range of isolates, adherence is not totally abolished, and antibodies affinity purified on a Pf332 repeat peptide do not inhibit cytoadherence. The monoclonal antibody may cross-react with an epitope in another molecule (43).

HRP-1

Carlson and colleagues (23) showed that antibodies to histidine-rich proteins can interfere with rosette formation and react with a 90,000-M_r polypeptide in rosetting parasites. The monoclonal antibody also precipitates a histidine-labeled molecule with an M_r of 28,000. As stated above, other molecules may also be important in the rosetting phenomenon.

Rosettins

If future studies confirm that rosettins are of parasite origin and are the targets of rosette-inhibiting antibodies, then rosettins could be components of a vaccine aimed at neutralizing the potentially important virulence characteristic of rosetting. The discussion above makes it clear that a number of different molecules could provide alternative receptor ligand coupling for this process. If rosettins are of modified host origin, then vaccination using this strategy may not be possible.

Modified Band 3

As a general principle, a vaccination strategy based on a modified host molecule is not likely to proceed unless it can be shown that deleterious cross-reactions to unaltered host tissues do not occur. It seems more likely that peptides corresponding to modified host molecules will be used as therapeutic agents rather than as vaccines.

Others

PfHRP2, a histidine- and alanine-rich protein, is released from IRBC, but some of it may remain on the outer surface of the IRBC membrane, as demonstrated by immunoelectron microscopy with a monoclonal antibody (42). Little is known of the role of this molecule in rosetting or cytoadherence, but the molecule is interesting because of promising results within monkey protection experiments (46).

Combination Vaccines

We have argued here that sequestration in vital organs may be the result of an interaction between specific ligands that occur only on specific parasites and molecules of the host surface that have been upregulated by cytokines. Neutralization of these cytokines in combination with immunity against the range of parasite neoantigens may amplify the protective benefit of a vaccine. Another requirement for successful vaccination would obviously be to include the range of different molecules or epitopes responsible for adherence to the wide range of receptors already defined. Inclusion of T-cell epitopes could be required to facilitate production of an effective immune response that is boosted by reinfection.

TESTING VACCINES BASED ON IMMUNE RESPONSES TO ANTIGENS ON THE SURFACES OF INFECTED CELLS

On theoretical grounds, vaccines based on immune responses to cell surface antigens are not likely to induce sterile immunity, and inappropriate immune responses could favor selection of null variants or of a more virulent parasite. Theoretically, antibody could enhance morbidity either directly, for example, by aggregation, or indirectly, by preventing interaction with cytotoxic cells of the host. Field-based studies in areas of endemicity provide clear evidence that cerebral malaria has its peak incidence long after a child's first experience with malaria, and there is good evidence, at least for *P. knowlesi*, that antibody is able to induce antigenic variation in the parasite (16). An optimal vaccine may include components to reduce replication, reduce cytoadherence (to reduce pathology), and neutralize the parasite products responsible for release of cytokines (so-called antidisease vaccines), an activity that would also prevent the upregulation of receptors for cytoadherence.

Passive transfer studies will probably be carried out before active immunization with antiadherence vaccines is attempted. Meticulous, lengthy observation would be required for detection of adverse effects, and large numbers of participants would be required before the beneficial effect of such a vaccine could be demonstrated.

SUMMARY AND CONCLUSIONS

The point of interaction between parasite and host is an obvious target for a vaccine, and compared with the time available for attack on sporozoites or merozoites, the time an IRBC is vulnerable is lengthy. Abnormal tissue sequestration probably results from interaction of IRBC expressing particular antigens with host cell receptors upregulated by particular cytokines, and a combined strategy of neutralizing the parasite ligand and the cytokines (or the parasite molecules responsible for their release) may be the most successful. Our current knowledge of the range of available host receptors and the characteristics of the parasite-dependent molecules on the surface of the IRBC is far from complete, but the discovery of the *var* gene family, which encodes the dominant adherence molecule PfEMP1; the emerging information about sequestrin (a CD36-binding ligand); and the discovery of band 3 peptides that may block adherence offer many new opportunities to investigate the adhesion process and embark on a process of designing vaccines that prevent cytoadherence.

Attempts at protective immunization will need to take into account the propensity for the response to be dominated by strain or phenotypic specificity and the impact that this propensity could have on favoring variation to a more pathogenic phenotype. It is likely that different receptors are important in different individuals at different times in different organs and that more than one receptor will need to be neutralized in order to induce a completely protective response. A vaccine based purely on sequestrin, for example, could be ineffective against parasites binding to CD36 by PfEMP1 or other ligands.

As the molecular mechanisms of this aspect of host-parasite interaction become clearer, vaccination strategies based on the prevention of cytoadherence may become a reality, particularly if an invariant binding site can be identified. Further studies relating these phenomena to pathogenicity must be undertaken, and efforts to understand the series of events and the final common pathways of binding that are susceptible to prevention or disruption by the host immune response must be made.

Acknowledgments. We acknowledge the help of Heather Saunders in preparing the manuscript and of John Reeder and Beverley Biggs in reviewing the text.

We are supported by the National Health and Medical Research Council of Australia and the John D. and Catherine T. MacArthur Foundation.

REFERENCES

1. **Aguiar, J. C., G. R. Albrecht, P. Cegielski, B. M. Greenwood, J. B. Jensen, G. Lallinger, A. Martinez, I. A. McGregor, J. N. Minjas, J. Neequaye, M. E. Patarroyo, J. A. Sherwood, and R. J. Howard.** 1992. Agglutination of *Plasmodium falciparum*-infected erythrocytes from East and West African isolates by human sera from distant geographic regions. *Am. J. Trop. Med. Hyg.* **47:**621–632.
2. **Aikawa, M.** 1988. Human cerebral malaria. *Am. J. Trop. Med. Hyg.* **39:**3–10.
3. **Aikawa, M., M. Iseki, J. W. Barnwell, D. Taylor, M. M. Oo, and R. J. Howard.** 1990. The pathology of human cerebral malaria. *Am. J. Trop. Med. Hyg.* **43:**30–37.
4. **Al-Yaman, F., B. Genton, D. Mokela, A. Raiko, S. Kati, S. Rogerson, J. Reeder, and M. Alpers.** 1995. Human cerebral malaria: lack of significant association between erythrocyte rosetting and disease severity. *Trans. R. Soc. Trop. Med. Hyg.* **89:**55–58.
5. **Asch, A. S., I. Liu, F. M. Briccetti, J. W. Barnwell, F. Kwakye-Berko, A. Dokun, J. Goldberger, and M. Pernambuco.** 1993. Analysis of CD36 binding domains: ligand specificity controlled by dephosphorylation of an ectodomain. *Science* **262:**1436–1440.
6. **Barnwell, J. W., C. F. Ockenhouse, and D. M. Knowles.** 1985. Monoclonal antibody OKM5 inhibits the in vitro binding of *Plasmodium falciparum*-infected erythrocytes to monocytes, endothelial, and C32 melanoma cells. *J. Immunol.* **135:**3494–3497.
7. **Baruch, D. I., B. L. Pasloske, H. B. Singh, X. Bi, X. C. Ma, M. Feldman, T. F. Taraschi, and R. J. Howard.** 1995. Cloning the *P. falciparum* gene encoding PfEMP1, a malarial variant antigen and adherence receptor on the surface of parasitized human erythrocytes. *Cell* **82:**77–87.
8. **Berendt, A. R., D. L. Simmons, J. Tansey, C. I. Newbold, and K. Marsh.** 1989. Intercellular adhesion molecule 1 is an endothelial cell adhesion molecule for *Plasmodium falciparum. Nature* (London) **341:**57–59.
9. **Berendt, A. R., G. D. H. Turner, and C. I. Newbold.** 1994. Cerebral malaria: the sequestration hypothesis. *Parasitol. Today* **10:**412–414.
10. **Biggs, B. A., R. F. Anders, H. E. Dillon, K. M. Davern, M. Martin, C. Petersen, and G. V. Brown.** 1992. Adherence of infected erythrocytes to venular endothelium selects for antigenic variants of *Plasmodium falciparum. J. Immunol.* **149:**2047–2054.
11. **Biggs, B. A., J. G. Culvenor, J. S. Ng, D. J. Kemp, and G. V. Brown.** 1989. *Plasmodium falciparum*: cytoadherence of a knobless clone. *Exp. Parasitol.* **69:**189–197.
12. **Biggs, B. A., L. Goozé, K. Wycherley, W. Wollish, B. Southwell, J. H. Leech, and G. V. Brown.** 1991. Antigenic variation in *Plasmodium falciparum. Proc. Natl. Acad. Sci. USA* **88:**9171–9174.
13. **Borst, P., and D. R. Greaves.** 1987. Programmed gene rearrangements altering gene expression. *Science* **235:**658–667.
14. **Bouharoun-Tayoun, H., P. Attanath, A. Sabchareon, T. Chongsuphajaisiddhi, and P. Druilhe.** 1990. Antibodies that protect humans against *Plasmodium falciparum* blood stages do not on their own inhibit parasite growth and invasion in vitro, but act in cooperation with monocytes. *J. Exp. Med.* **172:**1633–1641.
15. **Brown, J., and M. E. Smalley.** 1980. Specific antibody-dependent cellular cytotoxicity in human malaria. *Clin. Exp. Immunol.* **41:**423–429.
16. **Brown, K. N.** 1973. Antibody induced variation in malaria parasites. *Nature* (London) **242:**49–50.
17. **Brown, K. N., and I. N. Brown.** 1965. Immunity to malaria: antigenic variation in chronic infection of *Plasmodium knowlesi. Nature* (London) **208:**1286–1288.
18. **Brown, K. N., I. N. Brown, and L. A. Hills.** 1970. Immunity to malaria. I. Protection

against *Plasmodium knowlesi* shown by monkeys sensitized with drug-suppressed infections or by dead parasites in Freund's adjuvant. *Exp. Parasitol.* **28:**301–317.

19. **Brown, K. N., I. N. Brown, P. I. Trigg, R. S. Phillips, and L. A. Hills.** 1970. Immunity to malaria. II. Serological response of monkeys sensitized by drug-suppressed infection or by dead parasitized cells in Freund's complete adjuvant. *Exp. Parasitol.* **28:**318–338.
20. **Brown, K. N., and L. A. Hills.** 1974. Antigenic variation and immunity to *Plasmodium knowlesi*: antibodies which induce antigenic variation and antibodies which destroy parasites. *Trans. R. Soc. Trop. Med. Hyg.* **68:**139–142.
21. **Carlson, J.** 1993. Erythrocyte rosetting in *Plasmodium falciparum* malaria—with special reference to the pathogenesis of cerebral malaria. *Scand. J. Infect. Dis.* **86:**1–79.
22. **Carlson, J., H. Helmby, A. V. S. Hill, D. Brewster, B. M. Greenwood, and M. Wahlgren.** 1990. Human cerebral malaria: association with erythrocyte rosetting and lack of anti-rosetting antibodies. *Lancet* **336:**1457–1460.
23. **Carlson, J., G. Holmquist, D. W. Taylor, P. Perlmann, and M. Wahlgren.** 1990. Antibodies to a histidine-rich protein (PfHRP1) disrupt spontaneously formed *Plasmodium falciparum* erythrocyte rosettes. *Proc. Natl. Acad. Sci. USA* **87:**2511–2515.
24. **Carlson, J., and M. Wahlgren.** 1992. *Plasmodium falciparum* erythrocyte rosetting is mediated by promiscuous lectin-like interactions. *J. Exp. Med.* **176:**1311–1317.
25. **Celada, A., A. Cruchaud, and L. H. Perrin.** 1982. Opsonic activity of human immune serum on in vitro phagocytosis of *Plasmodium falciparum* infected red blood cells by monocytes. *Clin. Exp. Immunol.* **47:**635–644.
26. **Chaiyaroj, S. C., R. L. Coppel, S. Novakovic, and G. V. Brown.** 1994. Multiple ligands for cytoadherence are present on the surface of *Plasmodium falciparum* infected erythrocytes. *Proc. Natl. Acad. Sci. USA* **91:**10805–10808.
27. **Clark, I. A., and K. A. Rockett.** 1994. How important is mechanical blockage of blood vessels in human cerebral malaria? *Parasitol. Today* **10:**6.
28. **Cooke, B. M., S. Morris-Jones, B. M. Greenwood, and G. B. Nash.** 1993. Adhesion of parasitized red blood cells to cultured endothelial cells: a flow-based study of isolates from Gambian children with falciparum malaria. *Parasitology* **107:**359–368.
29. **Crandall, I., W. E. Collins, J. Gysin, and I. W. Sherman.** 1993. Synthetic peptides based on motifs present in human band 3 protein inhibit cytoadherence/sequestration of the malaria parasite *Plasmodium falciparum*. *Proc. Natl. Acad. Sci. USA* **90:**4703–4707.
30. **Crandall, I., K. M. Land, and I. W. Sherman.** 1994. *Plasmodium falciparum*: Pfalhesin and CD36 form an adhesion/receptor pair that is responsible for the pH-dependent portion of cytoadherence/sequestration. *Exp. Parasitol.* **78:**203–209.
31. **Crandall, I., and I. W. Sherman.** 1991. *Plasmodium falciparum* (human malaria)-induced modifications in human erythrocyte band 3 protein. *Parasitology* **102:**335–340.
32. **Culvenor, J. G., C. J. Langford, P. E. Crewther, R. B. Saint, R. L. Coppel, D. J. Kemp, R. F. Anders, and G. V. Brown.** 1987. *Plasmodium falciparum*: identification and localization of a knob protein antigen expressed by a cDNA clone. *Exp. Parasitol.* **63:**58–67.
33. **David, P. H., M. Hommel, L. H. Miller, I. J. Udeinya, and L. D. Oligino.** 1983. Parasite sequestration in *Plasmodium falciparum* malaria: spleen and antibody modulation of cytoadherence of infected erythrocytes. *Proc. Natl. Acad. Sci. USA* **80:**5075–5079.
34. **Forsyth, K. P., G. Philip, T. Smith, S. Kum, B. Southwell, and G. V. Brown.** 1989. Diversity of antigens expressed on the surface of the erythrocytes infected with mature *Plasmodium falciparum* parasites in Papua New Guinea. *Am. J. Trop. Med. Hyg.* **41:**259–265.
35. **Gustafson, K. S., G. M. Vercellotti, C. M. Bendel, and M. K. Hostetter.** 1991. Molecular mimicry in *Candidate albicans*. *Am. Soc. Clin. Invest.* **87:**1896–1902.
36. **Handunnetti, S. M., M. R. van Schravendijk, T. Hasler, J. W. Barnwell, D. E. Green-**

walt, and R. J. Howard. 1992. Involvement of CD36 on erythrocytes as a rosetting receptor for *Plasmodium falciparum*-infected erythrocytes. *Blood* **80:**2097–2104.

37. **Hasler, T., G. R. Albrecht, M. R. van Schravendijk, J. C. Aguiar, K. E. Morehead, B. L. Pasloske, C. Ma, J. W. Barnwell, B. Greenwood, and R. J. Howard.** 1993. An improved microassay for *Plasmodium falciparum* cytoadherence using stable transformants of Chinese hamster ovary cells expressing CD36 or intercellular adhesion molecule-1. *Am. J. Trop. Med. Hyg.* **48:**332–347.
38. **Helmby, H., L. Cavelier, U. Pettersson, and M. Wahlgren.** 1993. Rosetting *Plasmodium falciparum*-infected erythrocytes express unique strain-specific antigens on their surface. *Infect. Immun.* **61:**284–288.
39. **Ho, M., B. Singh, S. Looareesuwan, T. Davis, D. Bunnag, and N. J. White.** 1991. Clinical correlates of in vitro *Plasmodium falciparum* cytoadherence. *Infect. Immun.* **59:**873–878.
40. **Hommel, M., P. H. David, L. D. Oligino, and J. R. David.** 1982. Expression of strain-specific surface antigens on *Plasmodium falciparum*-infected erythrocytes. *Parasite Immunol.* **4:**409–419.
41. **Howard, R. J.** 1988. Malarial proteins at the membrane of *Plasmodium falciparum*-infected erythrocytes and their involvement in cytoadherence to endothelial cells. *Prog. Allergy* **41:**98–147.
42. **Howard, R. J., S. Uni, M. Aikawa, S. B. Aley, J. H. Leech, A. M. Lew, T. E. Wellems, J. Rener, and D. W. Taylor.** 1986. Secretion of a malarial histidine-rich protein (PfHRP II) from *Plasmodium falciparum*-infected erythrocytes. *J. Cell Biol.* **103:**1269–1277.
43. **Iqbal, J., P. Perlmann, and K. Berzins.** 1993. *Plasmodium falciparum*: analysis of the cytoadherence inhibition of the human monoclonal antibody 33G2 and of antibodies reactive with antigen Pf332. *Exp. Parasitol.* **77:**79–87.
44. **Jeffrey, G. M.** 1966. Epidemiological significance of repeated infections with homologous and heterologous strains and species of *Plasmodium. Bull. W.H.O.* **35:**873–882.
45. **Johnson, J. K., R. A. Swerlick, K. K. Grady, P. Millet, and T. M. Wick.** 1993. Cytoadherence of *Plasmodium falciparum*-infected erythrocytes to microvascular endothelium is regulatable by cytokines and phorbol ester. *J. Infect. Dis.* **167:**698–703.
46. **Knapp, B., E. Hundt, B. Enders, and H. A. Kupper.** 1992. Protection of *Aotus* monkeys from malaria infection by immunization with recombinant hybrid proteins. *Infect. Immun.* **60:**2397–2401.
47. **Leech, J. H., J. W. Barnwell, L. H. Miller, and R. J. Howard.** 1984. Identification of a strain-specific malarial antigen exposed on the surface of *Plasmodium falciparum*-infected erythrocytes. *J. Exp. Med.* **159:**1567–1575.
48. **Luse, S. A., and L. H. Miller.** 1971. Plasmodium falciparum malaria. Ultrastructure of parasitized erythrocytes in cardiac vessels. *Am. J. Trop. Med. Hyg.* **20:**655–660.
49. **MacPherson, G. G., M. J. Warrell, N. J. White, S. Looareesuwan, and D. A. Warrell.** 1985. Human cerebral malaria. A quantitative ultrastructural analysis of parasitized erythrocyte sequestration. *Am. J. Pathol.* **119:**385–401.
50. **Marsh, K., and R. J. Howard.** 1986. Antigens induced on erythrocytes by *P. falciparum*: expression of diverse and conserved determinants. *Science* **231:**150–153.
51. **Marsh, K., L. Otoo, R. J. Hayes, D. C. Carson, and B. M. Greenwood.** 1989. Antibodies to blood stage antigens of *Plasmodium falciparum* in rural Gambians and their relation to protection against infection. *Trans. R. Soc. Trop. Med. Hyg.* **83:**293–303.
52. **Mattei, D., and A. Scherf.** 1992. The Pf332 gene of *Plasmodium falciparum* codes for a giant protein that is translocated from the parasite to the membrane of infected erythrocytes. *Gene* **110:**71–79.
53. **Nash, G. B., B. M. Cooke, J. Carlson, and M. Wahlgren.** 1992. Rheological properties of rosettes formed by red blood cells parasitized by *Plasmodium falciparum. Br. J. Haematol.* **82:**757–763.

54. **Ockenhouse, C. F., and J. D. Chulay.** 1988. *Plasmodium falciparum* sequestration: OKM5 antigen (CD36) mediates cytoadherence of parasitized erythrocytes to a myelomonocytic cell line. *J. Infect. Dis.* **157:**584–588.
55. **Ockenhouse, C. F., M. Ho, N. N. Tandon, G. A. Van-Seventer, S. Shaw, N. J. White, G. A. Jamieson, J. D. Chulay, and H. K. Webster.** 1991. Molecular basis of sequestration in severe and uncomplicated *Plasmodium falciparum* malaria: differential adhesion of infected erythrocytes to CD36 and ICAM-1. *J. Infect. Dis.* **164:**163–169.
56. **Ockenhouse, C. F., F. W. Klotz, N. N. Tandon, and G. A. Jamieson.** 1991. Sequestrin, a CD36 recognition protein on *Plasmodium falciparum* malaria-infected erythrocytes identified by anti-idiotype antibodies. *Proc. Natl. Acad. Sci. USA* **88:**3175–3179.
57. **Ockenhouse, C. F., T. Tegoshi, Y. Maeno, C. Benjamin, M. Ho, K. E. Kan, Y. Thway, K. Win, M. Aikawa, and R. R. Lobb.** 1992. Human vascular endothelial cell adhesion receptors for *Plasmodium falciparum*-infected erythrocytes: roles for endothelial leukocyte adhesion molecule 1 and vascular cell adhesion molecule 1. *J. Exp. Med.* **176:**1183–1189.
58. **Oquendo, P., E. Hundt, J. Lawler, and B. Seed.** 1989. CD36 directly mediates cytoadherence of *Plasmodium falciparum* parasitized erythrocytes. *Cell* **58:**95–101.
59. **Panton, L. J., J. H. Leech, L. H. Miller, and R. J. Howard.** 1987. Cytoadherence of *Plasmodium falciparum*-infected erythrocytes to human melanoma cell lines correlates with surface OKM5 antigen. *Infect. Immun.* **55:**2754–2758.
60. **Playfair, J. H. L., J. Taverne, C. A. W. Bate, and J. B. de Souza.** 1990. The malaria vaccine: anti-parasite or anti-disease? *Immunol. Today* **11:**25–27.
61. **Pongponratn, E., M. Riganti, B. Punpoowong, and M. Aikawa.** 1991. Microvascular sequestration of parasitized erythrocytes in human falciparum malaria: a pathological study. *Am. J. Trop. Med. Hyg.* **44:**168–175.
62. **Reeder, J. C., S. J. Rogerson, F. Al-Yaman, R. F. Anders, R. L. Coppel, S. Novakovic, M. P. Alpers, and G. V. Brown.** 1994. Diversity of agglutinating phenotype, cytoadherence and rosette-forming characteristics of *Plasmodium falciparum* isolates from Papua New Guinean children. *Am. J. Trop. Med. Hyg.* **51:**45–55.
63. **Ringwald, P., F. Peyron, J. P. Lepers, P. Rabarison, C. Rakotomalala, M. Razanamparany, M. Rabodonirina, J. Roux, and J. Le Bras.** 1993. Parasite virulence factors during falciparum malaria: rosetting, cytoadherence, and modulation of cytoadherence by cytokines. *Infect. Immun.* **61:**5198–5204.
64. **Roberts, D. D., J. A. Sherwood, S. L. Spitalnik, L. J. Panton, R. J. Howard, V. M. Dixit, W. A. Frazier, L. H. Miller, and V. Ginsburg.** 1985. Thrombospondin binds falciparum malaria parasitized erythrocytes and may mediate cytoadherence. *Nature* (London) **318:**64–66.
65. **Roberts, D. J., A. G. Craig, A. R. Berendt, R. Pinches, G. Nash, G. Marsh, and C. I. Newbold.** 1992. Rapid switching to multiple antigenic and adhesive phenotypes in malaria. *Nature* (London) **357:**689–692.
66. **Rogerson, S. J., S. C. Chaiyaroj, K. Ng, J. C. Reeder, and G. V. Brown.** 1995. Chondroitin sulfate A is a cell surface receptor for *Plasmodium falciparum*-infected erythrocytes. *J. Exp. Med.* **182:**15–20.
67. **Rogerson, S. J., J. C. Reeder, F. Al-Yaman, and G. V. Brown.** 1994. Sulfated glycoconjugates as disrupters of *Plasmodium falciparum* erythrocyte rosettes. *Am. J. Trop. Med. Hyg.* **51:**198–203.
68. **Rowe, A., J. Obeiro, C. I. Newbold, and K. Marsh.** 1995. *Plasmodium falciparum* rosetting is associated with malaria severity in Kenya. *Infect. Immun.* **63:**2323–2326.
69. **Ruangjirachuporn, W., B. A. Afzelius, H. Helmby, A. V. S. Hill, B. M. Greenwood, J. H. Carlson, K. Berzins, P. Perlmann, and M. Wahlgren.** 1992. Ultrastructural analysis

of fresh *Plasmodium falciparum*-infected erythrocytes and their cytoadherence to human leukocytes. *Am. J. Trop. Med. Hyg.* **46:**511–519.

70. **Sabchaeron, A., T. Burnouf, D. Ouattara, P. Attanath, H. Bouharoun-Tayoun, P. Chantavanich, C. Foucault, T. Chongsuphajaisiddhi, and P. Druilhe.** 1991. Parasitologic and clinical human response to immunoglobulin administration in falciparum malaria. *Am. J. Trop. Med. Hyg.* **45:**297–308.
71. **Schofield, L., S. Novakovic, P. Gerold, R. Schwarz, M. McConville, and S. D. Tachado.** Glycosylphosphatidylinositol toxin of *Plasmodium* upregulates ICAM-1, VCAM-1 and E-selectin expression in vascular endothelial cells and increases leukocyte and parasite cytoadherence *via* tyrosine kinase dependent signal transduction. *J. Immunol.*, in press.
72. **Sherman, I. W., J. R. Greenan, and P. De La Vega.** 1988. Immunofluorescent and immunoelectron microscopic localization of protein antigens in red cells infected with the human malaria *Plasmodium falciparum. Ann. Trop. Med. Parasitol.* **82:**531–545.
73. **Sherwood, J. A., D. D. Roberts, K. Marsh, E. B. Harvey, S. L. Spitalnik, L. H. Miller, and R. J. Howard.** 1987. Thrombospondin binding by parasitized erythrocyte isolates in *falciparum* malaria. *Am. J. Trop. Med. Hyg.* **36:**228–233.
74. **Smith, H., J. A. Nelson, C. G. Gahmberg, I. Crandall, and I. W. Sherman.** 1992. *Plasmodium falciparum*: cytoadherence of malaria-infected erythrocytes to human brain capillary and umbilical vein endothelial cells—a comparative study of adhesive ligands. *Exp. Parasitol.* **75:**269–280.
75. **Smith, J. D., C. E. Chitnis, A. G. Craig, D. J. Roberts, D. E. Hudson-Taylor, D. S. Peterson, R. Pinches, C. I. Newbold, and L. H. Miller.** 1995. Switches in expression of *Plasmodium falciparum var* genes correlate with changes in antigenic and cytoadherent phenotypes of infected erythrocytes. *Cell* **82:**101–110.
76. **Southwell, B., G. V. Brown, K. P. Forsyth, T. Smith, G. Philip, and R. F. Anders.** 1989. Field applications of agglutination and cytoadherence assays with *Plasmodium falciparum* from Papua New Guinea. *Trans. R. Soc. Trop. Med. Hyg.* **83:**464–469.
77. **Staunton, D. E., C. F. Ockenhouse, and T. A. Springer.** 1992. Soluble intercellular adhesion molecule 1-immunoglobulin G1 immunoadhesin mediates phagocytosis of malaria-infected erythrocytes. *J. Exp. Med.* **176:**1471–1476.
78. **Su, X., V. M. Heatwole, S. P. Wertheimer, F. Guinet, J. A. Herrfeldt, D. S. Peterson, J. A. Ravetch, and T. E. Wellems.** 1995. The large diverse gene family *var* encodes proteins involved in cytoadherence and antigenic variation of *Plasmodium falciparum*-infected erythrocytes. *Cell* **82:**89–100.
79. **Swerlick, R. A., K. H. Lee, T. M. Wick, and T. J. Lawley.** 1992. Human dermal microvascular endothelial but not human umbilical vein/endothelial cells express CD36 *in vivo* and *in vitro*. *J. Immunol.* **148:**78–83.
80. **Taylor, T. E., M. E. Molyneux, J. J. Wirima, A. Borgstein, J. D. Goldring, and M. Hommel.** 1992. Intravenous immunoglobulin in the treatment of paediatric cerebral malaria. *Clin. Exp. Immunol.* **90:**357–362.
81. **Trager, W., M. A. Rudzinska, and P. C. Bradbury.** 1966. The fine structure of *Plasmodium falciparum* and its host erythrocytes in natural malarial infections in man. *Bull. W.H.O.* **35:**883–885.
82. **Turner, G. D. H., H. Morrison, M. Jones, T. M. E. Davis, S. Looareesuwan, I. D. Buley, K. C. Gatter, C. I. Newbold, S. Pukritayakamee, B. Nagachinta, N. J. White, and A. R. Berendt.** 1994. An immunohistochemical study of the pathology of fatal malaria. Evidence for widespread endothelial activation and a potential role for intercellular adhesion molecule-1 in cerebral sequestration. *Am. J. Pathol.* **145:**1057–1069.
83. **Udeinya, I. J., and C. O. Akogyeram.** 1993. Induction of adhesiveness in human en-

dothelial cells by *Plasmodium falciparum*-infected erythrocytes. *Am. J. Trop. Med. Hyg.* **48:**488–495.

84. **Udeinya, I. J., L. H. Miller, I. A. McGregor, and J. B. Jensen.** 1983. *Plasmodium falciparum* strain-specific antibody blocks binding of infected erythrocytes to amelanotic melanoma cells. *Nature* (London) **303:**429–431.
85. **Udeinya, I. J., J. A. Schmidt, M. Aikawa, L. H. Miller, and I. Green.** 1981. *Falciparum* malaria-infected erythrocytes specifically bind to cultured human endothelial cells. *Science* **213:**555–557.
86. **Udomsangpetch, R., M. Aikawa, K. Berzins, M. Wahlgren, and P. Perlmann.** 1989. Cytoadherence of knobless *Plasmodium falciparum*-infected erythrocytes and its inhibition by a human monoclonal antibody. *Nature* (London) **338:**763–765.
87. **Udomsangpetch, R., J. Carlsson, B. Wåhlin, G. Holmquist, L. S. Ozaki, A. Scherf, D. Mattei, O. Mercereau-Puijalon, S. Uni, M. Aikawa, K. Berzins, and P. Perlmann.** 1989. Reactivity of the human monoclonal antibody 33G2 with repeated sequences of three distinct *Plasmodium falciparum* antigens. *J. Immunol.* **142:**3620–3626.
88. **Udomsangpetch, R., H. K. Webster, K. Pattanapanyasat, S. Pitchayangkul, and S. Thaithong.** 1992. Cytoadherence characteristics of rosette-forming *Plasmodium falciparum. Infect. Immun.* **60:**4483–4490.
89. **Wahlgren, M., V. Fernandez, C. Scholander, and J. Carlson.** 1994. Rosetting. *Parasitol. Today* **10:**73–79.
90. **Winograd, E., J. R. Greenan, and I. W. Sherman.** 1987. Expression of senescent antigen on erythrocytes infected with a knobby variant of the human malaria parasite *Plasmodium falciparum. Proc. Natl. Acad. Sci. USA* **84:**1931–1935.
91. **Winograd, E., and I. W. Sherman.** 1989. Characterization of a modified red cell membrane protein expressed on erythrocytes infected with the human malaria parasite *Plasmodium falciparum*: possible role as a cytoadherent mediating protein. *J. Cell Biol.* **108:**23–30.

Malaria Vaccine Development: A Multi-Immune Response Approach
Edited by Stephen L. Hoffman

Chapter 7

An Antitoxic Vaccine for Malaria?

J. H. L. Playfair

INTRODUCTION

In this chapter, the case is made for a vaccine that acts against the pathology of malaria rather than against the parasite. This chapter is slightly different from the other chapters, since much of the work to be described is still at the laboratory animal stage, and its application to human disease remains largely theoretical.

MALARIA AS A TOXIC DISEASE

Several lines of argument suggest that a malaria vaccine that operates against the disease rather than against the parasite itself might be worth considering. To begin with, there is a long literature devoted to the concept of toxins in clinical malaria and the development of antitoxic immunity in areas of endemicity, and standard vaccines against well-established toxins (e.g., tetanus and diphtheria) are among the most effective in existence. More recently, attention has focused on the possible toxic role of cytokines, particularly tumor necrosis factor (TNF) in severe *Plasmodium falciparum* malaria, and on the identification of the parasite components that induce cytokines. Lastly, there is the fact that solid immunity against the parasite is still, after decades of work, proving quite difficult to induce. This last point will not be pursued further in this chapter, which concentrates on the evidence that an antitoxic vaccine might eventually be feasible.

ANTITOXIC IMMUNITY

Fifty years ago and more, malaria was widely considered a toxic disease, and it was recognized that not all patients with the same parasitemias were equally ill. Thus, the simple presence of parasites was not sufficient to cause symptoms, which must therefore be produced by indirect means. It was also proposed that children could become immune to these toxic effects, and thus suffer less severe symptoms, at an earlier age than they acquired immunity to the parasite (26, 46, 47). More recent studies in The Gambia confirmed that clinical tolerance (defined as the ability to live asymptomatically with fairly dense parasitemia) developed by the age of 3 to 5 years despite the persistence of high parasitemias for up to 5 more years (36). No evidence has yet been produced that this tolerance is due to an immune response, and it is still possible that it represents a shift in cell populations (23), the presence of some natural blocking factor, or differences in, for example, macrophage activity. However, recent work in experimental models, reviewed below, does suggest that acquired immunity to symptom-inducing antigens, possibly acting to reduce the levels TNF or other cytokines, can occur and might form the basis of a vaccine.

In considering malaria from this viewpoint, it is necessary to distinguish several patterns of disease. Fever, the hallmark of malaria, is prominent in both *P. falciparum* and *Plasmodium vivax* infections, and quite good evidence links fever to TNF overproduction (32, 35; see also below). Anemia, another almost universal complication, is particularly a feature of the first year of life, whereas cerebral malaria, the major cause of mortality, tends to occur from about a year later and is not seen in *P. vivax* infection. Hypoglycemia, which is frequent in *P. falciparum* infection, is also absent in *P. vivax* infection. Thus, a single pathogenetic mechanism is hardly likely to be responsible for all the clinical effects of the disease.

CYTOKINES

Nevertheless, as Clark et al. were the first to point out (14, 15), there are theoretical arguments for an involvement of TNF in these and many other complications of *P. falciparum* malaria, which bear a striking resemblance to the toxic effects of high-dose TNF. This cytokine is the one most thoroughly studied so far, though there is also evidence that interleukin 6 (IL-6), for example, can be induced independently (41). There is a general consensus that elevated levels of TNF in plasma are associated with severity of disease and mortality in patients with cerebral malaria (25, 33, 34). In a mouse model that somewhat resembles cerebral malaria, anti-TNF antibody protected against cerebral disease but not against anemia, but so did antibodies against several other cytokines (24). In a small trial of anti-TNF antibody in patients with human malaria, a reduction in fever was the only significant effect noted (35). Fever and elevated TNF levels are also found in patients with *P. vivax* infection (32), so if TNF is really involved in cerebral malaria, some cofactor(s) must be, too; a different pattern of sequestration could clearly be one of these cofactors, or there may be synergy with other cytokines or mediators. Interestingly, additive or synergistic effects are also seen in the antiparasite activity of TNF: killing of *P. vivax* gametocytes in vitro appears to require TNF plus a complementary factor (37). Evidently, we have much to learn about both the pathogenetic and the protective activities of even this one cytokine, but it seemed to us worth trying to understand why it is overproduced in malaria patients.

WHAT INDUCES TNF IN PATIENTS WITH MALARIA?

In our first experiments, in which we used the mouse parasite *Plasmodium yoelii* 17X, we found that parasitized erythrocytes incubated overnight with peritoneal macrophages induced amounts of TNF comparable to those of bacterial lipopolysaccharide (LPS). Both lethal and nonlethal strains of the parasite were active. We then found that incubating the parasitized blood overnight yielded supernatants that themselves induced TNF (10). The human parasite *P. falciparum* behaved similarly, and either mouse macrophages or human blood monocytes could be used interchangeably (48). When injected into mice, these supernatants induced TNF sufficient to be detectable in the serum and were lethal in mice that had been made hypersensitive to TNF by pretreatment with D-galactosamine; this lethal effect was inhibited by anti-TNF antibody (11).

Evidently, then, there are molecules in or on parasitized erythrocytes that induce TNF from macrophages. One obvious possibility was that we were picking up contaminating bacterial LPS, but for reasons to be summarized later, we believe that this was not the case. However, the active material did resemble LPS in certain respects: it retained activity after boiling, digestion with proteases, and deamination by nitrous acid but was destroyed by lipase treatment, by some types of phospholipase C, by dephosphorylation with HF, and by deacylation with NaOH (12). In a two-phase lipid extraction, the material appeared in the chloroform-methanol rather than the water-methanol phase. When injected into healthy mice at subtoxic doses, it induced antibody that blocked its ability to induce TNF both in vitro and in vivo and was predominantly immunoglobulin M (IgM); we found no evidence of a memory response or a switch to IgG on boosting (7). In this assay, culture supernatants from all the species we have tested so far, including the mouse parasites *P. yoelii* and *Plasmodium berghei* and the human parasites *P. falciparum* and *P. vivax*, show essentially complete cross-blocking, a situation in striking contrast to the species, strain, and even variant specificity of the protein antigens commonly studied (8). We concluded that parasitized erythrocytes contain and can release (in vitro at least) phospholipid molecules that induce TNF and behave as T-independent antigens. We refer to these molecules as parasite toxic antigens (13). A similar induction of TNF from human macrophages by *P. falciparum* supernatants has been reported; as in the mouse, the toxic antigens were heat stable, and both knob-positive and knob-negative parasites were active (39). However, another group found that rosetting subcultures of *P. falciparum* induced more TNF than rosetting-negative ones (2).

To further characterize the active molecule, we compared it with standard phospholipids of the type found in erythrocyte membranes (phosphatidylcholine, phosphatidylserine, phosphatidylinositol [PI], and phosphatidylethanolamine) and with phosphatidic acid and cardiolipin. None of these molecules induced TNF from macrophages, but one of them, PI, completely blocked the ability of toxic antigens to induce TNF. Surprisingly, the phosphorylated monosaccharide inositol monophosphate (IMP) also blocked TNF induction, and so did preparations of toxic antigen treated with either lipase or HF; we refer to these as detoxified antigens (6). At this point, we concluded that a phosphate group and an inositol ring were somehow involved in the TNF-triggering part of the exoantigens. This hypothesis was further supported by the finding that antisera raised by simply immunizing mice with PI or IMP, but not the other phospholipids, blocked TNF release. However, phosphatidylcholine, phosphatidylserine, and phosphatidic acid incorporated into liposomes with or without cholesterol did induce blocking antibody, but the antibody could be absorbed out by phosphatidylcholine liposomes, while antibody raised to PI or IMP could not (6). From this we concluded that blocking antibodies are of two types: one specific for phosphate groups and one recognizing inositol as well.

Recently, a group working with *P. falciparum* demonstrated very similar properties in a preparation containing the glycosylphosphatidylinositol anchor from the well-known merozoite antigen MSP-1; they postulate a disruptive role for this molecule in intracellular signaling (44). In another study, a *P. falciparum* extract induced nitric oxide synthase in endothelial cells, which might be relevant to cerebral malaria; in this case, the active molecule did not appear to be a phospholipid (22). The well-known protein antigen Pf155/RESA also triggers TNF (38). Until these and other candidate molecules have been compared directly, we cannot be sure whether there is one major TNF-inducing molecule or several, but we believe the evidence suggests that the phospholipid(s) mentioned above represents the predominant element.

DISTINCTION FROM LPS

Contamination of laboratory experiments by bacterial LPS is extremely difficult to avoid, but our toxic antigens appear to differ from LPS in a number of important ways. First, their TNF-triggering activity is not blocked by the inclusion of polymyxin B in the cultures, and the blocking antisera, whether against the exoantigens themselves or against PI or IMP, do not block triggering by LPS. Second, they stimulate macrophages from the LPS-hyporesponsive C3H/HeJ mouse strain. Third, triggering by LPS is not abolished, as that of the toxic antigens is, by phospholipase C digestion; contrariwise deamination abolishes the activity of LPS but not of the antigens. Fourth, neither PI, IMP, nor detoxified antigens block triggering by LPS, which may even suggest that different macrophage receptors are involved. We therefore feel safe in concluding that LPS alone does not account for our results. However, we cannot eliminate the possibility that trace amounts of LPS, insufficient to stimulate on their own, are needed as adjuvant to the effects of the toxic antigens in the induction of blocking antibody (3, 45); if this is so, it would be in keeping with the actual physiology, since malaria patients probably have increased amounts of circulating LPS (21).

HYPOGLYCEMIA

During the experiments on toxicity of exoantigens in vivo, we noticed that injected mice often looked ruffled and ill. Investigating this, we found a dramatic drop in blood sugar lasting from 2 until at least 8 h after injection. By comparison with the normal midmorning blood sugar of 7.6 mmol/liter, the 4-h level averaged 3.4 mmol/liter, and levels below 2 mmol/liter were quite common. Since TNF (and other cytokines) can cause hypoglycemia, we attempted to inhibit this drop with repeated injections of a monoclonal anti-TNF antibody. Though this antibody protected D-galactosamine-pretreated mice from death, it had no significant effect on hypoglycemia. On the other hand, a second and third injection of exoantigen at 2-week intervals induced progressively less hypoglycemia in healthy mice, but this protection was not seen in mice with severe combined immunodeficiency (SCID mice). We concluded that antibody against the toxic antigens could protect against hypoglycemia. In line with the TNF experiments described earlier, we then showed that IMP mixed with the exoantigens blocked their hypoglycemic effect, as also did antibody raised against IMP or against a PI-bovine serum albumin conjugate (51). Subsequent experiments with adipocytes in vitro suggested that the parasite supernatants increase glucose uptake, but only in the presence of small amounts of insulin (52).

Our initial assumption that the same molecule(s) induces both TNF and hypoglycemia is called into question by the fact that we quite often find that antigen preparations induce TNF and not hypoglycemia, and vice versa (50). Moreover, we have occasionally detected low levels of TNF and hypoglycemia induction with supernatants of healthy erythrocytes, so it remains possible that the active phospholipids are healthy erythrocyte components that are increased in concentration or otherwise modified by the parasite (27, 53).

We next measured insulin levels and found them markedly raised at the time of hypoglycemia (18); parasite supernatants were also able to induce hyperinsulinemia. Surprisingly, they did so in mice made partially diabetic with streptozotocin, suggesting that a malaria product is a stronger stimulus to insulin secretion than glucose itself (17). Furthermore, the insulin-inhibiting drugs diazoxide, adrenalin, and somatostatin were all capable of preventing hypoglycemia during infection (19). We concluded that at least two mechanisms operate to induce hypoglycemia in mouse models of malaria: one that acts on glucose transport and one that acts directly on the pancreatic islets. The molecule or molecules responsible, like those that induce TNF, can reasonably be referred to as toxins.

ANEMIA

Anemia is recognized as one of the severest complications of malaria, particularly in very young children, and researchers generally agree that it cannot be accounted for solely by erythrocyte parasitization (1). In our nonlethal *P. yoelii* mouse model, the hematocrit falls to 20% of normal in 10 days, despite the fact that the parasite is restricted to reticulocytes. The same is true in SCID mice, ruling out a major role for antibody. From about day 8 of infection, ^{51}Cr-labeled healthy erythrocytes are eliminated from the circulation unusually rapidly. Neither this effect nor the anemia itself is prevented by prior immunization with toxic antigens or by anti-TNF antibody, though TNF depresses erythropoiesis (31) and mice transgenic for human TNF are profoundly anemic (49).

RELEVANCE TO THE INFECTION

Having shown that interesting and potentially toxic molecules are released by parasitized erythrocytes in vitro, we wished to establish whether these molecules are released during the infection. Here the evidence is far from complete. On the one hand, we cannot detect TNF in the blood of infected mice, nor do infected mice develop hypoglycemia until about 1 day before they die; these facts might argue against a significant release of toxic antigens in vivo. On the other hand, serum from infected mice does contain antibodies that block TNF induction by toxic antigens as well as antibodies that bind PI and other phospholipids, which at least suggests that the immune system has been exposed to these antigens. This paradox might be resolved if antibody produced early during infection was able to block both TNF induction and hypoglycemia; a detailed study in SCID mice would probably be necessary to test this. Another possible explanation is that normal serum contains nonantibody blocking factors that are capable of binding and neutralizing the antigens until their concentration becomes overwhelming. These factors would be analogous to the acute-phase proteins or the lipoproteins that have been reported to inactive LPS (20). However, although acute-phase proteins such as serum amyloid precursor and orosomucoid are raised during malaria, purified preparations of these murine acute-phase proteins or of human C-reactive protein did not block TNF induction by malaria toxins (unpublished data).

RELEVANCE TO PROTECTIVE IMMUNITY AND VACCINATION

To be valid as vaccine candidates, the malaria toxins clearly should be protective against disease. Unfortunately, mice, which usually either are self-curing or suffer rapidly fatal infection and are highly resistant to TNF compared to humans, are not ideal for this purpose. However, in a large group of genetically homogeneous F_1 hybrid mice immunized with toxic antigens and challenged with the uniformly lethal *P. yoelii* 13 days later, about 60% recovered after about 3 weeks with very high parasitemias (in some mice, more than 80%) but without looking particularly ill, which is exactly what antidisease immunity, if it existed in mice, would be expected to look like. Similar protection has been obtained in smaller groups of mice immunized with PI and IMP. Neither repeated boosting nor the use of adjuvants improves this level of protection, and though preliminary experiments with antigens coupled to protein carriers such as bovine serum albumin and keyhole limpet hemocyanin have yielded substantial increases in blocking antibody, most of which is IgG, the level of actual protection remains about the same (9). Perhaps IgM is the more effective isotype in this situation.

The above discussion may simply reflect the fact that our *P. yoelii* model is not really suitable for this type of protection study. For example, as noted above, both vaccinated and nonvaccinated mice become extremely anemic during infection, and it may be that mice are dying of anemia even though they are protected against other aspects of pathology. This situation is typical of the difficulties encountered when trying to study antidisease immunity to such a multifactorial disease. However, in a model of cerebral malaria using C57BL mice and *P. berghei*, immunization with exoantigens gave 60% protection against early cerebral mortality (16), and a somewhat similar partial protection with exoantigens was reported for a *P. falciparum* model in squirrel monkeys (43) and in cattle with babesiosis (27).

RELEVANCE TO HUMAN DISEASE

Our hypothesis, then, is that the antitoxic immunity observed in human populations may be due to the development of immunity against toxic antigens that is probably mediated by antibody that blocks the ability of the antigens to induce TNF, hypoglycemia, etc. This hypothesis will be hard to prove, but one could make certain predictions. "Tolerant" children with high parasitemia but mild or no symptoms and patients recovering from mild malaria ought to have blocking antibody in their blood. In immune adults who leave areas of endemicity, clinical immunity (T-independent IgM?) might be expected to wane more rapidly than immunity to the parasite itself (T-dependent IgG?), and this does often seem to be the case. If we are right about the conserved nature of the toxic antigens, then antitoxic immunity should show a wider degree of cross-protection than antiparasite immunity, but this is still controversial (30). Finally, antibody against the relevant antigens ought to correlate with clinical protection. One such exoantigen, originally named antigen 7 by its discoverers, meets this criterion; it is a TNF-inducing molecule found in *P. falciparum* culture supernatants that cross-reacts with mouse toxic antigens. Although antibodies to antigen 7 peak in Gambian children at the age of about 5 years, which is when clinical immunity develops, antibodies to most other exoantigens peak considerably later (28). Moreover, T-cell responses to antigen 7, assayed by gamma interferon release, correlate with clinical severity, which might be important in view of the synergy between gamma interferon and TNF-inducing molecules (42). However, in individual patients, the best correlation with clinical immunity was with antibody to another exoantigen. In a more recent study, IgM antibody that blocked TNF induction was found in two European patients, one with *P. falciparum* infection and one with *P. vivax* infection (5); the antibodies did not block LPS but were inhibited by PI, exactly like those found in mice. Subsequently, both IgM and IgG blocking antibodies have been found in Gambian patients, the levels being lower in severe than in mild infections and lowest in patients with cerebral malaria (4). This kind of analysis is further complicated by the fact that several of these antibodies may also affect parasitemia. Careful, substantial studies of TNF and other cytokine responses to a range of toxic antigens in the presence of sera from both immune and nonimmune patients will be necessary before the existence or otherwise of true antitoxin antibody-mediated immunity in humans can be established. But whether or not it exists in the natural state, the ability to induce it by immunization could be of great value.

RELEVANCE TO HUMAN VACCINES

Compared to the candidate vaccines already under study, the antidisease approach has both advantages and disadvantages. In its favor is the fact that it aims to deal with what really matters: disease—the obvious analogy being to tetanus and diphtheria. Also, the apparent lack of antigenic variation noted in the mouse experiments suggests that a whole series of species- and variant-specific antigens may not be needed. Moreover, by acting against cytokine-inducing antigens rather than against the cytokines themselves, as, for example, antibody to TNF or soluble TNF receptors do, the antidisease antibody might affect other cytokines induced at the same time, such as IL-1 or IL-6, and thus avoid the need for a complex cocktail of inhibitors. In addition, it might reduce the incidence of cytokine-unrelated complications such as hypoglycemia.

Against the antidisease approach is the lack of T-dependent antibody, with memory, IgG, etc., in the mouse experiments, which, if translated directly into the field, might necessitate a vaccine injection every year unless there was a boosting effect from repeated infections. Nevertheless, other nonprotein antigens, such as the meningococcal, pneumococcal, and haemophilus polysaccharides, are successful vaccines. Moreover, as noted above, humans do appear to make blocking IgG. Another theoretical danger is that the phospholipid antigens responsible for disease might turn out to be true autoantigens or at least to cross-react significantly with host molecules; the vaccine might then induce autoimmunity, though this has never been reported. We are currently concentrating on attempts to confer T dependence on the relevant antigen, which might possibly overcome both these problems, as suggested earlier (40).

Acknowledgments. The work from our laboratory described in this chapter was supported by generous grants from the Medical Research Council of Great Britain, the Wellcome Trust, and the European Community.

REFERENCES

1. **Abdalla, S., D. J. Weatherall, S. N. Wickramasinghe, and M. Hughes.** 1980. The anaemia of *P. falciparum* malaria. *Br. J. Haematol.* **46:**171–183.
2. **Allan, R., A. Rowe, and D. Kwiatkowski.** 1993. *Plasmodium falciparum* varies in its ability to induce tumor necrosis factor. *Infect. Immun.* **61:**4772–4776.
3. **Banerji, B., and C. R. Alving.** 1990. Antibodies to liposomal phosphatidylserine and phosphatidic acid. *Biochem. Cell Biol.* **68:**96–101.
4. **Bate, C. A. W., and D. Kwiatkoski.** Personal communication.
5. **Bate, C. A. W., and D. Kwiatkowski.** 1994. Inhibitory immunoglobulin M antibodies to tumour necrosis factor-inducing toxins in patients with malaria. *Infect. Immun.* **62:**3086.
6. **Bate, C. A. W., J. Taverne, H. J. Bootsma, R. C. S. Mason, N. Skalko, G. Gregoriadis, and J. H. L. Playfair.** 1992. Antibodies against phosphatidylinositol and inositol monophosphate specifically inhibit tumour necrosis factor induction by malaria exoantigens. *Immunology* **76:**35–41.
7. **Bate, C. A. W., J. Taverne, A. Dave, and J. H. L. Playfair.** 1990. Malaria exoantigens induce T-independent antibody that blocks their ability to induce TNF. *Immunology* **70:**315–320.
8. **Bate, C. A. W., J. Taverne, N. D. Karunaweera, K. N. Mendis, D. Kwiatkowski, and J. H. L. Playfair.** 1992. Serological relationship of tumour necrosis factor-inducing exoantigens of *Plasmodium falciparum* and *Plasmodium vivax*. *Infect. Immun.* **60:**1241–1243.
9. **Bate, C. A. W., J. Taverne, D. Kwiatkowski, and J. H. L. Playfair.** 1993. Malaria exoantigens coupled to carriers induce IgG antibody that blocks their ability to induce TNF. *Immunology* **79:**138–145.
10. **Bate, C. A. W., J. Taverne, and J. H. L. Playfair.** 1988. Malarial parasites induce TNF production by macrophages. *Immunology* **64:**227–231.
11. **Bate, C. A. W., J. Taverne, and J. H. L. Playfair.** 1989. Soluble malarial; antigens are toxic and induce the production of tumour necrosis in vivo. *Immunology* **66:**600–605.
12. **Bate, C. A. W., J. Taverne, and J. H. L. Playfair.** 1992. Detoxified exoantigens and phosphotidylinositol derivatives inhibit tumor necrosis factor induction by malarial exoantigens. *Infect. Immun.* **60:**1894–1901.
13. **Bate, C. A. W., J. Taverne, E. Roman, C. Moreno, and J. H. L. Playfair.** 1992. Tumour

necrosis factor induction by malarial exoantigens depends on phospholipid. *Immunology* **75:**129–135.

14. **Clark, I. A.** 1987. Cell mediated immunity in protection and pathology of malaria. *Parasitol. Today* **3:**300–305.
15. **Clark, I. A., G. Chaudri, and W. B. Cowden.** 1989. Roles of tumour necrosis factor in the illness and pathology of malaria. *Trans. R. Soc. Trop. Med. Hyg.* **83:**436–440.
16. **Curfs, J. H. A. J., C. C. Hermsen, J. H. E. T. Meuwissen, and W. M. C. Eling.** 1992. Immunization against cerebral pathology in *Plasmodium berghei*-infected mice. *Parasitology* **105:**7–14.
17. **Elased, K., J. B. de Souza, and J. H. L. Playfair.** 1995. Blood-stage malaria infection in diabetic mice. *Clin. Exp. Immunol.* **99:**440–444.
18. **Elased, K., and J. H. L. Playfair.** 1994. Hypoglycemia and hyperinsulinemia in rodent models of severe malaria infection. *Infect. Immun.* **62:**5157–5160.
19. **Elased, K. M., and J. H. L. Playfair.** Reversal of hypoglycaemia in murine malaria by drugs that inhibit insulin secretion. *Parasitology,* in press.
20. **Emancipator, K., G. Csako, and R. J. Elin.** 1992. In vitro inactivation of bacterial endotoxin by human lipoproteins and apolipoproteins. *Infect. Immun.* **60:**596–601.
21. **Felton, S. C., R. B. Prior, V. A. Spagna, and J. P. Kreier.** 1980. Evaluation of *Plasmodium berghei* for endotoxin by the limulus lysate assay. *J. Parasitol.* **66:**846–847.
22. **Ghigo, D., R. Todde, H. Ginsburg, C. Costamagna, P. Gautret, F. Bussolino, D. Ulliers, G. Giribaldi, E. Deharo, G. Gabrielli, G. Pescarmona, and A. Bopsia.** 1995. Erythrocyte stages of *Plasmodium falciparum* exhibit a high nitric oxide synthase (NOS) activity and release an NOS-inducing soluble factor. *J. Exp. Med.* **182:**677–688.
23. **Goonewardene, R., R. Carter, P. Gamage, G. Del Giudice, P. H. David, S. Howie, and K. Mendis.** 1990. Human T-cell proliferative responses to P. vivax antigens: evidence of immunosuppression following prolonged exposure to endemic malaria. *Eur. J. Immunol.* **20:**1387–1391.
24. **Grau, G. E., L. F. Fajardo, P.-F. Piguet, B. Allet, P.-H. Lambert, and P. Vassalli.** 1988. Tumor necrosis factor (cachectin) as an essential mediator in murine cerebral malaria. *Science* **237:**1210–1212.
25. **Grau, G. E., T. E. Taylor, M. E. Molyneux, J. J. Wirima, P. Vassalli, M. Hommel, and P. H. Lambert.** 1989. Tumor necrosis factor and disease severity in children with *falciparum* malaria. *N. Engl. J. Med.* **320:**1586–1591.
26. **Hill, R. B., F. J. C. Cambournac, and M. P. Simoes.** 1943. Observations on the course of malaria in children in an endemic region. *Am. J. Trop. Med.* **23:**147–162.
27. **Hsaio, L. L., R. J. Howard, M. Aikawa, and T. F. Taraschi.** 1991. Modification of host cell membrane lipid composition by the intra-erythrocytic human malaria parasite *Plasmodium falciparum. Biochem. J.* **274:**121–132.
28. **Jakobsen, P. H., E. M. Riley, S. J. Allen, S. O. Larsen, S. Bennett, S. Jepsen, and B. M. Greenwood.** 1991. Differential antibody response of Gambian donors to soluble *Plasmodium falciparum* antigens. *Trans. R. Soc. Trop. Med. Hyg.* **85:**26–32.
29. **James, M. A.** 1989. Application of exoantigens of Babesia and Plasmodium in vaccine development. *Trans. R. Soc. Trop. Med. Hyg.* **83:**67–72.
30. **Jeffery, G. M.** 1966. Epidemiological significance of repeated infections with homologous and heterologous strains and species of Plasmodium. *Bull. W.H.O.* **35:** 873–882.
31. **Johnson, R. A., T. A. Waddelow, J. Caro, A. Oliff, and G. D. Roodman.** 1989. Chronic exposure to tumour necrosis factor in vivo preferentially inhibits erythropoiesis in nude mice. *Blood* **74:**130–138.
32. **Karunaweera, N. D., G. E. Grau, P. Gamage, R. Carter, and K. N. Mendis.** 1992. Dy-

namics of fever and serum TNF levels are closely associated during clinical paroxysms in *P. vivax* malaria. *Proc. Natl. Acad. Sci. USA* **89:**3200–3203.

33. **Kern, P. C., J. Hemmer, J. Van Damme, H.-J. Gruss, and M. Dietrich.** 1989. Elevated tumor necrosis factor α and interleukin 6 serum levels as markers for complicated *Plasmodium falciparum* malaria. *Am. J. Med.* **57:**139–143.
34. **Kwiatkowski, D., A. V. S. Hill, I. Sambou, P. Twumasi, J. Castracane, K. R. Manogue, A. Cerami, D. R. Brewster, and B. M. Greenwood.** 1990. TNF concentration in fatal cerebral, non-fatal cerebral, and uncomplicated *Plasmodium falciparum* malaria. *Lancet* **336:**1201–1204.
35. **Kwiatkowski, D., M. E. Molyneux, P. Pointaire, N. Curtis, N. Klein, M. Smit, R. Allan, S. Stephens, G. E. Grau, P. Holloway, D. R. Brewster, and B. M. Greenwood.** 1993. Anti-TNF therapy inhibits fever in cerebral malaria. *Q. J. Med.* **86:**91–98.
36. **McGregor, I. A., H. M. Giles, J. H. Walters, A. H. Davies, F. A. Pearson.** 1956. Effects of heavy and repeated malarial infections on Gambian infants and children. *Br. Med. J.* **2:**686–692.
37. **Naotunne, T. D., N. D. Karunaweera, G. Del Giudice, M. Kularatne, G. E. Grau, R. Carter, and K. N. Mendis.** 1990. Cytokines kill malaria parasites during infection crisis: extracellular complementary factors are essential. *J. Exp. Med.* **17:**523–529.
38. **Picot, S., F. Peyron, P. Deloron, C. Boudin, B. Chumpitazi, G. Barbe, J. P. Vuillez, A. Donadille, and P. Ambroise-Thomas.** 1993. Ring-infected erythrocyte surface antigen (Pf155/RESA) induces tumour necrosis factor-alpha production. *Clin. Exp. Immunol.* **93:**184–188.
39. **Picot, S., F. Peyron, J.-P. Vuillez, G. Barbe, K. Marsh, and P. Ambroise-Thomas.** 1990. Tumor necrosis factor production by human macrophages stimulated in vitro by *Plasmodium falciparum. Infect. Immun.* **58:**214–216.
40. **Playfair, J. H. L., J. Taverne, C. A. W. Bate, and J. B. de Souza.** 1990. The malaria vaccine: anti-parasite or anti-disease? *Immunol. Today* **11:**5–27.
41. **Prada, J., J. Malinowski, S. Muller, U. Bienzle, and P. G. Kremsner.** 1995. Hemozoin differentially modulates the production of interleukin 6 and tumor necrosis factor in murine malaria. *Eur. Cytokine Netw.* **6:**109–112.
42. **Riley, E. M., P. H. Jakobsen, S. J. Allen, J. G. Wheeler, S. Bennett, S. Jepsen, B. M. Greenwood.** 1991. Immune response to soluble exoantigens of *Plasmodium falciparum* may contribute to both pathogenesis and protection in clinical malaria: evidence from a longitudinal, prospective study of semi-immune African children. *Eur. J. Immunol.* **21:**1019–1025.
43. **Ristic, M., and J. P. Kreier.** 1984. Malaria and babesiosis: similarities and differences, p. 3–33. *In* M. Ristic, P. Ambroise-Thomas, and J. P. Kreier (ed.), *Malaria and Babesioisis: Research Findings and Control Measures.* Martinus Nijhoff Publishers, Dordrecht, The Netherlands.
41. **Schofield, L., and F. Hackett.** 1993. Signal transduction in host cells by a glycosylphosphatidylinositol toxin of malaria parasites. *J. Exp. Med.* **177:**145–153.
45. **Schuster, B. G., M. Neidig, B. M. Alving, and C. R. Alving.** 1979. Production of antibodies against phosphocholine, phosphatidyl choline, sphingomyelin, and lipid A by injection of liposomes containing lipid A. *J. Immunol.* **122:**900–905.
46. **Sinton, J. A., and J. Harbhagwan Singh.** 1931. The numerical prevalence of parasites in relation to fever in chronic benign tertian malaria. *Indian J. Med. Res.* **18:**871–879.
47. **Taliaferro, W. H.** 1949. Immunity to the malaria infections, p. 935–965. *In* M. F. Boyd (ed.), *Malariology,* vol. II. The W. B. Saunders Co., Philadelphia.
48. **Taverne, J., C. A. W. Bate, D. A. Sarkar, A. Meager, G. A. W. Rook, and J. H. L. Play-**

fair. 1990. Human and murine macrophages produce TNF in response to soluble antigens of *Plasmodium falciparum*. *Parasite Immunol.* **12:**33–43.

49. **Taverne, J., N. Sheikh, J. B. de Souza, J. H. L. Playfair, L. Probert, and G. Kollias.** 1994. Anaemia and resistance to malaria in transgenic mice expressing human tumour necrosis factor. *Immunology* **82:**397–403.
50. **Taverne, J., N. Sheikh, K. Elased, and J. H. L. Playfair.** 1995. Malaria toxins: hypoglycaemia and TNF are induced by different components. *Parasitol. Today* **11:**462–463.
51. **Taylor, K., C. A. W. Bate, R. E. Carr, G. A. Butcher, J. Taverne, and J. H. L. Playfair.** 1992. Phospholipid-containing malaria exoantigens induce hypoglycaemia. *Clin. Exp. Immunol.* **90:**1–5.
52. **Taylor, K., R. Carr, J. H. L. Playfair, and E. D. Saggerson.** 1992. Malarial toxic antigens synergistically enhance insulin signalling. *FEBS Lett.* **311:**231–234.
53. **Vial, H. J., M.-L. Ancelin, J. R. Philippot, and M. J. Thuet.** 1990. Biosynthesis and dynamics of lipids in Plasmodium-infected mature mammalian erythrocytes. *Blood Cells* **16:**531–555.

Malaria Vaccine Development: A Multi-Immune Response Approach
Edited by Stephen L. Hoffman

Chapter 8

Transmission-Blocking Vaccines

David C. Kaslow

BIOLOGY OF THE SEXUAL STAGES OF THE MALARIA PARASITE

The biology of the sexual stages of the malaria parasite's life cycle has played a central role in the history of malaria research. A male gametocyte exflagellating in blood drawn from an infected 24-year-old patient of Dr. Laveran in Algeria on November 5, 1880, was the crucial observation that led to the discovery of the causal agent of malaria (54). In addition, the discovery that mosquitoes transmit malaria parasites, observations made by Ross (the presence of oocysts in infected mosquitoes) (80) and MacCallum (fertilization of an avian malaria parasite) (56) in 1897, has been critical to almost all successful malaria control programs implemented to date (for example, see reference 5). The concept of a transmission-blocking vaccine, however, came 60 years later, in 1958, when Huff et al. showed that vaccination with avian malaria parasites could reduce the infectivity of parasites in mosquitoes that fed on the parasitemic vaccinated chickens (33). Although much remains to be understood, our current knowledge of the biology of sporogonic development (from gametogenesis to zygote transformation to ookinete development into an oocyst to oocyst maturation) has proven useful in designing, developing, and testing various aspects of transmission-blocking vaccine candidate antigens. A brief review of the biology of gametocytes and sporogonic development is presented here. A more detailed review can be found elsewhere (8).

Gametocyte Differentiation

Various in vitro manipulations of cultures of *Plasmodium falciparum* parasites have been reported to induce differentiation into male and female gametocytes (also called microgametocytes and macrogametocytes, respectively) (48). However, the natural signal(s) responsible for the differentiation of a merozoite into a gametocyte has yet to be defined. The rare schizont that produces gametocytes appears to produce multiple gametocytes, although not all of the progeny of a single schizont seem committed to undergoing simultaneous gametocytogenesis (6, 35). Early *P. falciparum* gametocytes are morphologically similar to trophozoites and are rarely observed circulating in the human bloodstream; instead, they sequester themselves (88, 94). Although the surfaces of circulating gametocyte-infected erythrocytes appear to be devoid of unique antigenic markers (8), the same or a similar molecule(s) involved in sequestration of asexual-blood-stage parasites is undoubtedly also present on the surfaces of the early stages of gametocytes. Similar to the surface molecules responsible for cytoadherence of asexual-stage malaria parasites, the presumptive surface molecules that mediate sequestration of early gametocytes might be effective target antigens of a humoral immune response.

Emergence and Exflagellation

In vitro-cultured, synchronized *P. falciparum* gametocytes require on average 7 days to fully mature (37, 87). Once mature, male gametocytes are competent to undergo exflagellation, in which approximately eight threadlike motile male gametes form from a single male gametocyte (Fig. 1). Female gametocytes emerge from erythrocytes, probably as a result of the release from the parasite of substances that disrupt the integrity of

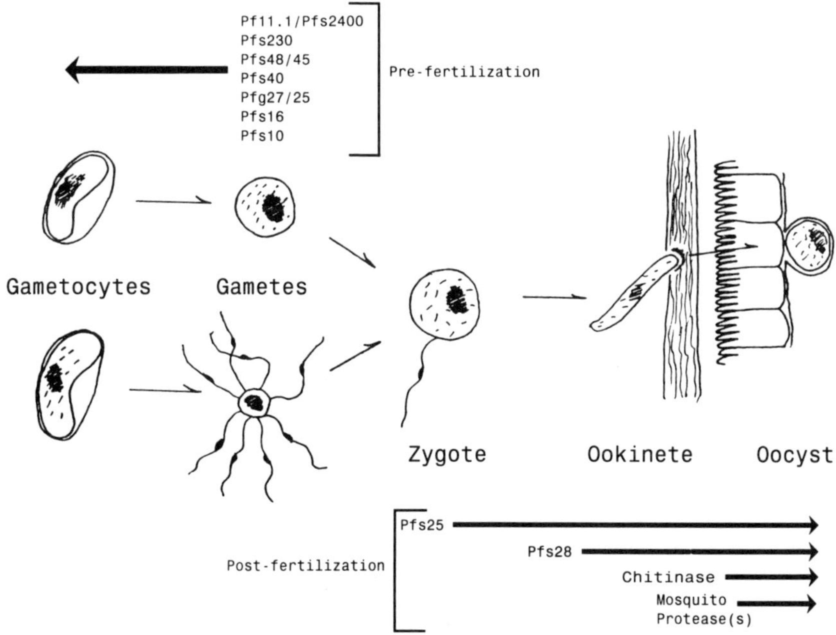

Figure 1. Developmental stages of malaria parasites in the mosquito midgut. Potential target antigens that appear in or on the parasite before fertilization are listed above the developmental stages; those that appear to block transmission after fertilization are listed below.

erythrocyte membranes (71). Once released, each female gametocyte forms a single nonmotile female gamete (Fig. 1). These processes normally do not occur within the bloodstream of the vertebrate host but readily occur immediately after ingestion of the parasite by the appropriate mosquito vector. The critical factors that trigger exflagellation and emergence are (i) a drop in temperature that occurs when the parasite goes from the homeothermic vertebrate host to the poikilothermic mosquito vector (8, 86); (ii) an increase in pH, probably mediated via bicarbonate, in going from the pH 7.4 of the host bloodstream to the pH 8.3 in the mosquito midgut (65); and (iii) to a lesser degree, a very low-molecular-weight, heat-stable factor(s) (mosquito exflagellation factor) that is present in extracts from the mosquito midgut (64).

Fertilization, Zygote Transformation, and Egress of Ookinetes from the Blood Meal

Within 10 to 60 min of emergence, a motile, haploid, male gamete fertilizes a nonmotile, haploid female gamete to form a diploid zygote (56, 85) (Fig. 1). Clearly, antibodies directed toward presumptive fertilization receptors involved in male recognition of female gametes or any heretofore undescribed "pheromones" involved in attracting males to female gametes could confer transmission-blocking activity. In addition, interference with the cell membrane fusion between the male and female gametes that occurs after initial recognition would certainly have a deleterious effect on further zygotic development.

During the 20- to 24-h transformation into the elongated, motile ookinete (Fig. 1), the spherical, nonmotile zygote extends a cytoplasmic protrusion containing the apical complex to form the irregularly shaped intermediate form, referred to as a retort. The antigens present on the surfaces of gametes and early zygotes are replaced by a series of new surface antigens (12) (Fig. 1). The functions of either set of these surface antigens are presently not known.

By 24 h, yet another set of antigens predominates on the surface (12, 52) of the now oblong, motile form, the ookinete. In addition to these new surface antigens, at least one new protein, a chitinolytic enzyme, is secreted from the ookinete. During this time, ookinetes must move from the midgut lumen to penetrate the peritrophic matrix (PM) that encapsulates the blood meal. The PM, which is composed of chitin (81), is a physical barrier that traps many ookinetes (57). Ookinetes that are able to traverse the PM appear to push aside the microvillus border of the midgut epithelium, leaving trails of antigen(s) behind as they move (57).

Penetration of the Midgut Epithelium and Oocyst Formation

Although early studies suggested that ookinetes traverse the midgut epithelium by an intracellular route (21), more recent electron microscopic studies of *P. falciparum* ookinetes suggest that the parasites disrupt desmosomal contacts between cells, perhaps by an enzymatic process, and take an intercellular route (58). Not all parasites, however, traverse the epithelium between cells; for example, *Plasmodium berghei* appears to take an intracellular route (58). The diploid ookinetes end their migration and become nonmotile oocysts when they come to reside between the epithelium and the basal lamina. The signals mediating loss of movement and morphologic changes from an oblong to a spherical form are not yet known but may well be components of the basal lamina. The sporozoites that subsequently develop within the oocyst have a haploid genome (36). Zygotes, ookinetes, and early oocysts are the only stages during the life cycle that have diploid genomes (36, 85).

DEVELOPMENT OF TRANSMISSION-BLOCKING VACCINES

Goal of Attacking Sexual-Stage Parasites

In addition to achieving the primary goal of preventing morbidity or mortality or in some cases inducing sterile immunity, effective antisporozoite and/or anti-asexual-parasite-stage (protective) vaccines may as a consequence of reducing parasitemia substantially reduce transmission. In contrast, the primary goal of anti-sexual- or anti-sporogonic-stage vaccines is to block parasite transmission. If achieved, this blocking could substantially reduce morbidity or mortality and might even in some cases lead to eradication of the parasite in a geographically isolated area. When combined with a protective vaccine or a chemotherapeutic agent, an anti-sexual- or anti-sporogonic-stage vaccine could block the spread of escape mutants resistant to the protective vaccine components or the antimalarial medication. By doing so, the transmission-blocking component would prolong the time that the protective components or antimalarial drug was effective.

A transmission-blocking vaccine might also convert a marginally effective antisporozoite vaccine, particularly any such vaccine that was effective only at low sporozoite inoculation rates, to a fully effective one.

Finally, by requiring all visitors and returnees to be fully immunized against sexual or sporogonic stages, a transmission-blocking vaccine might prevent the reintroduction of malaria parasites into areas where an eradication program has been successful.

Structure, Biology, and Immunology of Known Target Antigens

Target antigens of transmission-blocking immunity fall into three categories: (i) prefertilization target antigens, which are expressed either solely or predominantly in gametocytes and elicit antibodies that disrupt parasite development either by mediating complement-dependent lysis of gametes or by somehow interfering with emergence or fertilization; (ii) postfertilization target antigens, which are expressed solely or predominantly on zygotes or ookinetes and elicit antibodies that block the morphological transformation of the round, nonmotile zygote into the oblong, motile ookinete and/or block the egress of the ookinete from the blood meal to the midgut epithelial basal lamina; and (iii) late-midgut-stage target antigens such as parasite-produced chitinase, which is required for penetration of the chitin-containing PM that the midgut epithelium secretes around the blood meal, and a mosquito-produced protease that activates the parasite-produced chitinase.

Prefertilization (gametocyte or gamete) target antigens

The list of potential target antigens present in gametocyte-infected erythrocytes has grown from two (Pfs230 and Pfs48/45) in the mid-1980s to more than half a dozen in 1995. At least some of the genes encoding these potential target antigens have now been cloned (Table 1). Unfortunately, recombinant protein has not yet been successfully used to elicit transmission-blocking immunity for any of these proteins. Thus, the greatest challenge for this group of target antigens is preparation and/or delivery of protein in a vehicle that elicits transmission-blocking antibodies.

Pf11.1/Pfs2400. Within minutes of triggering gametogenesis, the erythrocyte membrane that surrounds a *P. falciparum* gametocyte lyses. Associated with this lysis is the movement of a RESA-like parasite protein from granules in the cytoplasm of a gametocyte adjacent to the parasitophorous vacuole through the erythrocyte cytoplasm to the erythrocyte membrane. Although the protein was originally thought to be Pf155/RESA because it was recognized by anti-Pf155/RESA monoclonal antibodies (MAbs), subsequent studies by Scherf et al. indicated that the gametocyte-specific protein is a megadalton polypeptide, referred to as Pf11.1, with RESA-like repeats (82).

Recently, Feng et al. found a MAb, 1A1, that recognizes the glutamate-rich repeat region of Pfs2400 and suppresses transmission in the presence of complement. This MAb does not substantially reduce the number of infected mosquitoes or completely block transmission (16). Transmission-blocking assays in the absence of complement were not re-

Table 1. Summary of potential transmission-blocking target antigens

Antigen	Epitope			Epitope conservation		H2-limited immune response	Sequence status		Recombinant protein expression[g]
	No.	CONF[a]	TBA[b]	Ab[c]	PCR[d]		AA[e]	DNA[f]	
Pf11.1/Pfs2400	?	?	+	?	?	?	?	+	Yes/?TBA
	I	C	+	Yes	?				
Pfs230						Yes	+	+	Yes/?TBA
	II	C	+	Yes	?				
	I	C	+	Yes	?				
	II	C	+	No	?				
Pfs48/45	III	C	+	Yes	?	Yes	+	+	Yes/?TBA
	IV	C	+	Yes	?				
	V	L	−	Yes	?				
Pfs40	?	?	?	?	?	Yes	+	+	Yes/−TBA
	I	L	+	?	?				
Pfg27/25						No	+	+	Yes/?TBA
	II	L	−	?	?				
Pfs28	?	?	+	?	Yes	?	−	+	Yes/+TBA
	I	C	+	?	Yes				
Pfs25						No	+	+	Yes/+TBA
	II	C	−	?	Yes				
	I	L	?	?	Yes				
Pfs16						?	−	+	Yes/?TBA
	II	L	?	?	Yes				

[a]Epitope conformation (CONF) was either linear (L) or reduction sensitive (C).
[b]TBA, transmission-blocking activity associated with antibodies to the epitope.
[c]Epitope conservation according to immunological data.
[d]Epitope conservation according to nucleotide sequence in PCR-amplified sequence.
[e]Amino acid sequence (AA) directly from peptide microsequencing.
[f]Nucleotide sequence available.
[g]TBA, transmission-blocking activity under study; −TBA, nonblocking; +TBA, transmission blocking.

ported (which may be irrelevant for 1A1, given that the isotype of the monoclonal antibody is immunoglobulin G1, but may be important when polyclonal sera to this epitope are tested). As mentioned below, perhaps polyclonal serum to this molecule will more favorably block transmission.

How antibodies to Pfs2400 suppress transmission is unknown. In this regard, it is interesting that the molecule is confined to the parasitophorous vacuolar membrane rather than the true surface of the gametocyte. These data, like those from Pfg27/25 (105), suggest that transmission-blocking antibodies can penetrate the membrane of a gametocyte-infected erythrocyte; however, not yet excluded is the possibility that the transmission-blocking effect is actually mediated by cross-reactivity to some other glutamate-rich, transmission-blocking antigen. Pfs230, a known target antigen of transmission-blocking antibodies (see below), became a likely suspect when Williamson et al. reported that Pfs230 contains glutamate-rich repeats as well as a stretch of glutamate (103). Whatever the mechanism of transmission suppression, the viability of Pf11.1 as a transmission-blocking vaccine candidate will be determined by whether polyclonal sera to either parasite-produced or recombinant Pf11.1 (rPf11.1) proteins block transmission.

Pfs230. Within 48 h of invading the erythrocyte, gametocytes of both sexes express a 363-kDa protein precursor that after emergence is processed to a 310-kDa protein (99, 104a); this 310-kDa protein is referred to as Pfs230 in *P. falciparum* because it was described originally as a 230-kDa protein. The precursor protein appears to be synthesized throughout the maturation of gametocytes, but only Pfs230 is present on the surfaces of newly emerged gametes, where, although no longer synthesized, it can readily be radiolabeled with iodine and recognized by transmission-blocking MAbs (70, 99). Within hours of fertilization, Pfs230 is no longer detectable. Although by immunoblot analysis some MAbs that recognize Pfs230 also immunoprecipitate Pfs48/45 (see below) when parasite surface molecules are extracted with nonionic detergents (51), pulse-chase experiments indicate that Pfs230 is not a precursor for any of the lower-molecular-weight target proteins (99). With the recent cloning of the genes encoding Pfs230 and Pfs48/45, it is now clear that these two proteins are encoded by separate genes.

At least some Pfs230 and Pfs48/45 molecules exist as a stable, membrane-bound complex in gametocytes and gametes (51). Pfs230 also exists in the gamete surface membrane in a free, uncomplexed form. Although Pfs230 and Pfg27 were originally presumed to have no antigenic or structural relationship with Pfs48/45, Wizel and Kumar recently reported that the three proteins have a common epitope that is recognized by transmission-blocking MAbs (105) (see the section on Pfg27/25 below). More recently, Pfs48/45 was found to share elements present in the larger 7-cysteine consensus sequence of Pfs230 (7a, 104b).

Pfs230 is hydrophilic, can be extracted from the surfaces of gametes in isotonic salt in the absence of any detergent (104a), and is not metabolically labeled with either carbohydrates or fatty acids. Williamson et al. resorted to immunoaffinity chromatography to microsequence tryptic peptides of Pfs230, and using the recently cloned cDNA that encodes Pfs230 as a hybridization probe of total cellular RNA, they found that Pfs230 is translated from a >12-kb sexual-stage-specific transcript (103).

The gene contains a 9.4-kb open reading frame that predicts a 363-kDa protein (103). A secretory signal sequence is predicted but a transmembrane region is not, a situation consistent with experimental data showing that Pfs230 is on the surface of the parasite but is not an integral membrane protein. The deduced amino acid sequence further predicts that Pfs230 is highly charged and contains eight copies of a 4-amino-acid repeat, four copies of an 8-amino-acid repeat, and multiple copies of a 7-cysteine motif (103) (Fig. 2). The mature protein is 50 kDa smaller than its 363-kDa precursor (Fig. 2). Since the amino acid repeats reside in the first 45 kDa of the protein, the mature protein may lack these repeats, which is consistent with unpublished data suggesting that Pfs230 lacks repetitive epitopes.

Although originally identified by Vermeulen et al. as a surface antigen (98), Pfs230 was thought not to be a target of transmission-blocking antibodies because of the failure of their MAbs to block transmission. Two subsequent studies, however, indicated that Pfs230 is indeed a target antigen: (i) Graves et al. reported that the presence of anti-Pfs230 antibodies in the sera collected from humans residing in malaria-endemic area was strongly associated with transmission-blocking activity (25), and (ii) Quakyi et al. subsequently found that two immunoglobulin G2a MAbs to Pfs230 could independently block transmission but only in the presence of complement and only if added prior to fertilization (70). Thus, all of the data reported so far are consistent with the notion that antibodies to Pfs230 require complement to block transmission. In contrast, *Plasmodium gallinaceum* MAbs to Pgs230 act synergistically to block fertilization, even in the absence of complement (49). It is now crucial to determine whether polyclonal antisera to either parasite-produced Pfs230 or recombinant Pfs230 require complement, because if the transmission-blocking activity of anti-Pfs230 antibodies is mediated solely by complement-dependent lysis, then inclusion of an appropriate adjuvant induces anti-Pfs230 antibodies with the proper isotype may be required.

MAbs against Pfs230 distinguish at least two B-cell epitopes, both of

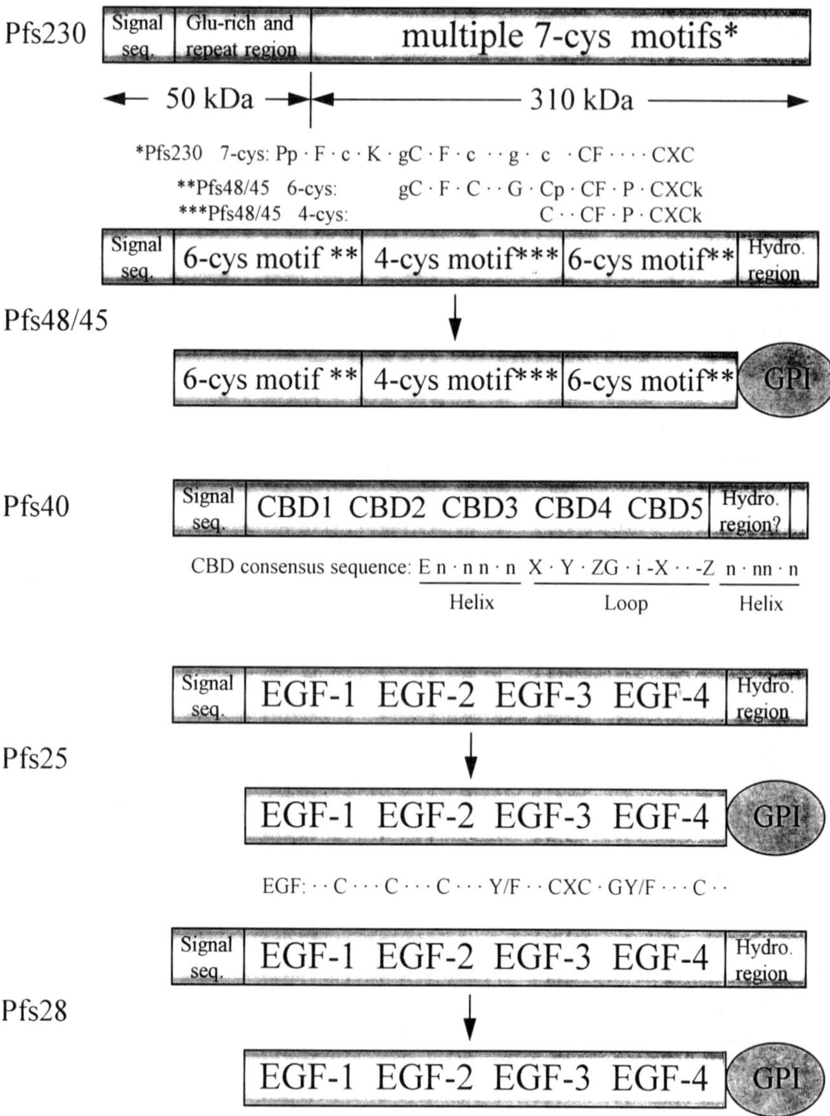

Figure 2. Motifs and domains identified in target antigens Pfs230, Pfs48/45, Pfs40, Pfs25, and Pfs28. For the consensus sequences of the 7-, 6-, and 4-cysteine motifs and EGF-like domain, uppercase letters designate amino acids present in almost all copies, lowercase letters indicate amino acids present in some copies, and dots represent any amino acid. For the consensus sequences of the calcium-binding domains (CBD), n indicates polar amino acid residues (L, I, V, F, or M); a dot indicates any residues; X, Y, Z, -X, and -Z are calcium-chelating residues (D, E, N, S, or T); and I indicates hydrophobic residues (I, L, or V).

which are conformation dependent (17). B-cell-epitope diversity has been studied by using two MAbs against Pfs230 in Malaysian isolates. In 42 of 45 isolates, all gametocytes gave a strong positive reaction; in the other 3 isolates, a minor population of gametocytes failed to react (17). Thus, Pfs230 appears to be generally well conserved. These findings are consistent with those of Williamson and Kaslow, who found that strain differences in the nucleotide sequence of the Pfs230 gene clustered in two regions: the repeat region that is cleaved off of the mature Pfs230 form and the second 7-cysteine motif (104). For the most part, the rest of the gene was conserved. These data, however, must be considered preliminary, because only two strains were compared throughout their complete open reading frames (albeit the two strains chosen were quite different: strain 3D7 is chloroquine sensitive and presumably from Africa, and strain 7G8 is chloroquine resistant and from Brazil).

With regard to the immunogenicity of Pfs230 in humans, studies of serum samples from individuals living in malaria-endemic Papua New Guinea revealed that 22% of the samples reduced or blocked transmission in the standard membrane feeding assay (25). The presence of antibodies to Pfs230 (found in 37% of the samples examined) strongly correlated with reduction of transmission, whereas antibodies to Pfs48/45 did not, in part because so few individuals had antibodies to Pfs48/45. Also from Papua New Guinea, 34% of relatively nonimmune adults experiencing acute attacks of malaria had antibodies that inhibited binding of a Pfs230 MAb (28). The prevalence of antibodies to Pfs230 increased with increasing past exposure to *P. falciparum*, peaking by the time three to five attacks had been experienced (28). Continued exposure, however, leads to acquired nonresponsiveness to Pfs230 (28).

Subsequent studies gave results similar to those of the Papua New Guinea study. In The Gambia, again in individuals frequently exposed to *P. falciparum*, 43% of immune serum samples from adults reacted with Pfs230 (72). In that study, no correlation between a specific HLA type and an antibody response was found.

Pfs48/45. Similar to Pfs230, Pfs48/45, a doublet consisting of a 48- and a 45-kDa surface protein of *P. falciparum*, is detectable from day 2 of gametocytogenesis through gametogenesis and fertilization (99). By pulse-chase experiments, neither Pfs48 nor Pfs45 has a higher-M_r precursor (unlike Pfs230), nor is one the precursor of the other (99). Both have a pI of 6.0, so the difference in mobility on sodium dodecyl sulfate-polyacrylamide gel electrophoresis (SDS-PAGE) has to be attributed to size differences rather than to charge (99). In fact, Pfs48 and Pfs45 are so antigenically similar that all five B-cell epitopes described to date are present in both of these antigens (17). Therefore, these antigens are almost always referred to collectively as Pfs48/45 (50).

As described above, Pfs48/45 forms a noncovalent but stable complex with Pfs230. Although the primary amino acid sequence for Pfs48/45 has been determined and analysis of the sequence predicts that Pfs48/45 has elements of the 7-cysteine motifs present in Pfs230 (Fig. 2), the sequence has not yet assisted in identification of the region(s) and mechanism of interaction between Pfs48/45 and Pfs230. How Pfs230 and Pfs48/45 interact may have significant practical applications, particularly in view of data that suggest that polyclonal antiserum to Pfs230 enhances binding of MAbs to Pfs48/45 (see below). Understanding the mechanism of this enhancement may help us in designing a more efficacious multiantigen vaccine.

MAbs to Pfs48/45 can block transmission in the absence of complement but must be present before fertilization. Three of the five known B-cell epitopes are reduction sensitive, while the other two appear to be linear. Transmission-blocking MAbs to Pfs48/45 published so far map to four of the five epitopes (thus, one of the target antigens is to a linear epitope; see also the section on Pfg27/25 below); however, not all of the MAbs that map to these four epitopes block transmission. Interestingly, binding of certain MAbs to one epitope enhances the binding of MAbs to another epitope, which perhaps explains how some MAbs that do not suppress transmission when tested alone do block transmission when mixed together (77). How anti-Pfs48/45 MAbs block transmission is not yet known. Because MAbs to Pfs48/45 prevent the formation of ookinetes, Vermeulen et al. proposed that Pfs48/45 plays a significant role in fertilization (98).

Although enhancement of transmission at low concentrations of transmission-blocking antibodies has not been reported for any of the MAbs developed against *P. falciparum*, progressively lowering the number of infectious gametocytes in the presence of the same concentration of MAb to Pfs48/45 inexplicably weakened the transmission-blocking activity of the MAb. Keeping the infectious gametocyte numbers constant and progressively decreasing the concentration of MAbs (68) weakened the transmission-blocking activity of the MAbs, as would be predicted.

The antigenic diversity of Pfs48/45 has now been well documented at least by antibody reactivity. Earlier studies with two MAbs against different epitopes of Pfs48/45 revealed that certain geographical isolates failed to react with either MAb (26). This failure could have been due to polymorphism, differential accessibility, or complete absence of Pfs48/45. A larger, more recent collaborative study confirmed antigenic diversity in at least two of the five epitopes defined in Pfs48/45 (17).

In the studies from Papua New Guinea, the presence of antibodies to Pfs48/45 did not correlate with transmission-blocking activity in the serum samples tested, partly because so few individuals had antibodies to Pfs48/45 (25). This lack of immune responsiveness to Pfs48/45 in individuals living in malaria-endemic regions was documented in a number of subsequent studies (11, 28, 72). For instance, only 9% of individuals from The Gambia who were tested reacted with Pfs48/45 (72). Subsequent studies suggested that antibodies to Pfs48/45 are present in up to 60% of serum samples from individuals after their first attack of *P. falciparum* malaria. With subsequent infections, the percentage of individuals responding may rise slightly, but with long-term exposure to *P. falciparum*, the antibody response is somehow downregulated (28). As with Pfs230, this nonresponsiveness in humans residing in malaria-endemic areas is not HLA linked (72).

Pfs40. Pfs40, a very acidic, 40-kDa radioiodinable surface protein in *P. falciparum* gametes and zygotes (75), was first identified as a potential target of transmission-blocking immunity by an immunogenetic approach. The limited immune response to this protein among congenic, major histocompatibility complex-disparate mouse strains has been interpreted as evidence that Pfs40 has a paucity of T-helper-cell epitopes as a consequence of immune selection (22, 38). Recognition of this molecule by only a limited number of hyperimmune adults from a malaria-endemic region of Ghana is also consistent with this interpretation (75). Thus, one possible explanation for the poor immunogenicity of Pfs40 is that it is indeed a target of natural transmission-blocking antibodies.

To characterize the biological role(s) of Pfs40 and the immunological response directed against the molecule, the protein was purified by two-dimensional gel electrophoresis, and three high-pressure-liquid-chromatography-purified tryptic fragments were microsequenced. Degenerate synthetic oligonucleotides based on the microsequence were used to clone a gametocyte cDNA that contained a 1,125-bp open reading frame that encodes a polypeptide of 374 amino acids with a predicted molecular mass of 43.5 kDa and a pI of 5.12 (75).

A striking feature of the deduced amino acid sequence of the cloned gene is five EF-hand calcium-binding domains predicted on the basis of homology with the calcium-binding domains of members of the EF-hand calcium-modulated protein superfamily (2, 13) (Fig. 2). ^{45}Ca autoradiography of total soluble gamete-zygote protein separated by two-dimensional gel electrophoresis was used to show that Pfs40 has calcium-binding activity (75). By SDS-PAGE separation and ^{45}Ca autoradiography of total gamete proteins, Pfs40 appears to be the single, predominant calcium-binding protein (75).

Unique among EF-hand calcium-binding proteins, Pfs40 has a presumptive secretory signal sequence and a carboxyl-terminal hydrophobic region. These features, along with its ability to be surface radiolabeled with ^{125}I, strongly suggest that Pfs40 is a noncytosolic, membrane-associated polypeptide (75). The only other reported exception to the cytosolic localization of EF-hand-containing calcium-binding proteins is in another parasite, *Toxoplasma gondii*, also of the subclass *Coccidia* (13).

Rawlings et al. immunized mice with rPfs40 produced in *Escherichia coli*. The resulting mouse serum recognized parasite-produced Pfs40 on Western blot (immunoblot) but did not immunoprecipitate ^{125}I-labeled Pfs40 extracted with nonionic detergent (75). ^{125}I-labeled Pfs40 extract treated with SDS, however, was immunoprecipitated by the anti-rPfs40 serum (75). Thus, despite binding calcium and lacking any cysteine residues, the rPfs40 induced only antibodies that recognized denatured parasite-produced Pfs40. Determining the role, if any, that Pfs40 plays in the calcium-mediated events of exflagellation and transmission-blocking immunity must await the production of properly folded rPfs40 that induces antiserum that recognizes nondenatured Pfs40.

Pfg27/25. Perhaps the most potent immunogen of gametocytes is an abundant (5 to 10% of total protein content) sexual-stage-specific protein doublet with an apparent M_r of 27,000 and 25,000 (9). Even before early gametocytes are morphologically distinguishable from trophozoites, *P. falciparum* gametocytes synthesize an abundance of these highly hydrophilic, 25- to 27-kDa cytoplasmic proteins, collectively referred to as Pfg27/25 (9). The protein appears in the cytoplasm of a gametocyte within 40 h after invasion of an erythrocyte by a committed merozoite (9). By two-dimensional gel electrophoresis, these sexual-stage-specific proteins are resolved into peptides with at least four isoelectric points, ranging from pH 5.0 to 6.0 (9). Unlike Pfs25, Pfg27/25 does not occur on the surfaces of gametes or zygotes (11); however, Wizel and Kumar found that in live parasites, the protein can be ^{125}I radiolabeled by the lactoperoxidase method (105).

The gene encoding Pfg27/25 has been cloned by immunoscreening and sequenced (1). Amino acid sequences from three tryptic peptides confirmed that the 651-bp open reading frame is the one translated in vivo (1). The deduced amino acid sequence predicts a highly hydrophilic 29.4-kDa polypeptide devoid of any secretory signal peptide, membrane anchor or transmembrane region, or N-linked glycosylation sites (1). Two cysteine residues are predicted; however, both of the B-cell epitopes described so far appear to be nonconformational (1). Furthermore, it is unlikely that the cysteines form an intermolecular disulfide bond, since the protein migrates on SDS-PAGE (nonreducing conditions) as a doublet of 25 and 27 kDa, the size expected for a monomer (9).

Recently, a reduction-insensitive, highly conserved B-cell epitope (referred to as C3 because it is continuous and nonconformational, cross-reactive, and conserved) in Pfg27/25 was identified. MAb 6B6, which recognizes this epitope, almost completely blocks transmission in the presence of added complement (105). In the absence of complement, transmission is only suppressed (19 of 42 mosquitoes examined were uninfected). The epitope for MAb 6B6 is highly conserved (105). This MAb also recognizes Pfs48/45 and Pfs230 (105).

Although Quakyi et al. (70) mentioned the cross-reactivity of anti-Pfs230 MAb 2B4 to a 70-kDa protein, the data they presented (see their Fig. 4) clearly indicate that an approximately 27-kDa protein is present in immunoprecipitations of Iodogen-labeled gamete whole-cell lysates. Interestingly, this cross-reactivity was not present on immunoblots of gamete extracts. These data suggest either that the 27-kDa protein forms a noncovalent complex with Pfs230 or that a cross-reactive epitope in the 27-kDa protein is destroyed during SDS-PAGE and immunoblotting. Evidence from Wizel and Kumar (105) indicates that SDS-PAGE-purified Pfs230 and SDS-PAGE-purified Pfg27/25 are immunoprecipitated by MAb 6B6 and thus favors the latter explanation, that a protein is destroyed. Similar to the situation for Pfs2400, the critical question is whether recombinant Pfg27/25 or a synthetic peptide derivative induce complete transmission-blocking activity in humans. Even if recombinant Pfg27/25 does not induce transmission-blocking immunity, the immune response to this protein is an important epidemiological tool for ascertaining prior exposure to gametocytes (11).

Pfs16. An alternative cloning strategy, subtractive hybridization of asexual-stage parasite cDNA from sexual-stage DNA, resulted in the isolation of a full-length gene that encodes a previously unidentified gametocyte and a sporozoite membrane-bound protein, referred to as Pfs16 (also known as Pfs17 and SHEBA) (15). Simultaneously, some or all of the genes encoding this protein were cloned by several other strategies (6, 61) (see below). The gene consists of a 471-bp open reading frame that encodes a polypeptide of 16.6 kDa (15). The deduced amino acid sequence predicts a hydrophobic region at the NH_2 terminus, which suggests a secretory signal sequence at this terminus, and a second hydrophobic region followed by a highly charged region at the COOH terminus, which suggests a transmembrane region followed by a cytoplasmic tail (15). The amino acid sequence does not contain any cysteine, tyrosine, or tryptophan residues or any predicted N-linked glycosylation sites.

This integral membrane protein of gametocytes and sporozoites has not yet been shown to be a target of transmission-blocking antibodies. Although it is membrane-bound, Pfs16 appears to be associated with the parasitophorous vacuolar membranes of gametocytes rather than the surfaces of gametes (15). Interestingly, Pfs16 does appear on the surfaces of sporozoites (15). Pfs16 has limited antigenic diversity (60), despite the presence of antibodies to Pfs16 in individuals living in malaria-endemic areas (27). This limited diversity may be a clue that Pfs16 is not a target of transmission-blocking antibodies. The role that Pfs16 plays in sporozoite invasion of hepatocytes is beyond the scope of this chapter.

Preliminary evidence suggests that two nonconformational B-cell epitopes exist in Pfs16 (93). Where these epitopes map relative to the two areas found to be polymorphic (region I: Ile-85 to Leu-85 in strains Dd2 and GH1, and Ser-90 to Asn-90 in strain Dd2; region II: insertion of Asp-Lys between Gly-139 and Asp-140 in strain GH1) in Pfs16 (60) has not yet been reported. Currently, Pfs16 is not being actively developed as a transmission-blocking vaccine because of a paucity of conclusive data demonstrating that it is a target of transmission-blocking immunity.

Pfs10. Very little is known about Pfs10, a surface-iodinated gametocyte-gamete protein. Pfs10 seems to be immunologically related to a 40-kDa protein or to Pfs40 in that MAbs that recognize a 40-kDa protein by immunoprecipitation also recognize Pfs10, and whenever mice immunized with gametes produce antibodies that immunoprecipitate Pfs40, Pfs10 is always coprecipitated (76a, 98). The MAbs that recognize Pfs10 on immunoblot do not immunoprecipitate the radioiodinated Pfs10 and do not block oocyst formation in the absence of complement; however, these MAbs do show a positive immunofluorescence on live gametes (98). Antisera to rPfs40 that immunoprecipitate radiolabeled Pfs40 and recognize Pfs40 on immunoblot do not immunoprecipitate Pfs10 or recognize it on immunoblot (75). Because evidence from other sexual-stage antigens (see sections on Pfs25 and Pgs28 below) suggests that polyclonal antibodies are qualitatively better than MAbs in blocking transmission (42), until high-quality monospecific polyclonal serum to Pfs10 is tested for transmission-blocking activity, Pfs10 should not be discounted as a potential target antigen.

Fertilization receptors. None of the specific potential target antigens mentioned above have proven to be sex specific; nevertheless, male-specific and female-specific surface proteins have been identified by surface labeling purified male and female gametes of *P. gallinaceum* (47), respectively. By analogy to other fertilization systems, some of these sex-specific surface proteins may be unique receptors on male gametes that bind to ligands on female gametes, and also by analogy, the moieties involved in binding may not all be proteins (for a review, see Wassarman [102]). Antibodies directed against the receptor(s) and ligand(s) involved in fertilization could interfere with zygote formation and thus prevent oocyst development. If the interacting molecules are proteins, then they would be prime vaccine candidates, especially if used together in a transmission-blocking vaccine. Mutants selected by use of such a transmission-blocking vaccine would probably not complement one another.

Postfertilization (zygote-early ookinete) target antigens

Surface and secreted proteins expressed predominantly by the malaria parasite after fertilization constitute a privileged class of antigens. Although antibodies artificially induced by sensitizing the vertebrate host to late-sexual-stage antigens can bind to these antigens and potentially interfere with normal sporogonic development, these proteins will not have been under selective pressure by the vertebrate immune system. Therefore, the parasite will not have had the advantage of millennia of evolution to select for some mechanism by which to evade the immune response to these antigens. The obvious and major disadvantage of these proteins that are not expressed by the parasite when it is in the vertebrate host is that the antibody response to them will not be boosted after a natural infection. As vaccine delivery technology advances, this disadvantage may no longer be of great importance. But because natural boosting has been considered important and because producing substantial quantities of late-sexual-stage parasites, especially of *P. falciparum*, has been difficult, few of the antigens expressed during late sporogonic development have been tested as potential vaccine candidates or even identified as potential target antigens as fully as gametocyte and gamete antigens have with the exception of two major surface proteins: P25 (Pfs25 in *P. falciparum* and its analogs in other *Plasmodium* spp.) and P28 (Pfs28 in *P. falciparum* and its analogs in other *Plasmodium* spp.).

The prototype P25 is Pfs25, a 25-kDa transmission-blocking target antigen present in *P. falciparum* (44); its analog, Pgs25, is present in the avian malaria parasite *P. gallinaceum* (29, 46). The prototypes of P28 are Pfs28, a 28-kDa protein recently identified in *P. falciparum* (13a); Pgs28, a 28-kDa protein present in *P. gallinaceum* (29); and Pbs21, a 21-kDa protein in *Plasmodium berghei* (95). Both P25 and P28 have now been shown to be targets of transmission-blocking antibodies in the avian and falciparum malaria model systems (14, 29). The 25- and 28-kDa antigens have similar structures, and both are cysteine-rich, lipid-anchored proteins (14, 18, 19, 46) (Fig. 2).

Pfs25. The gene that encodes Pfs25 was the first malaria sexual-stage-antigen gene to be cloned; hence, more is known about the expression, structure, antigenicity, and immunogenicity of this molecule than about any of the other known target antigens. The gene is transcribed in gametocytes (44), and stored mRNA, in addition to newly synthesized mRNA, is used for Pfs25 protein synthesis early in gametogenesis (19). Accumulating evidence suggest that Pfs25 is expressed, but in very limited amounts, in gametocytes (39a, 97, 99). The deduced amino acid sequences predict that the protein contains an amino-terminal secretory signal sequence and a short hydrophobic carboxy-terminal sequence that signals the transfer of the mature protein to a glycosylphosphatidylinositol (GPI) anchor (44). A striking feature of the protein is the presence of four epidermal growth factor (EGF)-like domains (44) (Fig. 2). The third EGF-like domain contains at least a portion of the major B-cell epitope (69, 90, 91, 97). Interestingly, the variant that has been identified in laboratory isolates (43) and the variants that were found in field isolates (106) all map to this third EGF-like domain. Whether these variants arose because of immunoselection is suspect, because antibodies to Pfs25 have yet to be detected in sera from humans living in areas of endemicity.

Although Pfs25 has been detected in maturing gametocytes from days 2 through 8 (99), expression of Pfs25 occurs predominantly after emergence and exflagellation (19). By 3 h after exflagellation, Pfs25 has replaced Pfs230, Pfs48/45, and Pfs40/10 as the predominant surface protein of zygotes and is continually expressed in ookinetes and even oocysts. It is an extremely cysteine-rich protein with a pI of 5.6 and incorporates mannose, glucosamine, and palmitic and myristic acids (99), probably via a GPI anchor (44).

Cloned by synthetic oligonucleotides designed from a tryptic peptide sequence, the gene consists of 654 bp of coding region on a 1.4-kb sexual-stage-specific transcript. This coding region predicts a 217-amino-acid polypeptide with a putative secretory signal sequence, a hydrophobic carboxyl terminus consistent with transfer to a GPI anchor, and four potential glycosylation sites (47). The most striking feature of the deduced amino acid sequence, however, is the similarity of cysteine (11% of all of the amino acid residues) spacing in Pfs25 to that in EGF and other EGF-like domains in a number of eukaryotic extracellular proteins (47). As evidence that these disulfide bonds are crucial for the conformational integrity of Pfs25, reduction of the cystines in Pfs25 destroys all epitopes recognized by transmission-blocking MAbs (18, 39a). Therefore, formation of the appropriate disulfide bonds is predicted to be critical in eliciting transmission-blocking immunity (see below).

Attempts to produce recombinant Pfs25 have focused on expression systems likely to recreate the native conformation by allowing disulfide bond formation to occur readily. Mammalian cells infected with recombinant vaccinia virus that express Pfs25 specifically bind transmission-blocking MAbs (61). After several inoculations with this recombinant virus, mice develop antibodies that completely block transmission (42). The polyclonal sera from these mice were qualitatively, not just quantitatively, superior to MAbs in blocking transmission (42). A polypeptide analog of Pfs25 secreted from yeast cells also induces in mice as well as in monkeys antibodies that completely block transmission (3, 45). Various formulations of Pfs25 have been tried in attempts to induce transmission-blocking immunity after a single immunization. At present, a slow-release particle containing Pfs25 looks particularly promising (4).

Two B-cell epitopes in Pfs25 have been defined: one is recognized by a series of MAbs that block transmission, and the other is a target of MAbs that do not suppress oocyst development (98). Both of these epitopes are reduction sensitive and thus conformation dependent. Fries et al. reported that these epitopes are also dependent on the presence of linked fatty acids but not carbohydrate groups (18). Subsequent studies using a new MAb, 4B7, and a recombinant protein lacking these lipid and sugar moieties suggest that neither lipids nor sugars are required to induce or bind transmission-blocking antibodies (3). MAb 4B7, selected because it recognizes reduced Pfs25, recognizes both nonreduced and reduced recombinant yeast products and blocks transmission (3).

In studies of 45 Malaysian isolates, the region reacting with nonsuppressive MAbs appears to be invariant; however, variability in reactivity has been reported at the transmission-blocking target epitope, suggesting structural diversity (17). In striking contrast, DNA sequence data from Pfs25 amplified by PCR from eight isolates from different geographical areas suggest that the nucleotide sequence is very highly conserved, with only a single nucleotide substitution that results in a conservative amino acid substitution (Ala to Gly) in one of eight laboratory strains (43). Follow-up studies with parasites collected in the field from Papua New Guinea (a high-transmission area) and Brazil (a low-transmission area) confirm that the gene lacks significant antigenic diversity. In the low-transmission area, not a single nucleotide variant was observed in the 14 isolates studied, and only one nonsilent change (two silent changes) was observed in the 20 isolates from the high-transmission area (106). The conservative substitution (Val to Ala) situated 12 amino acids carboxyl to the one described earlier (43) was found in 11 of 20 isolates.

The lack of significant antigenic diversity along with the lack of naturally occurring antibodies to Pfs25 in hyperimmune sera from individuals residing in malaria-endemic regions and the lack of immunological nonresponsiveness in major histocompatibility complex-congenic mouse strains suggests that Pfs25 has not been subjected to significant immune selective pressure (38). The most logical explanation for the lack of any apparent immune selection is that Pfs25 is not expressed while the parasite is in the human host. This explanation is at odds with observations by Vermeulen et al., who detected Pfs25 in gametocytes as early as day 2 of gametocytogenesis (99), and by Kaslow et al., who detected a transcript for Pfs25 in gametocytes (44). Whether both of these in vitro findings are simply artifacts of culturing and handling parasites is not known. The *Aotus* monkey model (see below) may be useful in sorting out these apparent discrepancies.

Pfs28, Pgs28, and Pbs21. Between 3 and 5 h after gametogenesis, *P. gallinaceum* zygotes begin synthesizing a 28-kDa surface protein, Pgs28 (52). By the time transformation into ookinetes is complete, Pgs28 has replaced Pgs25 as the predominant surface protein (52). Pgs28 incorporates glucosamine, mannose, and palmitic acid (52), which suggests that the protein is anchored to the plasma membrane via GPI (14) (Fig. 2).

Polyclonal antisera to Pgs28 block both in vitro transformation of zygotes into ookinetes and in vivo development of ookinetes into oocysts (14); therefore, the blocking activity appears to occur during at least two developmental steps in sporogony. These data lead to further characterization of the molecular structure of this protein (14), whose deduced amino acid sequence was strikingly similar to those of Pfs25 and Pgs25 (the *P. gallinaceum* analog of Pfs25) (44,46).

The gene encoding Pgs28 consists of a 666-bp open reading frame in a 1.5-kb RNA transcript (14). The deduced amino acid sequence of Pgs28 is strikingly similar to that of Pgs25, consisting of a putative secretory signal sequence, four EGF-like domains, and a COOH-terminal hydrophobic region without a cytoplasmic tail (14) (Fig. 2). Although two sites within the EGF-like domains are potential substrates for N-linked glycosylation, as mentioned above, the mannose and glucosamine that are incorporated into Pgs28 are probably present in a GPI anchor (14). Using PCR technology, Duffy and Kaslow recently identified the analogous gene in *P. falciparum* and showed that antiserum raised against yeast-secreted recombinant Pfs28 elicits transmission-blocking activity in laboratory animals (13a). The genes encoding Pfs25 and Pfs28 are physically linked on chromosome 10 (13a).

Recently, Paton et al. published the deduced amino acid sequence of Pbs21, the predominant surface antigen of *P. berghei* ookinetes (67). Pbs21 elicits a potent transmission-blocking immune response in the murine model (95). Although Pbs21 was originally thought to be the protein analogous to Pfs25 (67, 95), a comparison of this family of proteins led Duffy and me to conclude that Pbs21 is analogous to Pgs28 (39). Because Pfs28, Pgs28, and Pbs21 induce transmission-blocking immunity, an intensive search for the analogous proteins in *Plasmodium vivax* may lead to a malaria transmission-blocking vaccine candidate for the most prevalent form of human malaria.

Other target antigens. Other postfertilization target antigens probably exist, and their identification should be pursued in the hope that they lack the problems that may plague the known prefertilization target antigens as components of a transmission-blocking vaccine. Because the postfertilization target antigens are presumably not expressed while in the human host, immunoselective pressure has never been applied to these antigens. Preexisting antigenic diversity and/or poor immunogenicity should therefore not be present.

Late-midgut-stage (late ookinete-oocyst) target antigens

Studying the basic biology of the parasite during its development in the mosquito recently led to the discovery of two other potential target antigens, a parasite-produced chitinase and a mosquito-produced protease.

Chitinase. After a mosquito ingests a blood meal, a chitin-containing PM forms around the food bolus, and this PM must be traversed during the parasite's egress from the lumen of the midgut. Sieber et al. found that as the parasite penetrates the PM, a dense material appears external to the apical end of the parasite and is associated with disruption of the normal laminated structure of the PM (84). These observations suggest that the parasite may secrete a chitinase that digests the peritrophic membrane in the mosquito midgut during penetration (84). Subsequently, Huber et al. (32) found that, starting 15 h after exflagellation, ookinetes contain a chitinase, and by 24 h, the chitinase activity is secreted into the culture supernatant. Furthermore, Huber et al. (32) determined that in vitro, the PM could be disrupted by *Streptomyces griseus* chitinase.

As discussed in detail below, Shahabuddin et al. found that inhibition of chitinase in vivo blocks parasite infectivity to mosquitoes and thus parasite transmission (83).

Mosquito-produced protease. The protease-rich environment of the PM and the periphery of the blood meal seem to be hostile places for most proteins. To examine the stability of parasite chitinase during proteolysis, purified mosquito midgut protease was added to parasite chitinase, and the resulting reaction mixture was assayed for chitinase activity. Instead of decreasing chitinase activity, mosquito midgut protease increased chitinase activity at least threefold (83). Activation of parasite-produced chitinase by a mosquito-produced protease may be necessary for the parasite's penetration of the chitinous PM.

The major proteases produced by the mosquito midgut are trypsins, which cleave peptide bonds C terminal to arginine and lysine residues. To examine the nature of the peptide cleavage responsible for activation of parasite chitinase, Shahabuddin et al. treated parasite-produced chitinase with various purified proteases and assayed for chitinase activity (83). Lys-C, which is a thermolabile, trypsinlike enzyme that specifically cleaves peptide bonds COOH terminal to lysine residues, increased chitinase activity 12- to 13-fold (83). The other proteases had little to no effect on chitinase activity. These data suggest that malaria parasite chitinase undergoes a postsecretory modification by mosquito midgut protease that activates the enzyme by cleaving it COOH terminal to a lysyl residue(s).

Preliminary data suggest that inhibition of the protease-mediated activation of parasite-produced chitinase may block transmission (82a). As discussed below, the major drawback of developing such an anti-mosquito protease-based vaccine would be the resulting selection of resistant mosquitoes. Clearly, this is more apt to happen if such a vaccine decreases the reproductive fitness of the mosquito vector.

Anti-mosquito antigen vaccines. With an eye toward making a global transmission-blocking vaccine against a wide range of vector-borne diseases, including arboviruses, a number of investigators have begun exploring immunity to mosquito midgut antigens as a means of blocking transmission. Ideally, such a transmission-blocking vaccine would have no deleterious effect on the fecundity of the vector but would completely block transmission, presumably by (i) interfering with initial attachment to the PM or midgut epithelium, (ii) blocking a mosquito-produced factor required by the parasite or virus (see the section on mosquito-produced protease above), (iii) inhibiting the egress of the pathologic agent from the midgut lumen to the hemolymph, or (iv) preventing the agent from being passed back to the human host by blocking binding to or passage through the salivary glands. The major advantage of an anti-mosquito-antigen-based vaccine would be the chance that such a vaccine would block transmission of all species of malaria parasites and potentially other arthropod-vector-borne diseases that required the mosquito antigen for development.

Ramasamy et al. looked at the transmission-blocking effect of rodent antibodies to mosquito midgut components (74). In one study, rabbit antimosquito antibodies significantly reduced the susceptibility of *Aedes aegypti* mosquitoes to infectivity with Ross River virus and Murray Valley encephalitis virus (74), although in the latter case, the effect was inconsistent (in one experiment, the control and immunized transmission rates were 79 and 37%, respectively, and in a second experiment, the rates were 62% in controls and 68% in immunized rabbits). As discussed below, transmission-blocking activity against murine malaria parasites has also been observed (53, 73). Interestingly, the mortality rate of mosquitoes was lower when they fed on mice that had received immunizations with midguts from mosquitos fed only sugar-water than when they fed on control mice. Mice that received immunizations of midguts from mosquitos fed blood did not develop significant transmission-blocking immunity. Clearly, further work is necessary to sort out how antibodies to midgut antigens block transmission and which target antigen(s) elicits the transmission-blocking effect.

Of practical importance, the complete gametocytogenic phase of the *P. falciparum* life cycle (34) and the complete sporogonic phases of the *P. gallinaceum* and *P. falciparum* life cycles can now be carried out in vitro (100, 101). Interestingly, these recent studies of sporogonic development in vitro showed that neither peritrophic membrane, epithelial cells, nor basal-lamina constituents are absolute requirements for the maturation of oocysts or the formation of sporozoites. However, whether antibodies to epithelial cells or basal-lamina components will block sporogonic development is not yet known (see the section on mosquito midgut antigens below).

Immune Response

Using the facile avian malaria model, Huff et al. showed that chickens immunized with *P. gallinaceum* parasites develop humoral immunity to transmission (33). These studies were confirmed by a series of studies in the 1970s in which purified sexual-stage parasite rather than a mixture of asexual- and sexual-stage parasites was used as the immunogen (7, 29, 30, 70). With the advent of MAb technology and the development of a mosquito membrane feeding assay, three *P. falciparum* target antigens of transmission-blocking antibodies and their analogs in *P. gallinaceum* were identified in the mid-1980s (78, 98). The first target antigen was cloned in 1988 (44). Since that time, more than half a dozen other potential target antigens have been identified and cloned. Studies of the immune response to parasite-produced and recombinantly produced proteins and a search for and identification of variant B- and T-cell epitopes in these target antigens have now begun in earnest.

IMMUNE RESPONSES TO SEXUAL-STAGE ANTIGENS IN HUMANS

Antigenic Diversity

Similar to other infectious agents, parasites may evade the human immune response through natural selection of mutants that express target antigens that have variant or even no B- or helper-T-cell epitopes. Indeed, antigenic diversity appears to be present in some of the target antigens of prefertilization target antigens. Because the nucleotide sequences for most of the target antigens have only recently become available, antigenic diversity has been studied primarily by studying the immunofluorescence reactivity of MAbs with laboratory or field isolates. Graves et al. found that Pfs48/45 has at least one variant epitope and that the transmission-blocking effects of anti-Pfs48/45 MAbs are strain specific (26). Two subsequent studies, one using field isolates from Malaysia (17) and the other using laboratory isolates (10), confirmed the earlier findings. Foo et al. also reported limited antigenic diversity in a minor population of gametocytes at one of two Pfs230 B-cell epitopes and one of two Pfs25 B-cell epitopes (17). Now that the genes for many of the target antigens have been cloned and sequenced, the exact structural basis of this diversity can be determined. Whether including multiple variant forms of each target antigen will be required to elicit an effective transmission-blocking antibody response has not yet been addressed adequately enough for us to reach a firm conclusion.

When one is considering the selective pressure for vaccine-resistant parasites that will result from the use of transmission-blocking vaccines, it is important to note that sporogonic development represents a significant bottleneck in the parasite's life cycle. First, most of the mature gametocytes circulating in the bloodstream are never ingested by mosquitoes, since only a small fraction of the total blood volume of the host is ingested by mosquitoes feeding on the host; second, of those gametocytes that are actually ingested by the mosquito, only 1 in 100 to 1,000, or less, develops into an oocyst (20); and third, in nature, most mosquitoes that feed on an infectious host do not become infected (e.g., the probability in the Mandang area of Papua New Guinea of a mosquito feeding on any human host and becoming infected is 0.013; if the host is infected *and* has infectious gametocytes, the probability is 0.38) (24). Therefore, in an individual with a low gametocytemia (e.g., 2 gametocytes per mm^3 or 10^7 total circulating gametocytes in 5 liters of blood) who is subjected to 50 mosquito bites per night (the average number of mosquito bites per night in the Sri Lanka study was fewer than 1) (20), with 50% of the mosquitoes becoming infected (in an area of high transmission in Papua New Guinea, the probability of infection when a mosquito bites an infectious host was 38%) (24) with an average of five or fewer oocysts per mosquito (in the Gold Coast region, Muirhead-Thompson found that 78% of infected mosquitoes had five or fewer oocysts per gut [62]; see also below), approximately 100 oocysts will develop. Roughly then, fewer than 1 in every 100,000 gametocytes (10^2 of 10^7) that circulate in the bloodstream will typically end up as a mature oocyst in a mosquito. The selective pressure of a transmission-blocking vaccine is thus on the 10^2 gametocytes that manage to be transmitted rather than the 10^7 gametocytes that circulate in the host.

Because most mature gametocytes are not ingested by mosquitoes and have a half-life of only 2 to 3 days (89), most gametocytes are probably cleared and destroyed by the reticuloendothelial system. Which gametocyte antigens are presented and how they are presented to the human immune system during this clearance is not known.

Boosting after Natural Infection

Because the prefertilization target antigens are expressed while the parasite is in the human host, boosting of the immune response may occur after natural infection. This could provide long-lasting transmission-blocking immunity after just a single immunization (or a few immunizations). Unfortunately, the evidence from field studies to support this "boosting hypothesis" is scant. Laboratory studies that have demonstrated a boosting response following natural infection have been limited to immunization of animals that had not previously had a malaria infection (31). Whether a prior natural infection will irreversibly perturb the balance of the immune response to a subunit vaccine such that an ineffective response results has not been adequately studied. Furthermore, prefertilization target antigens suffer from the potential disadvantages of antigenic diversity or variation (immunoselection on the B-cell epitope) and poor immunogenicity (immunoselection on helper-T-cell epitopes). With the advent of slow-release particle technology and other novel delivery systems, the notion that boosting following natural infection is a requirement for developing an effective transmission-blocking vaccine may well be obsolete.

Immunogenicity to Prefertilization Target Antigens

Two lines of evidence from early studies suggest that poor immunogenicity might be a significant problem for prefertilization target antigens, i.e., those target antigens that elicit an immune response in humans previously exposed to gametocytes. One line of evidence comes from immunogenicity studies by Good et al. of congenic mouse strains disparate at the *Ir* gene locus (22). All of the ^{125}I-radiolabeled surface proteins of early gametes appeared to elicit a limited antibody response, leading Good et al. to conclude that these antigens are inherently poor immunogens because of a paucity of helper-T-cell epitopes. However, in a subsequent study in which a different adjuvant was used and a subunit vaccine was available (75), Rawlings and Kaslow reached a different conclusion (76). The limited immunogenicity appeared to be both adjuvant dependent and antigen dependent (76). In the latter study, poor immunogenicity was not inherent in the antigen but instead was a function of the antigen being in the milieu of a whole parasite (76).

The second line of evidence for poor immunogenicity comes from studies of sera collected from humans living in malaria-endemic areas. As reviewed above, nonresponsiveness to gametocyte target antigens of transmission-blocking antibodies has been observed repeatedly in humans living in malaria-endemic regions. The mechanism(s) of this nonresponsiveness has not been fully elucidated. In addition, several key questions remain to be answered. (i) What proportion of semi-immune adults and children will respond to vaccination with subunit gametocyte target antigens by eliciting antibodies that block transmission? (ii) Does a prior exposure to gametocytes during a natural infection irreversibly divert the immune response so that subsequent immunization is ineffective? (That is, because the nonimmune individual's immune systems have not been primed previously with whole gametocytes, will nonimmune individuals respond more favorably than semi-immune individuals to immunization with a transmission-blocking vaccine?) (iii) Will boosting of an effective transmission-blocking immune response after vaccination occur with subsequent natural infection in previously nonimmune or semi-immune individuals?

Immune responses to gametocytes in humans naturally exposed to malaria

A variety of field studies of humans with malaria transmission-blocking immunity and laboratory-based studies of transmission-blocking activity have now been published. Laboratory studies indicate that the two major mechanisms of transmission-blocking immunity are (i) nonspecific mediators, such as cytokines (tumor necrosis factor and gamma interferon), that cripple the transmissibility of gametocytes (63) and (ii) human antibodies that specifically recognize sexual-stage parasite surface proteins and block development of the parasite in the mosquito midgut (25). As discussed below, several field studies of semi-immune adults or school-age children have shown that antibodies to transmission-blocking target antigen are present in some individuals exposed to malaria parasites. Conclusions based on the results of these studies are limited by the number of localities examined (to date, mainly Papua New Guinea, The Gambia, and Sri Lanka) and the absence of data for either in vivo or in vitro transmission-blocking activity in these samples (with one exception [25]). Furthermore, it has not been determined conclusively whether the source of any transmission-blocking effect in a serum sample is due to antibodies against prefertilization target antigens or to some other ill-defined humoral factor. Nevertheless, we can draw from these studies several important conclusions, some of which may help in the rational design of transmission-blocking vaccine strategies.

Antibody responses to prefertilization target antigens in semi-immune adults and children and nonimmune adults

By whatever mechanisms, some individuals living in malaria-endemic regions do develop some form of transmission-blocking immunity. In a study done in Papua New Guinea, a strong correlation was found between the presence of antibodies against gamete surface proteins and the suppression of *P. falciparum* infectivity in mosquitoes (25). However, sera from only 9 individuals of 41 examined consistently reduced infectivity, and antibodies to Pfs230 but not Pfs48/45 correlated with reduction of transmission (25). In the hyperendemicity region of Madang (Papua New Guinea), less than 50% of adults had antibodies to Pfs230 and Pfs48/45 (23). Carter et al. confirmed these results in a subsequent study of sera collected in Papua New Guinea in which Pfg27/25 (a highly immunogenic internal protein) was used as a marker for prior exposure to gametocytes (11). Antibodies to Pfs25 were not found in any of the 47 serum samples analyzed. Graves et al. found that in children residing in an area of hyperendemicity in Papua New Guinea, the antibody response to Pfs48/45 was generally comparable to that in immune adults from areas of hyperendemicity; however, the antibody response to Pfs230 was considerably higher (27). Approximately 85% of the children had anti-Pfs230 and anti-Pfs16 antibodies (27). In that study, the antibody response was reassessed 4 weeks after treatment. The number of individuals responsive to Pfs230 dropped significantly after treatment (down to the level seen in immune adults), whereas the number of individuals responsive to Pfs48/45 was highly dependent on the epitope examined (27). The antibody titers to Pfs16 did not change significantly.

Similar to adults living in areas of hyperendemicity, a minority of individuals exposed to malaria for the first time make antibodies that compete with MAbs for binding to Pfs230 or Pfs48/45. Graves et al. studied a population of Papua New Guinea highlanders and found that only 23% had antibodies to Pfs230 (28). For Pfs48/45, the number of individuals who responded in the competitive enzyme-linked immunosorbent assay (ELISA) was dependent on which MAb was used. The responses ranged from 6 to 65%. With each subsequent malaria parasite infection, the antibody response continued to rise. Ong et al. also found a 60% response rate to Pfs48/45 in individuals recovering from an initial attack of malaria (66). Convalescent-phase serum was not analyzed in the study by Ong et al.

Studies of sera collected from individuals residing in areas of hyperendemicity in The Gambia, West Africa, also revealed that less than 60% of adults develop antibodies to Pfs230 and only 9% develop antibodies to Pfs48/45 (72, 79). Again, antibody response to Pfs25 was completely absent. Interestingly, in one of these studies, the lack of responsiveness to Pfs230 was remarkably stable over time and did not appear to be related to prior malaria exposure, age, or the class II major histocompatibility complex genotype (79). Studies of twins suggested that environmental rather than genetic factors predominate (79). Unanswered questions include whether these unresponsive individuals will respond at all to a subunit Pfs230 vaccine and, if they do, whether boosting of the primary immune response from a subunit vaccine will occur with each subsequent natural infection. With the exception of the Graves et al. study (25), transmission-blocking activity was not measured in these studies.

Studies of naturally acquired transmission-blocking immunity in Sri Lanka have focused mainly on *P. vivax*. In *P. falciparum*, polymorphisms of antigen M_r are not a common feature, but poor immunogenicity is, whereas in *P. vivax* antigen, M_r polymorphisms are common (69). Therefore, transmission-blocking-immunity data in Sri Lanka should be extrapolated to transmission-blocking immunity for *P. falciparum* with reasonable skepticism. Nevertheless, a prominent feature of the immune response to malaria in Sri Lanka is the lack of immunological memory for boosting after successive infections unless the interval between infections is less than 4 months (for a review, see reference 59). Mendis and Carter concluded that transmission-blocking immunity is largely T-cell independent and that the target antigens may be lipids, carbohydrates, or other nonprotein moieties (59).

PROGRESS TOWARD DEVELOPING VACCINES AGAINST SPECIFIC ANTIGENS

The emphasis in development of a malaria transmission-blocking vaccine has been on humoral-mediated responses, particularly antibody-dependent ones. In part, this focus derives from the now-classic work done in the 1950s, in which chickens immunized with whole parasites were shown to develop humoral-mediated transmission-blocking immunity (33), and the 1980s, in which membrane feeding mosquitoes with MAbs to potential target antigens allowed investigators to search for and identify target antigens of antibody-mediated transmission-blocking immunity (78, 98).

In comparison, the cellular aspects of transmission-blocking immunity are largely unexplored. More work needs to be done to further develop the findings of cytokine-mediated inhibition of gametocyte infectivity. Because of the absence of any known target antigens of a cell-mediated transmission-blocking immune response, very little work appears to be directed toward development of vaccines designed to elicit T-cell-mediated immunity.

With the exception of research on Pfs25, most of the research directed toward evaluation of an antibody-mediated transmission-blocking vaccine is still early and unpublished; therefore, in this chapter, Pfs25 is used as a case study of malaria transmission-blocking vaccine development. Many of the issues that confront investigators who are developing transmission-blocking vaccines based on other target antigens have been faced during development of a Pfs25-based vaccine.

Mechanisms of Antibody-Dependent Protective Immune Responses

MAbs to Pfs48/45, Pfs28, or Pfs25 are sufficient alone to completely block oocyst development in mosquitoes (see above). For antibodies to recombinant Pfs25 and Pfs28, neither complement nor any cellular component of the immune system is required (13a, 40), although, as mentioned above, cytokines, complement, and cellular elements could conceivably enhance the transmission-blocking effect. The only described exceptions to the generalization that antibodies alone are sufficient are for Pfg27/25 and Pfs230, for which accumulating evidence suggests that complement may be required (70, 105). Antibody-dependent cell-mediated cytotoxicity has not been described adequately enough to determine what role it may play in mediating transmission-blocking activity.

The proposed mechanisms of MAb-dependent transmission-blocking activity are described above. Studies of the mechanisms of monospecific polyclonal-antibody-dependent transmission-blocking activity have to date been limited to Pfs25/Pgs25 (P25) and Pfs28/Pgs28/Pbs21 (P28). In both cases, polyclonal serum appears to mediate blocking at multiple stages of parasite development within the midgut. Although MAbs to P25 and P28 do not significantly decrease the transformation of zygotes into ookinetes (55, 84), polyclonal sera do disrupt this developmental process in vitro and/or in vivo (14, 55). In model systems in which in vitro-cultured ookinetes are available, polyclonal sera to P25 and P28 inhibit the infectivity of mature ookinetes in mosquitoes. Preliminary evidence also suggests that polyclonal sera may have a postookinete effect as well (55). Thus, the multiplicity of blocking mechanisms ascribed to polyclonal sera compared to those ascribed to MAbs may explain the enhanced potency of the former.

Current Status of Efforts To Design, Produce, and Evaluate Vaccines To Induce Antibodies to Target Antigens in Experimental Model Systems

Work to determine the importance of a particular antibody isotype has proved useful, especially with regard to complement fixation. Since serum complement is active immediately after ingestion of the blood meal but inactive within 8 to 10 h (8) thereafter, it is not surprising that transmission-blocking target antigens that require complement for full activity are present on the surface of the parasite before fertilization.

Antibodies to target antigens (P25 and P28) expressed later in sporogony do not appear to require the presence of complement to be fully transmission blocking. With this in mind, the greatest effort has been and should continue to be directed toward identifying and recreating the relevant B-cell epitopes that mediate transmission-blocking immunity. For those antigens expressed while the parasites circulate in the vertebrate host, work directed toward identifying relevant helper T cells might be useful in modulating the immune response to ensure a rapid and sustained boost in the immune response after a natural infection.

Mainly for historical reasons, identification and construction of relevant B-cell epitopes have been quite limited: although the gene for Pfs25 was cloned more than half a dozen years ago, the exact and complete B-cell epitope(s) that mediates transmission-blocking immunity is not yet known. The difficulty has been that the relevant B-cell epitope(s) is disulfide bond dependent, and recreating the appropriate disulfide bonds and/or conformational structures for both synthetic peptides and recombinant protein has been difficult.

Construction of B-cell epitopes

With a few exceptions, the known target epitopes of transmission-blocking MAbs are disulfide bond dependent. The exceptions include all of the potential target epitopes in Pfg27/25 (105) (at present, it is unclear whether these target epitopes directly mediate transmission blocking or simply cross-react with target epitopes in Pfs48/45 and/or Pfs230) and a single reduction-insensitive transmission-blocking target epitope in Pfs48/45 (92). Although the sequence of the latter epitope has not yet been described, there is hope that this epitope could be useful in a conventional linear-synthetic-peptide-based vaccine. The other described epitopes may require sophisticated syntheses to create conformationally constrained synthetic peptides that elicit transmission-blocking antibodies. Recently, published evidence for Pfs25 suggests that this latter approach may be feasible (81b, 90, 91).

Two distinct B-cell epitopes have been described for Pfs25, but only one of them has been tentatively mapped. Using PEPSCAN technology, in which overlapping linear peptides attached to pins in a microtiter plate format are screened for reactivity with MAbs of interest, a B-cell epitope that is a target of several transmission-blocking MAbs mapped to the B loop of the third EGF-like domain of Pfs25 (69, 97) (Fig. 3). Interestingly, the only site (Ala in position 130 instead of Gly as described in the original sequence published for the 3D7 strain) of antigenic diversity described in laboratory isolates was in the Dd2 strain of *P. falciparum* (43). The mutation maps to just this region (the B loop) of the third EGF-like domain. Whether the diversity in this region has practical importance is unclear; however, MAbs to Pfs25 that map to this region block transmission of Dd2 (Ala-130) parasites as well as 3D7 (Gly-130) parasites.

Linear synthetic peptides and mimeotopes from the B loop of the third EGF-like domain of Pfs25 elicit antibodies that recognize parasite-produced Pfs25, but despite inducing high-titer antibodies, they fail to

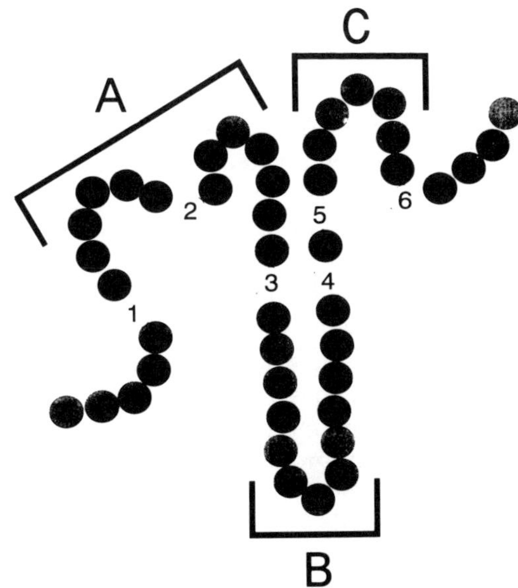

Figure 3. Presumed disulfide bond pattern and loop structure of an EGF-like domain. Loops are designated A, B, and C.

elicit transmission-blocking activity (69, 97). Recently, Stura et al. constructed conformationally constrained synthetic peptides. These peptides bind transmission-blocking MAbs to Pfs25 better than linear peptides and have been used in X-ray crystallography studies (90, 91). Although the evidence is still preliminary and antibody titers were low compared to those required by MAbs (>100 μg/ml) to block transmission, preliminary data suggest that these conformationally constrained peptides induce antibodies more similar to those produced with yeast-produced Pfs25 than those produced with linear peptides (81a). Furthermore, conformationally constrained peptides delivered with multiple-antigen peptides appear to elicit antibodies that suppress oocyst development (81a). Whether antibody titers sufficiently high to completely block transmission can be achieved with conformationally constrained synthetic peptides is currently being studied.

Carrier proteins and T-helper epitopes

Because success in using either recombinant protein or synthetic peptides to elicit a transmission-blocking antibody response has not been described other than for rPfs25 and rPfs28 expressed in eukaryotic cells (13a, 41, 42), the major emphasis has been on identifying and recreating the B-cell epitopes that are targets of transmission-blocking antibodies rather than on identifying the associated helper-T-cell epitopes. For those antigens expressed solely while the parasites develops in the vector (i.e., that are not expected to benefit from boosting following a natural infection), there is at present little practical reason to pursue identification of T-cell epitopes, especially for antigens such as Pfs25 and Pfs28, which are highly immunogenic (13a, 41, 42). If the antigen is a poor immunogen and additional T-cell help is required to induce transmission-blocking antibodies, including a heterologous helper-T-cell epitope, e.g., one from tetanus toxoid, may provide as much benefit as identifying and including a parasite-encoded one.

In contrast, for transmission-blocking vaccine candidates in which boosting following a natural infection is expected or at least is theoretically possible, identification of T-cell epitopes may be critically important. Elimination of immunodominant T- or B-cell epitopes that may be diverting the immune response away from eliciting transmission-blocking antibodies may prove effective. Furthermore, if the diversity already described (e.g., for Pfs230) maps to helper-T-cell epitopes, then inclusion of a variety of variant sequences in a single construct may broaden the immunogenicity of the vaccine. As described above, in experimental immunization models and in humans, poor immunogenicity of the antigens expected to benefit from boosting following a natural infection suggests that some immunoselection at helper-T-cell epitopes has occurred. Other than general boosting of transmission-blocking activity in naive animals immunized by injection with whole sexual-stage parasites and later challenged with an infection (31), there is scant evidence for boosting following a natural infection. In fact, some of the data recently obtained from field studies of humans living in malaria-endemic regions suggest that boosting does not occur naturally (59). Thus, whether the primary immune response to a subunit vaccine will be boosted following a natural infection remains to be determined. Although studies in which heterologous helper-T-cell epitopes are fused to potential B-cell epitopes are in progress, no studies on any of the potential transmission-blocking vaccine candidates have been published.

Adjuvants

Preliminary evidence suggests that the immune response to target antigens of the early sexual stages will be adjuvant dependent. For example, in intact gametes, Pfs40 is a poor immunogen when adjuvanted with complete Freund's adjuvant; however, this poor immunogenicity can be overcome with RIBI RAS adjuvant (76). Furthermore, modulation of poor immunogenicity may not be inherent in the protein itself (i.e., *cis* acting) but instead may be mediated by some *trans*-acting factor in whole gametes (76). If such a *trans*-activating factor actually exists, it does not have a global effect on the immunogenicity of all gametocyte antigens; for instance, in humans residing in malaria-endemic areas, Pfg27/25 is highly immunogenic, making it a good marker of prior exposure to gametocytes, whereas many individuals with antibodies to Pfg27/25 do not have antibodies to Pfs230 or Pfs48/45.

Adjuvants may also be important in eliciting the proper antibody isotype required to mediate transmission-blocking antibodies. For example, accumulating evidence suggests that the transmission-blocking activity of MAbs to Pfs230 is complement dependent (70). Although gametocytes are normally resistant to the presence of homologous complement, when they are mixed with some anti-Pfs230 MAbs in the presence of complement, transmission is inhibited if not completely blocked. The same MAb fed with infectious gametocytes in the absence of complement does not block transmission; likewise, complement in the absence of MAb does not block transmission (70). Thus, the parasite that normally resists the alternative pathway of complement activation may be made sensitive to the classical pathway by the addition of the right antibody, both antigen dependent and isotype dependent. Parenthetically, these data suggest that the point of resistance to lysis by complement occurs rather early in the complement activation pathway, at a point prior to C3 activation. A similar complement-dependent blocking effect appears to be present, albeit less dramatically, in the case of the Pfs230–Pfs48/45–Pfg27/25 cross-reacting MAbs (105).

The late-sexual-stage antigens, e.g., Pfs25, may require a carrier or adjuvant. Soluble rPfs25 does not induce complete transmission-blocking antibodies (39a). In contrast, the protein analogous to Pfs28 in the murine malaria model, *P. berghei*, does not appear to require an adjuvant: acetone-precipitated parasite-produced protein elicits complete transmission-blocking activity (95). A great deal more work is required to optimize the delivery of all of the transmission-blocking target antigens. One area that holds some hope of meeting the criteria for an ideal transmission-blocking vaccine (administered by a single shot or orally highly stable, immunogenic with long-lasting effective antibodies, and cheap) is microencapsulation-microsphere technology. Preliminary data with slow-release particles consisting of polylactide-polyglycolide polymers loaded with approximately 5.8% rPfs25 indicate that long-lasting transmission-blocking antibodies may be achievable starting within 2 to 3 weeks of a single injection (4). Other delivery systems, such as injection of naked DNA and highly attenuated live vectors, are also being explored.

Current Status of Efforts To Design, Produce, and Evaluate Vaccines To Induce Antibodies to Target Antigens in Humans

Construction of B-cell epitopes

In the absence of convincing data to support development of a synthetic-peptide-based transmission-blocking vaccine, there currently are no plans for designing, producing, or evaluating such a vaccine for use in human trials. With few exceptions, the target B-cell epitopes of transmission-blocking antibodies are disulfide bond dependent. Thus, linear peptides may have limited utility in eliciting transmission-blocking immunity. As discussed above, synthetic-peptide vaccines in which conformationally constrained synthetic peptides are covalently attached to heterologous T-cell epitopes are currently under active investigation. Whether vaccines that elicit an effective immune response to a single or just a few B-cell epitopes (rather than a large array of B-cell epitopes) will quickly be rendered useless by selection of escape mutants will need to be determined.

Carrier proteins and T-cell help

At present, the only formulations slated to be tested in humans in the near future are a yeast-secreted form of Pfs25 (TBV25H) adsorbed to alum and a highly attenuated vaccinia virus (NYVAC-7) that expresses Pfs25 in addition to six antigens from other stages of the life cycle (discussed further in chapter 3). Both formulations have been made using current good manufacturing practices. Neither formulation relies on carrier proteins or heterologous T-cell help to elicit transmission-blocking antibodies. Both formulations include all four EGF-like domains; however, the yeast-derived product is truncated after the last Cys of the fourth EGF-like domain and terminates with a 6-histidine tag used for initial purification by metal affinity chromatography (41). After this initial purification, the yeast product is desalted and further purified by size-exclusion gel chromatography. The highly purified material (>98% purity) is then adsorbed to alum and put into vials at 250 μg of TBV25H per dose. This material, to be administered intramuscularly at 0, 1, and 2 months, is due to go into a human phase I trial in the United States in the near future.

Because transmission-blocking vaccines may not afford immediate direct benefit to vaccine recipients, safety and toxicity are of the utmost concern. As long as these vaccines are completely safe and nontoxic, even a marginal transmission-blocking effect may warrant their use as public health measures to control the transmission of malaria parasites. Therefore, the initial emphasis of evaluating transmission-blocking vaccines previously shown to induce blocking antibodies in animals is on suitable preclinical animal studies followed by human trials to assess safety and toxicity. A preliminary study in *Aotus* monkeys of yeast-secreted Pfs25 formulated with a potent adjuvant, MF75.2, indicated that Pfs25 is safe and nontoxic in this model (40). Further toxicology studies using clinical-grade Pfs25 adjuvanted with alum in several other animal species are currently in progress.

Adjuvants

Preliminary evidence from studies in rodents and monkeys suggests that alum-adjuvanted Pfs25-based vaccines are sufficient to induce transmission-blocking antibodies but that the response is short-lived (41, 45). The data suggest that instead of enhancement of immunogenicity by an adjuvant, what is required is a delivery system that continues to stimulate an immune response for a prolonged period. If immunogenicity studies with alum-adjuvanted vaccine in humans also suggest that the immunity is short-lived, alternative formulations will need to be considered.

One such alternative currently being studied is slow-release particles. A prototype slow-release particle consisting of microspheres of polylactide-polyglycolide polymers loaded with approximately 5% (by weight) Pfs25 given as a single parenteral injection has shown promise in mouse studies (4, 39a). Whether such a formulation will induce similar results in primates and humans remains to be determined. Ideally, a transmission-blocking vaccine would be extremely stable and inexpensive and would induce long-lasting immunity after a single administration, preferably by a nonparenteral route. Such a vaccine appears to be a long way off.

Current Status of Field Trials of Transmission-Blocking Vaccines

As part of a safety study in humans (phase I), immunogenicity can be easily determined. A first approximation of the in vivo utility of a transmission-blocking vaccine may be extrapolated from membrane feeding data. Briefly, in this assay, serum samples mixed with in vitro-cultured gametocytes are fed to starved, laboratory-reared mosquitoes through a natural or artificial (Parafilm) membrane. Six to eight days after a blood meal, the midguts of female mosquitoes are dissected, stained with Mercurochrome, and scored for infectivity by counting malaria parasite oocysts. At present, only rather indirect proof exists that such a membrane feeding technique is an adequate predictor of the in vivo efficacy of a transmission-blocking vaccine. Ultimately, the membrane feeding assay needs to be validated as a predictive test for each of the specific target antigens. The rationale for such a statement derives from recent observations that the mechanism by which antibodies mediate transmission-blocking activity may differ according to the target antigen; e.g., transmission-blocking antibodies to Pfs230 may mediate blocking solely by a complement-dependent mechanism, whereas transmission-blocking antibodies to Pfs25 may mediate blocking by multiple mechanisms. How useful the *Aotus* monkey model will be in predicting in vivo efficacy in humans is also not yet known.

Use of the currently available malaria challenge model (challenge by the bite of laboratory-reared infected mosquitoes) is unethical for testing the in vivo efficacy of transmission-blocking vaccines in humans. To challenge human volunteers with a natural infection and allow the infection to proceed to the point at which infectious gametocytes are produced would put undue risk on volunteers, particularly because there would be no potential direct benefit from that volunteer receiving the transmission-blocking vaccine. Therefore, at present, the prevailing consensus is that in vivo efficacy studies must be done in humans residing in malaria-endemic areas, who would normally get a natural infection. Because wild mosquitoes caught in many malaria-endemic areas also transmit other human pathogens, such as arboviruses, the mosquitoes used to feed directly on volunteers must be reared in a laboratory long enough to ensure that they are free from any other human pathogens.

Selecting and preparing a site for field testing of a transmission-blocking vaccine have begun. The resources required to test a transmission-blocking vaccine, particularly the requirement for mosquitoes, are unique. For TBV25H (transmission-blocking vaccine based on Pfs25 with a carboxy histidine tag [45]), the goals of the first field studies will be to establish the safety of the vaccine and to provide the principle that a subunit transmission-blocking vaccine will elicit an immune response in humans that is efficacious in vivo. One design being considered for proving the principle is the so-called three-pronged mosquito feed: (i) laboratory-reared mosquitoes purposely fed on naturally infected volunteers or trapped wild mosquitoes that have fortuitously fed on these volunteers are quantitatively assayed for infectivity (e.g., scoring oocysts in the dissected midguts of mosquitoes 6 to 8 days post-blood meal, ELISAs of squashed mosquitoes for the presence of circumsporozoite protein, PCR of stage-specific rRNA, or a variety of other means); (ii) at about the same time that mosquitoes are fed directly on the volunteer, blood collected from the volunteer is washed free of any serum antibodies and then fed via a membrane feeding apparatus to laboratory-reared mosquitoes to establish the inherent infectivity of the circulating gametocytes in that donor at the time of the direct feed; and (iii) serum (or purified antibodies from the serum) collected from the volunteer are mixed with known infectious gametocytes and fed via a membrane feeding apparatus to laboratory-reared mosquitoes to establish that transmission-blocking activity is in fact due to antibodies. Ideally, the study would be randomized, placebo controlled, and blinded and would be performed at several sites that differed in rate and seasonality of transmission and in prevalence and incidence of parasitemia.

Estimating sample sizes for phase IIb studies

Before phase IIb field studies commence, the sample size required to detect a statistically significant result must be estimated. Some of the variables that need to be considered in determining the sample size include (i) the efficacy of the vaccine; (ii) the proportion of the target population that will become infected with malaria parasites, develop sufficient gametocytes, and actually transmit the parasites to mosquitoes in the absence of antivaccine antibodies; and (iii) the design, power, and significance level of the statistical test to be applied.

A rough estimate of the efficacy of the vaccine should be determined in the in vitro efficacy (membrane feeding) arms of phase I trials of that vaccine. Because the number of individuals enrolled in phase I studies tends to be small (8 to 10 individuals), the estimate of efficacy will be crude. With the obvious precautionary note that immunogenicity may vary in different genetic backgrounds, the estimate of efficacy may be more accurately determined by combining the efficacy determined in phase I studies done at the multiple venues of endemicity at which phase IIb studies will be done. Of course, this estimate of efficacy may not accurately reflect the efficacy in vivo, and this problem with accuracy is one of the primary reasons for doing a phase II study, i.e., to test how well membrane feeding efficacy correlates with in vivo efficacy.

In addition to assessing the infrastructure of a testing site (e.g., capacity to rear and dissect mosquitoes, presence of a validated clinical laboratory for safety studies, personnel available to monitor volunteers at the field site), the phase IIb site selection process assesses the transmissibility of malaria at the site. This value, which incorporates rate of infection, rate of gametocyte production, and prevalence of inherent transmission-blocking activity preexisting in the population, is required to determine the sample size for a phase IIb study. At first glance, sites with the highest transmission rates would be thought to be the best sites for phase IIb testing in adults. In fact, this may not be the case. For instance, the Madang area of Papua New Guinea is often cited as an area with intense transmission of malaria, yet less than 2% of the adult population has a detectable gametocytemia, and of those with smear-positive gametocytemia, less than 40% reliably transmit the parasite to laboratory-reared mosquitoes. Estimated sample sizes required in this setting (Table 2) would be quite large. In contrast, in Sri Lanka, where transmission intensity is often thought to be lower than in Papua New Guinea, the transmissibility rate is approximately 8%. Estimated sample sizes required in this setting would be quite reasonable.

The power and significance levels of the statistical test to be applied also influence the sample size. Commonly accepted values for type I and II errors are 0.05 and 0.10, respectively; these values were used in calculating the sample sizes given in Table 2. Despite the suspicion usually associated with using one-tail tests instead of two-tail tests, the appropriate test of significance in a phase IIb study of a transmission-blocking vaccine can be one sided. Clearly, the investigators designing the clinical trial specify the hypothesis that the trial is designed to test. If the investigators seek only to determine whether the test vaccine is superior to a placebo (that is, the practical consequences of the test vaccine being equal to or worse than placebo are the same), then a one-sided test of significance is appropriate. In practice and in the case of TBV25H, clinical tri-

Table 2. Sample sizes for phase IIb transmission-blocking trials based on infectivity rates of gametocytemic volunteers and vaccine efficacy[a]

% Efficacy	Sample size with gametocytemia of:						
	1%	2%	5%	10%	15%	20%	25%
				90% infectivity			
94	1,200	600	240	120	80	60	48
89	1,600	800	320	160	107	80	64
78	2,000	1,000	400	200	133	100	80
67	2,400	1,200	480	240	160	120	96
56	3,400	1,700	680	340	227	170	136
44	5,000	2,500	1,000	500	333	250	200
33	7,800	3,900	1,560	780	620	390	312
				70% infectivity			
93	2,000	1,000	400	200	133	100	80
86	2,400	1,200	480	240	160	120	98
71	3,600	1,800	720	360	240	180	144
57	6,200	3,100	1,240	620	413	310	248
43	10,600	5,300	2,120	1,060	707	530	424
29	22,200	11,100	4,440	2,220	1,480	1,110	888
14	40,800	20,400	8,160	4,080	2,720	2,040	1,632
				50% infectivity			
90	3,600	1,800	720	360	240	180	144
80	5,000	2,500	1,000	500	333	250	200
60	9,400	4,700	1,880	940	627	470	376
40	22,200	11,100	4,440	2,220	1,480	1,110	888
20	89,000	44,500	17,800	8,900	5,933	4,450	3,560
				30% infectivity			
83	9,400	4,700	1,880	940	627	470	378
67	14,800	7,400	2,960	1,480	987	740	592
33	67,600	33,800	13,520	6,760	4,507	3,380	2,704
				20% infectivity			
75	17,800	8,900	3,060	1,780	1,187	890	712
50	46,400	23,200	9,280	4,640	3,093	2,320	1,856

[a]Calculations are based on Haseman's tables (one-tailed significance test at a 5% level and a power of 90%). Boxed areas indicate population sizes generally considered feasible to conduct a phase IIb transmission-blocking trial.

als undertaken to document superiority of a test vaccine over placebo (i.e., required for Food and Drug Administration approval for licensure) are one-sided tests.

SUMMARY AND CONCLUSIONS

Accumulating evidence suggests that malaria transmission-blocking vaccines will be viable components of public health programs to control the spread of malaria and of triple-stage (pre-erythrocytic-, erythrocytic-, and sexual-sporogonic-stage) malaria vaccines. Undoubtedly, a transmission-blocking vaccine will be first used as a component of a multimodality approach to malaria control, either in combination with bed nets and drugs as a public health measure or in combination with vaccines from other stages of the life cycle as a protective measure for individuals in a population. In the latter case, the transmission-blocking component will serve to prevent the spread of "vaccine escape mutants" and to enhance the potency of the protective components by reducing transmission in a region (and thus reducing the sporozoite load to which an individual is exposed). Transmission-blocking vaccines may also serve as adjuncts to chemotherapy campaigns, especially to prevent the spread of the drug-resistant mutants that will no doubt arise when a new drug is introduced.

With the isolation of more than a dozen genes encoding sexual-stage-specific proteins, the challenges that remain are to produce recombinant proteins or synthetic peptides that induce transmission-blocking antibodies, deliver these immunogens in a form that induces long-lasting immunity, and combine them in a formulation that is safe and nontoxic in humans. In this chapter, a yeast-produced Pfs25-based subunit vaccine has been described as one example of a transmission-blocking vaccine for malaria. Much more needs to be done on Pfs25 and the other potential target antigens. Mustering the necessary expertise that resides in the biotechnology industrial complex and the financial support required to expedite the development of such a transmission-blocking vaccine may be the greatest challenge we have faced to date.

REFERENCES

1. **Alano, P., S. Premawansa, M. S. Bruce, and R. Carter.** 1991. A stage specific gene expressed at the onset of gametocytogenesis in *Plasmodium falciparum*. *Mol. Biochem. Parasitol.* **46:**81–88.
2. **Bairoch, A., and J. A. Cox.** 1990. EF-hand motifs in inositol phospholipid-specific phospholipase. *FEBS Lett.* **269:**454–456.

3. **Barr, P. J., K. M. Green, H. L. Gibson, I. C. Bathurst, I. A. Quakyi, and D. C. Kaslow.** 1991. Recombinant Pfs25 protein of Plasmodium falciparum elicits malaria transmission-blocking immunity in experimental animals. *J. Exp. Med.* **174:**1203–1208.

4. **Bathurst, I. C., P. J. Barr, D. C. Kaslow, D. H. Lewis, T. J. Atkins, and M. E. Rickey.** 1992. Development of a single injection transmission-blocking malaria vaccine using biodegradable microspheres. *Proc. Int. Symp. Release Bioact. Membr.* **19:**120–121.

5. **Bradley, G. H.** 1966. A review of malaria control and eradication in the United States. *Mosq. News* **26:**462–470.

6. **Bruce, M. C., D. A. Baker, P. Alano, N. C. Rogers, P. M. Graves, G. A. Targett, and R. Carter.** 1990. Sequence coding for a sexual stage specific protein of Plasmodium falciparum. *Nucleic Acids Res.* **18:**3637. (Erratum, **18:**4991.)

7. **Carter, R., and D. H. Chen.** 1976. Malaria transmission blocked by immunisation with gametes of the malaria parasite. *Nature* (London) **263:**57–60.

7a. **Carter, R., A. Coulson, S. Bhatti, B. J. Taylor, and J. F. Elliott.** 1995. Predicted disulfide-bonded structures for three uniquely related proteins of *Plasmodium falciparum*, Pfs230, Pfs 48/45 and Pf12. *Mol. Biochem. Parasitol.* **71:**203–210.

8. **Carter, R., and P. M. Graves.** 1988. Gametocytes, p. 253–305. *In* W. H. Wernsdorfer and I. McGregor (ed.), *Malaria: Principles and Practice of Malariology*. Churchill Livingstone, Edinburgh.

9. **Carter, R., P. M. Graves, A. Creasey, K. Byrne, D. Read, P. Alano, and B. Fenton.** 1989. Plasmodium falciparum: an abundant stage-specific protein expressed during early gametocyte development. *Exp. Parasitol.* **69:**140–149.

10. **Carter, R., P. M. Graves, D. B. Keister, and I. A. Quakyi.** 1990. Properties of epitopes of Pfs 48/45, a target of transmission blocking monoclonal antibodies, on gametes of different isolates of Plasmodium falciparum. *Parasite Immunol.* **12:**587–603.

11. **Carter, R., P. M. Graves, I. A. Quakyi, and M. F. Good.** 1989. Restricted or absent immune responses in human populations to Plasmodium falciparum gamete antigens that are targets of malaria transmission-blocking antibodies. *J. Exp. Med.* **169:**135–147.

12. **Carter, R., and D. C. Kaushal.** 1984. Characterization of antigens on mosquito midgut stages of Plasmodium gallinaceum. III. Changes in zygote surface proteins during transformation to mature ookinete. *Mol. Biochem. Parasitol.* **13:**235–241.

13. **Cesbron-Delauw, M. F., B. Guy, G. Torpier, R. J. Pierce, G. Lenzen, J. Y. Cesbron, H. Charif, P. Lepage, F. Darcy, J. P. Lecocq, and A. Capron.** 1989. Molecular characterization of a 23-kilodalton major antigen secreted from *Toxoplasma gondii*. *Proc. Natl. Acad. Sci USA* **86:**7537–7541.

13a. **Duffy, P. E., and D. C. Kaslow.** Unpublished data.

14. **Duffy, P. E., P. Pimenta, and D. C. Kaslow.** 1993. Pgs28 belongs to a family of epidermal growth factor-like antigens that are targets of malaria transmission-blocking antibodies. *J. Exp. Med.* **177:**505–510.

15. **Elliott, J. F., G. R. Albrecht, A. Gilladoga, S. M. Handunnetti, J. Neequaye, G. Lallinger, J. N. Minjas, and R. J. Howard.** 1990. Genes for *Plasmodium falciparum* surface antigens cloned by expression in COS cells. *Proc. Natl. Acad. Sci. USA* **87:**6363–6387.

16. **Feng, Z., R. N. Hoffmann, R. S. Nussenzweig, M. Tsuji, H. Fujioka, M. Aikawa, T. H. Lensen, T. Ponnudurai, and L. G. Pologe.** 1993. Pfs2400 can mediate antibody-dependent malaria transmission inhibition and may be the Plasmodium falciparum 11.1 gene product. *J. Exp. Med.* **177:**273–281.

17. **Foo, A., R. Carter, C. Lambros, P. Graves, I. Quakyi, G. A. Targett, T. Ponnudurai, and G. Lewis, Jr.** 1991. Conserved and variant epitopes of target antigens of trans-

mission-blocking antibodies among isolates of Plasmodium falciparum from Malaysia. *Am. J. Trop. Med. Hyg.* **44:**623–631.
18. **Fries, H. C., M. B. Lamers, M. A. Smits, T. Ponnudurai, and J. H. Meuwissen.** 1989. Characterization of epitopes on the 25 kD protein of the macrogametes/zygotes of Plasmodium falciparum. *Parasite Immunol.* **11:**31–45.
19. **Fries, H. C., M. B. Lamers, J. van Deursen, T. Ponnudurai, and J. H. Meuwissen.** 1990. Biosynthesis of the 25-kDa protein in the macrogametes/zygotes of Plasmodium falciparum. *Exp. Parasitol.* **71:**229–235.
20. **Gamage-Mendis, A. C., J. Rajakaruna, R. Carter, and K. Mendis.** 1991. Infectious reservoir of *Plasmodium falciparum* and *Plasmodium vivax* malaria in an endemic region of Sri Lanka. *Am. J. Trop. Med. Hyg.* **45:**479–487.
21. **Garnham, P. C. C., R. G. Bird, and J. R. Baker.** 1962. Electron microscope studies of motile stages of malaria parasites. III. The ookinetes of *Haemamoeba* and *Plasmodium. Trans. R. Soc. Trop. Med. Hyg.* **56:**116.
22. **Good, M. F., L. H. Miller, S. Kumar, I. A. Quakyi, D. Keister, J. H. Adams, B. Moss, J. A. Berzofsky, and R. Carter.** 1988. Limited immunological recognition of critical malaria vaccine candidate antigens. *Science* **242:**574–577.
23. **Graves, P. M., K. Bhatia, T. R. Burkot, M. Prasad, R. A. Wirtz, and P. Beckers.** 1989. Association between HLA type and antibody response to malaria sporozoite and gametocyte epitopes is not evident in immune Papua New Guineans. *Clin. Exp. Immunol.* **78:**418–423.
24. **Graves, P. M., T. R. Burkot, R. Carter, J. A. Cattani, M. Lagog, J. Parker, B. J. Brabin, F. D. Gibson, D. J. Bradley, and M. P. Alpers.** 1988. Measurement of malarial infectivity of human populations to mosquitoes in the Madang area, Papua, New Guinea. *Parasitology* **96:**251–263.
25. **Graves, P. M., R. Carter, T. R. Burkot, I. A. Quakyi, and N. Kumar.** 1988. Antibodies to Plasmodium falciparum gamete surface antigens in Papua New Guinea sera. *Parasite Immunol.* **10:**209–218.
26. **Graves, P. M., R. Carter, T. R. Burkot, J. Rener, D. C. Kaushal, and J. L. Williams.** 1985. Effects of transmission-blocking monoclonal antibodies on different isolates of *Plasmodium falciparum. Infect. Immun.* **48:**611–616.
27. **Graves, P. M., A. Doubrovsky, and P. Beckers.** 1991. Antibody responses to *Plasmodium falciparum* gametocyte antigens during and after malaria attacks in schoolchildren from Madang, Papua New Guinea. *Parasite Immunol.* **13:**291–299.
28. **Graves, P. M., A. Doubrovsky, R. Carter, S. Eida, and P. Beckers.** 1990. High frequency of antibody response to Plasmodium falciparum gametocyte antigens during acute malaria infections in Papua New Guinea highlanders. *Am. J. Trop. Med. Hyg.* **42:**515–520.
29. **Grotendorst, C. A., N. Kumar, R. Carter, and D. C. Kaushal.** 1984. A surface protein expressed during the transformation of zygotes of *Plasmodium gallinaceum* is a target of transmission-blocking antibodies. *Infect. Immun.* **45:**775–777.
30. **Gwadz, R. W.** 1976. Malaria: successful immunization against sexual stage of *Plasmodium gallinaceum. Science* **193:**1150–1151.
31. **Gwadz, R. W., and L. C. Koontz.** 1984. *Plasmodium knowlesi*: persistence of transmission-blocking immunity in monkeys immunized with gamete antigens. *Infect. Immun.* **44:**137–140.
32. **Huber, M., E. Cabib, and L. H. Miller.** 1991. Malaria parasite chitinase and penetration of the mosquito peritrophic membrane. *Proc. Natl. Acad. Sci. USA* **88:**2807–2810.
33. **Huff, C. G., D. F. Marchbank, and T. Shiroishi.** 1958. Changes in infectiousness of malarial gametocytes. II. Analysis of the possible causative factors. *Exp. Parasitol.* **7:**399–417.

34. **Ifediba, T., and J. P. Vanderberg.** 1981. Complete *in vitro* maturation of *Plasmodium falciparum. Nature* (London) **294:**364.
35. **Inselburg, J.** 1983. Gametocyte formation by the progeny of single *Plasmodium falciparum* schizonts. *J. Parasitol.* **69:**584–591.
36. **Janse, C. J., P. F. J. van der Klooster, H. J. van der Kaay, M. van der Ploeg, and J. P. Overdulve.** 1986. DNA synthesis in *Plasmodium berghei* during asexual and sexual development. *Mol. Biochem. Parasitol.* **20:**173–182.
37. **Jensen, J. B.** 1979. Observations on gametogenesis in *Plasmodium falciparum* from continuous culture. *J. Protozool.* **26:**129–132.
38. **Kaslow, D. C.** 1990. Immunogenicity of Plasmodium falciparum sexual stage antigens: implications for the design of a transmission blocking vaccine. *Immunol. Lett.* **25:**83–86.
39. **Kaslow, D. C.** 1993. Transmission-blocking immunity against malaria and other vector-borne diseases. *Curr. Opin. Immunol.* **5:**557–565.
39a. **Kaslow, D. C.** Unpublished data.
40. **Kaslow, D. C., I. C. Bathurst, D. B. Keister, G. H. Campbell, S. Adams, C. L. Morris, J. S. Sullivan, P. J. Barr, and W. E. Collins.** 1993. Safety, immunogenicity and in vitro efficacy of a muramyl tripeptide-based malaria transmission-blocking vaccine in an *Aotus nancymai* monkey model. *Vaccine Res.* **2:**95–103.
41. **Kaslow, D. C., I. C. Bathurst, T. Lensen, T. Ponnudurai, P. J. Barr, and D. B. Keister.** 1994. *Saccharomyces cerevisiae* recombinant Pfs25 adsorbed to alum elicits antibodies that block transmission of *Plasmodium falciparum. Infect. Immun.* **62:**5576–5580.
42. **Kaslow, D. C., S. N. Isaacs, I. A. Quakyi, R. W. Gwadz, B. Moss, and D. B. Keister.** 1991. Induction of Plasmodium falciparum transmission-blocking antibodies by recombinant vaccinia virus. *Science* **252:**1310–1313.
43. **Kaslow, D. C., I. A. Quakyi, and D. B. Keister.** 1989. Minimal variation in a vaccine candidate from the sexual stage of Plasmodium falciparum. *Mol. Biochem. Parasitol.* **32:**101–103.
44. **Kaslow, D. C., I. A. Quakyi, C. Syin, M. G. Raum, D. B. Keister, J. E. Coligan, T. F. McCutchan, and L. H. Miller.** 1988. A vaccine candidate from the sexual stage of human malaria that contains EGF-like domains. *Nature* (London) **333:**74–76.
45. **Kaslow, D. C., and J. Shiloach.** 1994. Production, purification and immunogenicity of a malaria transmission-blocking vaccine candidate: TBV25-H expressed in yeast and purified using nickel-NTA agarose. *Bio/Technology* **12:**494–499.
46. **Kaslow, D. C., C. Syin, T. F. McCutchan, and L. H. Miller.** 1989. Comparison of the primary structure of the 25 kDa ookinete surface antigens of Plasmodium falciparum and Plasmodium gallinaceum reveal six conserved regions. *Mol. Biochem. Parasitol.* **33:**283–287.
47. **Kaushal, D. C., and R. Carter.** 1984. Characterization of antigens on mosquito midgut stages of Plasmodium gallinaceum. II. Comparison of surface antigens of male and female gametes and zygotes. *Mol. Biochem. Parasitol.* **11:**145–156.
48. **Kaushal, D. C., R. Carter, L. H. Miller, and G. Krishna.** 1980. Gametocytogenesis by malaria parasites in continuous culture. *Nature* (London) **286:**490–492.
49. **Kaushal, D. C., R. Carter, J. Rener, C. A. Grotendorst, L. H. Miller, and R. J. Howard.** 1983. Monoclonal antibodies against surface determinants on gametes of Plasmodium gallinaceum block transmission of malaria parasites to mosquitoes. *J. Immunol.* **131:**2557–2562.
50. **Kocken, C. H., J. Jansen, A. M. Kaan, P. J. Beckers, T. Ponnudurai, D. C. Kaslow, R. N. Konings, and J. G. Schoenmakers.** 1993. Cloning and expression of the gene coding for the transmission blocking target antigen Pfs48/45 of Plasmodium falciparum. *Mol. Biochem. Parasitol.* **61:**59–68.

51. **Kumar, N.** 1987. Target antigens of malaria transmission blocking immunity exist as a stable membrane bound complex. *Parasite Immunol.* **9:**321–335.
52. **Kumar, N., and R. Carter.** 1985. Biosynthesis of two stage-specific membrane proteins during transformation of Plasmodium gallinaceum zygotes into ookinetes. *Mol. Biochem. Parasitol.* **14:**127–139.
53. **Lal, A. A., M. E. Schrieffer, J. B. Sacci, I. F. Goldman, V. Louis-Wileman, W. E. Collins, and A. F. Azad.** 1994. Inhibition of malaria parasite development in mosquitoes by anti-mosquito-midgut antibodies. *Infect. Immunol.* **62:**316–318.
54. **Laveran, A.** 1880. Note sur un noveau parasite trouve dans le sang du plusiers malades atteints de fievre palustre. *Bull. Acad. Natl. Med. Paris* **9:**1235.
55. **Lensen, A. H., G. J. Van Gemert, M. G. Bolmer, J. F. Meis, D. Kaslow, J. H. Meuwissen, and T. Ponnudurai.** 1992. Transmission blocking antibody of the Plasmodium falciparum zygote/ookinete surface protein Pfs25 also influences sporozoite development. *Parasite Immunol.* **14:**471–479.
56. **MacCallum, W. G.** 1898. On the haematozoan infections in birds. *J. Exp. Med.* **3:**117–136.
57. **Meis, J. F., and T. Ponnudurai.** 1987. Ultrastructural studies on the interaction of Plasmodium falciparum ookinetes with the midgut epithelium of Anopheles stephensi mosquitoes. *Parasitol. Res.* **73:**500–506.
58. **Meis, J. F., G. Pool, G. J. van Gemert, A. H. Lensen, T. Ponnudurai, and J. H. Meuwissen.** 1989. Plasmodium falciparum ookinetes migrate intercellularly through Anopheles stephensi midgut epithelium. *Parasitol. Res.* **76:**13–19.
59. **Mendis, K. N., and R. Carter.** 1991. Transmission blocking immunity may provide clues that antimalarial immunity is largely T-independent. *Res. Immunol.* **142:**687–690.
60. **Moelans, I. I., C. H. Klaassen, D. C. Kaslow, R. N. Konings, and J. G. Schoenmakers.** 1991. Minimal variation in Pfs16, a novel protein located in the membrane of gametes and sporozoites of Plasmodium falciparum. *Mol. Biochem. Parasitol.* **46:**311–313.
61. **Moelans, I. I., J. F. Meis, C. Kocken, R. N. Konings, and J. G. Schoenmakers.** 1991. A novel protein antigen of the malaria parasite *Plasmodium falciparum*, located on the surface of gametes and sporozoites. *Mol. Biochem. Parasitol.* **45:**193–204.
62. **Muirhead-Thompson, R. C.** 1957. The malarial infectivity of an African village population to mosquitoes (*Anopheles gambiae*): a random xenodiagnostic survey. *Am. J. Trop. Med. Hyg.* **6:**971–979.
63. **Naotunne, T. S., N. D. Karunaweera, G. Del Giudice, M. U. Kularatne, G. E. Grau, R. Carter, and K. N. Mendis.** 1991. Cytokines kill malaria parasites during infection crisis: extracellular complementary factors are essential. *J. Exp. Med.* **173:**523–529.
64. **Nijhout, M. M.** 1979. *Plasmodium gallinaceum*: exflagellation stimulated by a mosquito factor. *Exp. Parasitol.* **48:**75–80.
65. **Nijhout, M. M., and R. Carter.** 1978. Gamete development in malaria parasites: bicarbonate-dependent stimulation by pH in vitro. *Parasitology* **76:**39–53.
66. **Ong, C. S., K. Y. Zhang, S. J. Eida, P. M. Graves, C. Dow, M. Looker, N. C. Rogers, P. L. Chiodini, and G. A. Targett.** 1990. The primary antibody response of malaria patients to Plasmodium falciparum sexual stage antigens which are potential transmission blocking vaccine candidates. *Parasite Immunol.* **12:**447–456.
67. **Paton, M. G., G. C. Barker, H. Matsuoka, J. Ramesar, C. J. Janse, A. P. Waters, and R. E. Sinden.** 1993. Structure and expression of a post-transcriptionally regulated malaria gene encoding a surface protein from the sexual stages of Plasmodium berghei. *Mol. Biochem. Parasitol.* **59:**263–275.
68. **Ponnudurai, T., G. J. van Gemert, T. Bensink, A. H. Lensen, and J. H. Meuwissen.** 1987. Transmission blockade of Plasmodium falciparum: its variability with gametocyte numbers and concentration of antibody. *Trans. R. Soc. Trop. Med. Hyg.* **81:**491–493.

69. **Premawansa, S., J. S. Peiris, K. L. Perera, G. Ariyaratne, R. Carter, and K. N. Mendis.** 1990. Target antigens of transmission blocking immunity of *Plasmodium vivax* malaria. Characterization and polymorphism in natural parasites isolates. *J. Immunol.* **144:**4376–4383.
70. **Quakyi, I. A., R. Carter, J. Rener, N. Kumar, M. F. Good, and L. H. Miller.** 1987. The 230-kDa gamete surface protein of Plasmodium falciparum is also a target for transmission-blocking antibodies. *J. Immunol.* **139:**4213–4217.
71. **Quakyi, I. A., Y. Matsumoto, R. Carter, R. Udomsangpetch, A. Sjolander, K. Berzins, P. Perlmann, M. Aikawa, and L. H. Miller.** 1989. Movement of a falciparum malaria protein through the erythrocyte cytoplasm to the erythrocyte membrane is associated with lysis of the erythrocyte and release of gametes. *Infect. Immun.* **57:**833–839.
72. **Quakyi, I. A., L. N. Otoo, D. Pombo, L. Y. Sugars, A. Menon, A. S. DeGroot, A. Johnson, D. Alling, L. H. Miller, and M. F. Good.** 1989. Differential non-responsiveness in humans of candidate *Plasmodium falciparum* vaccine antigens. *Am. J. Trop. Med. Hyg.* **41:**125–134.
73. **Ramasamy, M. S., and R. Ramasamy.** 1990. Effect of anti-mosquito antibodies on the infectivity of the rodent malaria parasite Plasmodium berghei to Anopheles farauti. *Med. Vet. Entomol.* **4:**161–166.
74. **Ramasamy, M. S., M. Sands, B. H. Kay, I. D. Fanning, G. W. Lawrence, and R. Ramasamy.** 1990. Anti-mosquito antibodies reduce the susceptibility of Aedes aegypti to arbovirus infection. *Med. Vet. Entomol.* **4:**49–55.
75. **Rawlings, D. J., and D. C. Kaslow.** 1992. A novel 40-kDa membrane-associated EF-hand calcium-binding protein in Plasmodium falciparum. *J. Biol. Chem.* **267:**3976–3982.
76. **Rawlings, D. J., and D. C. Kaslow.** 1992. Adjuvant-dependent immune response to malarial transmission-blocking vaccine candidate antigens. *J. Exp. Med.* **176:**1483–1487. (Erratum, **177:**576, 1993.)
76a. **Rawlings, D. J., and D. C. Kaslow.** Unpublished data.
77. **Rener, J., R. Carter, Y. Rosenberg, and L. H. Miller.** 1980. Anti-gamete monoclonal antibodies synergistically block transmission of malaria by preventing fertilization in the mosquito. *Proc. Natl. Acad. Sci. USA* **77:**6797–6799.
78. **Rener, J., P. M. Graves, R. Carter, J. L. Williams, and T. R. Burkot.** 1983. Target antigens of transmission-blocking immunity on gametes of *Plasmodium falciparum. J. Exp. Med.* **158:**976–981.
79. **Riley, E. M., S. Bennett, A. Jepson, M. Hassan-King, H. Whittle, O. Olerup, and R. Carter.** 1994. Human antibody responses to Pfs230, a sexual stage-specific surface antigen of *Plasmodium falciparum*: non-responsiveness is a stable phenotype but does not appear to be genetically regulated. *Parasite Immunol.* **16:**55–62.
80. **Ross, R.** 1898. Report on the cultivation of Proteosoma, Labbe, in grey mosquitoes. *Indian Med. Gaz.* **33:**401–408.
81. **Rudin, W., and H. Hecker.** 1989. Lectin-binding sites in the midgut of the mosquitoes *Anopheles stephensi* Liston, and *Aedes egypti* L. (Diptera: Culicidae). *Parasitol. Res.* **75:**268–279.
81a. **Satterthwait, A. C., and D. C. Kaslow.** Unpublished data.
81b. **Satterthwait, A. C., E. A. Stura, and D. C. Kaslow.** Unpublished data.
82. **Scherf, A., R. Carter, C. Petersen, P. Alano, R. Nelson, M. Aikawa, D. Mattei, L. Pereira da Silva, and J. Leech.** 1992. Gene activation of Pf11-1 of *Plasmodium falciparum* by chromosome breakage and healing: identification of gametocyte-specific protein with a potential role in gametogenesis. *EMBO J.* **11:**2293–2301.
82a. **Shahabuddin, M., and D. C. Kaslow.** Unpublished data.
83. **Shahabuddin, M., T. Toyoshima, M. Aikawa, and D. C. Kaslow.** 1993. Transmis-

sion-blocking activity of a chitinase inhibitor and activation of malarial parasite chitinase by mosquito protease. *Proc. Natl. Acad. Sci. USA* **90:**4266–4270.

84. **Sieber, K. P., M. Huber, D. Kaslow, S. M. Banks, M. Torii, M. Aikawa, and L. H. Miller.** 1991. The peritrophic membrane as a barrier: its penetration by Plasmodium gallinaceum and the effect of a monoclonal antibody to ookinetes. *Exp. Parasitol.* **72:**145–156.
85. **Sinden, R. E., and R. H. Hartley.** 1985. Identification of the meiotic division of malarial parasites. *J. Protozool.* **32:**742.
86. **Sinden, R. E., and M. E. Smalley.** 1976. Gametocytes of Plasmodium falciparum: phagocytosis by leucocytes in vivo and in vitro. *Trans. R. Soc. Trop. Med. Hyg.* **70:**344–345.
87. **Smalley, M. E.** 1976. *Plasmodium falciparum* gametocytogenesis *in vitro. Nature* (London) **264:**271–272.
88. **Smalley, M. E., S. Abdalla, and J. Brown.** 1980. The distribution of *Plasmodium falciparum* in the peripheral blood and bone marrow of Gambian children. *Trans. R. Soc. Trop. Med. Hyg.* **75:**103–105.
89. **Smalley, M. E., and R. E. Sinden.** 1977. Plasmodium falciparum gametocytes: their longevity and infectivity. *Parasitology* **74:**1–8.
90. **Stura, E. A., A. S. Kang, R. Stefanko, D. C. Kaslow, and A. C. Satterthwait.** 1994. Crystallization, sequence and preliminary crystallographic data from transmission-blocking Fab 4B7 with cyclic peptides from the Pfs25 protein of *Plasmodium falciparum* malaria. *Acta Cryst. D* **D50:**535–542.
91. **Stura, E. A., A. C. Satterthwait, J. Clavo, R. Stefanko, J. P. Langeveld, and D. C. Kaslow.** 1994. Crystallization of an intact monoclonal antibody (4B7) against *Plasmodium falciparum* malaria parasites with peptides from the Pfs25 protein. *Acta Cryst. D* **D50:**556–562.
92. **Targett, G. A., P. G. Harte, S. Eida, N. C. Rogers, and C. S. Ong.** 1990. Plasmodium falciparum sexual stage antigens: immunogenicity and cell-mediated responses. *Immunol. Lett.* **25:**77–81.
93. **Targett, G. A. T., P. G. Harte, S. Eida, N. C. Rogers, and C. S. L. Ong.** 1990. *Plasmodium falciparum* sexual stage antigens: immunogenicity and cell-mediated responses. *Immunol. Lett.* **25:**77–82.
94. **Thomson, J. G., and A. Robertson.** 1935. The structure and development of *Plasmodium falciparum* gametocytes in the internal organs and peripheral circulation. *Trans. R. Soc. Trop. Med. Hyg.* **29:**31–40.
95. **Tirawanchai, N., L. A. Winger, J. Nicholas, and R. E. Sinden.** 1991. Analysis of immunity induced by the affinity-purified 21-kilodalton zygote-ookinete surface antigen of *Plasmodium berghei. Infect. Immun.* **59:**36–44.
96. **Van Amerongen, A., P. J. Beckers, H. H. Plasman, W. M. M. Schaaper, R. W. Sauerwein, J. H. E. T. Meuwissen, and R. H. Meloen.** 1992. Peptides reactive with a transmission-blocking antibody against *Plasmodium falciparum* Pfs25: 2000-fold affinity increase by PEPSCAN-based amino acid substitutions. *Pept. Res.* **5:**269–274.
97. **Van Amerongen, A., R. W. Sauerwein, P. J. Beckers, R. H. Meloen, and J. H. Meuwissen.** 1989. Identification of a peptide sequence of the 25 kD surface protein of *Plasmodium falciparum* recognized by transmission-blocking monoclonal antibodies: implications for synthetic vaccine development. *Parasite Immunol.* **11:**425–428.
98. **Vermeulen, A. N., T. Ponnudurai, P. J. A. Beckers, J. Verhave, M. A. Smits, and J. H. E. Meuwissen.** 1985. Sequential expression of antigens on sexual stage of *Plasmodium falciparum* accessible to transmission-blocking antibodies in the mosquito. *J. Exp. Med.* **162:**1460–1476.

99. **Vermeulen, A. N., J. van Deursen, R. H. Brakenhoff, T. H. Lensen, T. Ponnudurai, and J. H. Meuwissen.** 1986. Characterization of Plasmodium falciparum sexual stage antigens and their biosynthesis in synchronised gametocyte cultures. *Mol. Biochem. Parasitol.* **20:**155–163.

100. **Warburg, A., and I. Schneider.** 1993. In vitro culture of the mosquito stages of Plasmodium falciparum. *Exp. Parasitol.* **76:**121–126.

101. **Warburg, A., M. Touray, A. U. Krettli, and L. H. Miller.** 1992. Plasmodium gallinaceum: antibodies to circumsporozoite protein prevent sporozoites from invading the salivary glands of Aedes aegypti. *Exp. Parasitol.* **75:**303–307.

102. **Wassarman, P. M.** 1987. The biology and chemistry of fertilization. *Science* **255:**553–560.

103. **Williamson, K. C., M. D. Criscio, and D. C. Kaslow.** 1993. Cloning and expression of the gene for Plasmodium falciparum transmission-blocking target antigen, Pfs230. *Mol. Biochem. Parasitol.* **58:**355–358.

104. **Williamson, K. C., and D. C. Kaslow.** 1993. Strain polymorphism of *Plasmodium falciparum* transmission-blocking target antigen Pfs230. *Mol. Biochem. Parasitol.* **62:**125–128.

104a. **Williamson, K. C., and D. C. Kaslow.** Unpublished data.

104b. **Williamson, K. C., D. B. Keister, O. Muratova, and D. C. Kaslow.** Recombinant Pfs230, a *Plasmodium falciparum* gametocyte protein, induces antisera that reduce infectivity of *Plasmodium falciparum* in mosquitos. *Mol. Biochem. Parasitol.*, in press.

105. **Wizel, B., and N. Kumar.** 1991. Identification of a continuous and cross-reacting epitope for Plasmodium falciparum transmission-blocking immunity. *Proc. Natl. Acad. Sci. USA* **88:**9533–9537.

106. **Ya-Ping, S., M. P. Alpers, M. M. Povoa, and A. A. Lal.** 1992. Single amino acid variation in the ookinete vaccine antigen from field isolates of *Plasmodium falciparum*. *Mol. Biochem. Parasitol.* **50:**179–180.

Malaria Vaccine Development: A Multi-Immune Response Approach
Edited by Stephen L. Hoffman

Chapter 9

Development and Field-Testing of the Synthetic SPf66 Malaria Vaccine

R. Amador, J. J. Aponte, and M. E. Patarroyo

INTRODUCTION

According to recent statistics released by the World Health Organization (WHO), nearly 2 billion people inhabit malaria-endemic areas throughout the Third World, and the disease affects 300 million people. As a direct consequence of infection, 1.2 million children, most of them younger than 5 years of age, die each year (34).

The scarce economic resources and lack of road and communication infrastructure in zones of endemicity mean that traditional methods of control are not being properly implemented. To make matters worse, even if traditional methods of control were being adequately implemented, they are becoming less effective as parasite- and insect-resistant malaria strains arise in many parts of the world.

A heat-stable, inexpensive vaccine is becoming an essential component of future control and eradication programs, and the responsibility for the development of such vaccines lies increasingly with the affected societies.

A vaccine has been promoted as an essential component of an effective control and eradication measure, especially in Third World countries. The first attempts at producing a vaccine against the disease were begun in the 1940s, and for the next 40 years, the main focus of this research was the study of candidate antigens from the sporozoite stages. Only with the advent of in vitro culture techniques in the mid-1970s and early 1980s was the study of antigens from asexual blood stages begun, and only recently did studies with antigens from the hepatic or sexual stages begin (26).

The enormous cost in lives and in days of labor lost added to the cost of treatment of the patient makes malaria a considerable economic burden. The estimated annual direct and indirect cost of malaria in Africa was $800 million in 1987. This figure is expected to rise to more than $1.8 billion by 1995 (34). Parasite and insect vector control measures with insecticides and chloroquines were at first thought to be an effective remedy but were soon discovered to be enhancing the problem rather than alleviating it.

With these considerations in mind, our group began research on the immunology of malaria nearly a decade ago, and this research has led to the development of an inexpensive synthetic vaccine that is currently undergoing major field trials in South America and has involved many other Latin American research groups. The SPf66 vaccine, as it has been named, is the first generation of multiantigenic, multistage, synthetic peptide vaccines and the first malaria vaccine to be safely inoculated in large groups of people.

The road travelled so far has not been easy, and in this short review, we take the opportunity to recount some of our experiences so that perhaps they will serve to encourage others to advance further on this same path.

SPf66 MALARIA VACCINE DEVELOPMENT

A malaria infection starts with the inoculation of sporozoites by infected mosquitoes. The sporozoites infect the hepatocytes and become merozoites, which then infect erythrocytes, leading to a huge amplification of the parasite population and the production of clinical signs and symptoms. The parasite then differentiates to form gametocytes, the stage capable of infecting mosquitoes. *Plasmodium falciparum* is essentially an intracellular parasite; the extracellular period is too short to ensure adequate antigen presentation to the immune system. Possible points of attack by the immune system are various, and different research groups have focused their efforts on different stages of infection to produce vaccine candidates.

The first attempts at inducing an immunologic state of protection against malaria by using attenuated sporozoites (8) gave a clear indication that induced immunity in susceptible hosts is a potential method of controlling the disease. Several subunit vaccines based on fragments of the circumsporozoite protein, i.e., based on a repeat sequence of a 4-amino-acid sequence (Asn-Ala-Asn-Pro; NANP) have been developed (12). Human experimental trials have been carried out for two of these vaccines; in both cases, infection was slightly delayed in most of the volunteers, and although high antibody titers were elicited, the vaccines offered limited protection.

As we know, the 90% of the 300 to 500 million clinical cases of malaria per year worldwide occur in Africa. Therefore, the priority of research must be to prevent the deaths due to malaria of 1 million African children under 5 years of age every year. A more desirable vaccine in this scenario could be an asexual erythrocytic-stage vaccine capable of reducing severe and complicated malaria and malaria-related mortality in this risk group (34).

As knowledge of the parasite's biology has become more and more sophisticated, characterization of the whole organism by means of characterizing its component antigens has been undertaken. The biological role of each antigen is slowly being unravelled, a process that has been accelerated by the application of new technologies. The best candidates for malaria vaccine design are proteins that are required for parasite survival from multiple or alternative steps in the invasion process and that have low mutation rates and conserved epitopes. Additionally, these proteins have to be targets of a protective immune response. Recombinant DNA cloning and sequencing techniques combined with serological studies have made possible precise mapping of epitopes to the level of amino acid residues.

With this new knowledge in hand, laboratories the world over have made novel attempts at developing subunit vaccines. In order to select epitopes from proteins with demonstrated invasion properties to be included in a malaria synthetic vaccine, our approach has been to identify the epitopes with conserved structures that bind the erythrocyte (13). This has been achieved by analyzing the binding kinetics of radiolabeled synthetic peptides containing sequences derived from these proteins and their analogs. Amino acids that serve as contact residues for a dominant number of antibodies within a polyclonal serum cannot be replaced without altering recognition.

Because of the suspicion that parasites grown in vitro must shift their antigenic composition in order to adapt to culture conditions, our group decided to use fully infective parasites re-collected from malaria patients living in zones of endemicity. *P. falciparum* terminal cultures were initiated from infected individuals' blood samples, and the harvested parasites were subsequently lysed with sodium dodecyl sulfate (SDS). The lysate was run on 10% SDS-polyacrylamide gel electrophoresis (PAGE) gels, and after dialysis in RPMI medium, 200 to 400 μg each of the proteins with molecular masses of 155, 115, 105, 90, 83, 60, 55, 50, 40, 35, 30, and 23 kDa was isolated in a purified form. The purity of the isolated molecules was tested by SDS-PAGE and Western blots (immunoblots) with hyperimmune sera and by high-pressure liquid chromatography (21).

The N-terminal amino acid sequence of each of the 12 isolated proteins was analyzed, and synthetic peptides were constructed from the sequences by following the solid-phase peptide synthesis methodology described by Merrifield (14). The immunogenicities of peptides were tested in the *Aotus* monkey model. Several peptides were identified on the basis of recognition by monkey sera and cross-reactivity with native sequences. Four antigens that elicited immune and protective or delayed response in studies with Colombian *Aotus* monkeys were selected.

A vaccine that targets a single epitope of the parasite is inadequate as an inducer of complete protection from the disease. A multicomponent malaria vaccine that includes multiple antigens from various stages is ideal. The SPf66 vaccine developed by our group is a chimeric molecule consisting of four synthetic epitopes (21). Three are erythrocytic-stage epitopes that include MSP-1 and were selected from amino acid sequences in isolated proteins, and one was derived from the repeat domain of the circumsporozoite protein, the PNANP sequence (21). To establish a final Drug Master file and the minimal guidelines for peptide biologics, which do not exist at present, promoted by WHO-Tropical Disease Research-Pan American Health Organization and in collaboration with the National Institute of Biological Standards and Controls, the Instituto de Immunología is involved in the fine characterization of the monomeric and bulk SPf66 molecule.

EXPERIMENTAL AND FIELD-TESTING OF THE SPf66 VACCINE

Phase 0. *Aotus* Monkey Model

Each one of the 12 isolated and purified antigens was inoculated three times in two or three *Aotus* monkeys for each antigen. The monkeys were challenged by intravenous inoculation of 5×10^6 *P. falciparum* (FVO strain)-infected erythrocytes taken from naive donors. The results of this experiment showed that three of the antigens (the 155-, 55-, and 35-kDa proteins) presented a delay in the appearance of parasitemias, and one (the 83-kDa protein) induced sterilizing immunity. The synthesized peptides were coupled to bovine serum albumin and inoculated five times with Freund's complete adjuvant in another group of monkeys, which were then challenged with live parasite. Of the 40 different synthesized peptides, 8 induced a partial immune response (22).

The next step in this series of experiments was to test different mixtures of the peptides, using two or three different peptides in each mixture. Of the six monkeys immunized with a mixture of three peptides (83.1, 55.1, and 35.1), three developed low parasitemia levels that peaked between days 10 and 15, significantly later than in the control groups, and went on to recover spontaneously. The remaining three animals never developed parasitemias. With this evidence in hand, researchers developed a hybrid synthetic polymeric molecule and tested it in *Aotus* monkeys. This molecule, called SPf66, elicited protective immunity in four of eight animals (25). This experiment has been repeated several times with larger groups of animals and has shown similar results (15).

The *Aotus* monkey model system remains the best available for simulating human *P. falciparum* infection (10). These monkeys, unlike other animal models, are highly susceptible to malaria, and in them, the disease is lethal unless complete protection is elicited by the vaccine preparation. However, because of the lack of availability of this primate, many researchers have adopted other animal models and extrapolated their results to research in human subjects.

Genetic restrictions are involved in the immune response to malaria, and the same occurs in an outbred *Aotus* population. Consequently, the importance of testing a future vaccine in a larger number of animals cannot be understated. Because these monkeys are captured in family groups, it is important to know the origin of each animal so that members of the same family will be placed in different research groups.

Aotus monkeys from the Colombian Amazon were chosen for toxicity and safety assays in animals. Pyrogenicity and sterility tests for each batch of vaccine were performed in rabbits. Immediately before and 8 days after each immunization, the animals were bled for analysis of renal and hepatic function, blood count, and blood chemistry. No significant differences were found between groups immunized with the SPf66 vaccine and control groups, into which only the adjuvant was inoculated. At 15 days after the application of each dose, one animal from each group was sacrificed for anatomopathologic studies. Different histological stains were used for samples of brain, liver, kidney, lung, and gonads. No histological alteration was found in these samples. After these studies, the monkeys were held in captivity in a different region of the country, where they were able to procreate freely. The progeny of these monkeys were anatomically and behaviorally normal, without any evidence of teratogenicity caused by the vaccine.

Phase I Trials

The objectives of phase I trials are to assess human local and systemic tolerance as well as the immune response to antigens that have been shown to be safe during the preclinical studies.

To select the individual volunteers who will participate in this phase of the study, four important considerations are taken into account. First, the individuals must be readily accessible for the initial and booster shots and all subsequent monitoring. Second, they must be healthy young males. Third, the individuals must live in non-malaria-endemic areas and must not have had previous exposure to the parasite. Fourth and most important, volunteers have to be free to participate in the studies. We decided to work with soldiers from the Colombian military forces. Initially, 109 soldiers volunteered for the study. No incentives such as money or promotions were offered. The volunteers were examined from the physical and mental health standpoints (examinations included urinalysis, blood chemistry, blood cell count, and antimerozoite antibody titers [by indirect immunofluorescence assay]), and 30 individuals were chosen for further testing. Of these 30, 13 were finally selected for testing of the vaccine. The volunteers were divided into five groups: (i) two volunteers received three 2-mg doses of SPf66 on days 0, 60, and 80; (ii) three volunteers received 2-mg doses of the same protein on days 0 and 60; (iii) four volunteers received three 2-mg doses of SPf105 on days 0, 20, and 45; (iv) three controls received saline solution on days 0, 20, and 45; and (v) one volunteer served as a naive receptor for passing on the *P. falciparum* strain. The vaccines were challenged by intravenous inoculation of 10^6 infected erythrocytes (20).

The results showed that the vaccine was well tolerated, with only minor local reactions in a few cases. None of the volunteers presented with fever or showed significant changes in blood cell count, blood chemistry, or urinalysis on day −1 or on days 1, 3, and 5 after each immunization. Autoimmunity tests (rheumatoid factor, antinuclear antibodies, Coombs test, and antimyocardial fiber antibodies) were consistently negative.

To study the dynamics of the cellular and humoral immune responses, blood samples were taken the day before each immunization, 15 days later, and 1 day prior to challenge. Antibody titers were determined by enzyme-linked immunosorbent assay (ELISA), using the synthetic vaccines as antigens. A strong humoral response against the vaccine was found.

Proliferation assays of peripheral blood mononuclear cells using either the SPf66 molecule or a sonicate of purified schizonts showed stimulation indices below 3 before the first vaccination. After each vaccination and prior to challenge, the stimulation indices varied from 0.61 to 35.1.

Phase II

Phase IIA

During phase IIA, the efficacy of protective immunity conferred by the vaccine is assessed through experimental challenge of nonimmune individuals originating from non-malaria-endemic areas. The experiments for this phase of the study must be performed under strict medical supervision and must include access to an intensive care unit. Our group chose the Central Military Hospital in Bogotá, an institution accepted by the military forces and the Ethical Committee of the Colombian Ministry of Public Health, for performing this assay.

The volunteer soldiers were intravenously inoculated with 10^6 wild-strain *P. falciparum*-infected erythrocytes, each of which can produce 16 merozoites that will be released into the bloodstream. Twice a day, parasitemia levels were assayed for each vaccine recipient, and those with levels above 0.5% were treated. During intravenous challenge, it is more important to know the clinical behavior of the infecting strain than to use a quantitative parasitemia cutoff point in order to ascertain when to treat patients. It permit us to monitor the volunteers for a longer time to observe the delay in the disease. All the individuals were free to retire from the study whenever they chose to do so, as in fact one individual participating in trials with the SPf66 vaccine did.

Of the five individuals immunized with the SPf66 vaccine, three were protected, one retired in midstudy, and one required drug treatment, showing a trend to protection (20).

Phase IIB

In keeping with WHO guidelines (5), the efficacy of the vaccine was then assessed through natural exposure of the volunteers to the parasite in areas of endemicity. For this study, 399 male volunteers aged 18 to 21 years and belonging to the Colombian military forces were chosen (4). Of these, 185 were vaccinated with the SPf66 hybrid polymer, and the other 214 served as controls. The soldiers, who had had little previous exposure to the parasite, were on patrolling missions within a wide area of endemicity along Colombia's southwest coast. This area presents an annual parasitological index of 12% ± 2%, with a *P. falciparum* incidence of 80 to 90%.

Vaccinated volunteers were carefully monitored, and any that developed febrile symptoms were evacuated to the Tumaco Regional Military Health Centre (RMHC) for clinical and laboratory tests. The blood smears of these individuals were blindly and independently read by the staff members of Tumaco RMHC and the Servicio de Erradicación de la Malaria (SEM) in Tumaco. Both of these results were reread by staff at SEM's Central Laboratory in Bogotá, who either confirmed or discarded the original diagnosis.

To evaluate the safety of the vaccine, a thorough clinical examination was performed on all the volunteers during the first hour after each immunization and then again 24 and 48 h after the immunizations. Blood chemistry and autoimmune tests were performed for each vaccine 10 days after each immunization; no statistically significant differences were found when the vaccinated and control groups were compared. The protective efficacy of the vaccine was calculated as 82.3% against *P. falciparum* and 60.6% against *Plasmodium vivax*, with wide confidence intervals (CI), because of the small sample.

The immunoglobulin G (IgG) antibody production kinetics against the SPf66 vaccine were established in a parallel study (27). These antibodies increased their levels of 15 to 30 days after the second and third immunization, clearly showing a boosting effect after the third dose. The antibodies also recognized native parasite proteins, as assessed by Western blot. However, the response was not the same in all individuals. Following the second immunization, a characteristic pattern of the IgG response against the vaccine was observed by Falcon assay screening test (FAST)-ELISA; the pattern could be used to place the vaccines in three different groups. The first group, the high responders, presented antibody titers between 1:1,600 and 1:25,600. A second group, the intermediate responders showed antibody titers between 1:200 and 1:800, and a third group, the low responders, contained individuals whose antibody titers never increased above 1:100.

A further study showed a bimodal distribution of these antibodies in the vaccinated population, suggesting genetic control of the immune response to this protein (23). For this reason, HLA A, B, DR, and DQ typing was performed on 105 vaccine and 47 control individuals. The distribution of DR and DQ alleles in the high responders was similar to that in the control group. However, the population of HLA DR4 individuals within the low-responder group (68%) was significantly increased compared to that in the control group (36.2%) and the high-responder group

(25%). Using oligotyping methods after amplification of the DR4 B-1 exon, we proceeded to subtype 20 DR4 volunteers classified as high, intermediate, or low responders. No correlation was found between a specific DR4 subtype and the humoral immune response of the vaccinees (16).

Consequently, our group decided to investigate whether the genetic restriction to the vaccine was present at the T-cell-receptor level. Twenty T-cell clones were analyzed; 12 of them corresponded to volunteers classified as high responders, and the remaining 8 came from low responders. Specific and selective preferences were shown for the variable beta (VB) arrangements of the T-cell responder in each group. The VB has been associated with the antigen recognition process, and several reports have shown preferential associations of certain VB genes in response to specific antigens. Our results show that the T-cell clones of high responders are associated preferentially with the VB 8 rearrangement but also with the VB 2, 5, 6, and 7 genes. On the other hand, T-cell clones from low responders involved none of the VB genes associated with the high responders but expressed mainly the VB 10 genes and, at a lower frequency, the VB 3 and 11 genes (17). The molecular mechanisms involved in this genetic restriction are currently being studied by our group.

The next series of phase IIB field trials carried out by our group was once again performed on young, healthy, male volunteer soldiers originating from non-malaria-endemic areas who were brought to patrol in malaria-endemic areas after immunization. The aim of these trials was to define the number of doses required, the best interval between application of doses, the best vaccine concentration, and the best adjuvant (24). The results of the Tumaco A, B, C, and D trials show that under our conditions, the best immunization schedule for adults consists of three doses of the vaccine (on days 0, 30, and 180), with each dose containing 2 mg of the synthetic SPf66 polymer adsorbed onto 1 mg of alum hydroxide.

A pilot study was performed in children aged 1 to 14 years living in the malaria-endemic area of Tumaco, Colombia (19). Of the 292 children receiving the three doses of the vaccine, 94.6% showed no reaction. The rest, 12 children, presented with slight local reactions that did not persist in subsequent checkups. No delayed clinical or immune markers of autoimmunity were detected in the 62 children whom it was possible to evaluate 1 and 5 years after the first dose. Sera obtained from these children 20 days after the third immunization showed that 93.7% of them developed high titers, ranging from 1:100 to 1:12,800, to the vaccine. In both studies, reactivity to native proteins, as assessed by Western blot, was high, especially to the 135-, 115-, and 83-kDa proteins. The results of these studies establish beyond a reasonable doubt the safety and immunogenicity of the SPf66 vaccine.

Phase III

The previous field trials had been carried out with a small number of individuals originating in non-malaria-endemic areas; the next field trial was carried out under more representative conditions, with individuals living in areas of endemicity. For this trial, the town of Majadas, state of Bolivar, in eastern Venezuela was chosen as the site for study by the research group from the Universidad Central de Venezuela and the Ministry of Health. *P. falciparum* is endemic in this area, with an annual parasitological index of 10% and a parasite formula of 25%. Following ethical considerations, which weighed very heavily with us, we decided not to exclude any individual over 11 years of age (except for pregnant women) who manifested a desire to receive the vaccine. Of the townspeople, 976 (27.6%) were immunized three times with SPf66, and 938 were left completely unvaccinated. Both groups were homogeneous for all characteristics considered. The aims of this study were to evaluate any change in the epidemiological characteristics of the vaccinated population compared with those of the nonvaccinated population and to attribute the difference to the effects of vaccination. To compare the incidence of malaria after vaccination in the vaccinated and the nonvaccinated groups, after adjusting for a dissimilar malaria risk at the baseline, ratios were calculated for the incidence rate during the 12 months after the third vaccine dose in relation to that during an equivalent calendar period just before vaccination. The after/before vaccination incidence ratio of each group was used to derive vaccine efficacy. The protective efficacy against *P. falciparum* for those receiving the complete series of three vaccination doses (18) was calculated to be 55.1%. The vaccine protective effect against *P. vivax* was estimated as 41%, evidence that needs to be corroborated in future studies in areas where *P. vivax* is endemic.

Parallel to this study in Venezuela, our group in Colombia began a large trial with universal vaccination, randomized by localities and not by individuals, although pregnant women were excluded. For this trial, 9,957 persons, all older than 1 year and inhabiting the Tumaco and Francisco Pizarro municipalities on the southern Pacific coast of Colombia, were administered three doses of vaccine under the previously defined optimal schedule (3). Of the vaccinees, 95.7% showed no adverse reactions to the vaccine. In the remaining 4.3% of the individuals, local induration and erythema were the most frequent reactions. Among a randomly selected group of vaccinees, anti-SPf66 antibody titers were measured by ELISA, which showed that 55% of this group presented titers above 1:1,600. It is interesting that in this study, women of child-bearing age showed the largest number of adverse reactions. We believe the interplay between the endocrine and immune systems in women caused the undesired side effects. This issue needs to be addressed more carefully in future trials.

In a meeting between our research group and a committee from the WHO in Bogotá in mid-1990, the necessity of initiating double-blind placebo trials under different malaria epidemiological conditions was emphasized (9). These trials were initiated independently in cooperation with other internationally recognized research groups in Colombia (31), Venezuela, Ecuador (28), Brazil, Tanzania (29), The Gambia (11), and Thailand; they involved 500 to 3,000 individuals per trial. Variable efficacy is observed for any vaccine applied to different populations or even to different subgroups within a population because of inherent differences in the immune statuses of the individuals, the pattern of transmission and intensity of the disease, the different parasite strains, etc. For this reason, replicating trials under different epidemiological conditions is essential.

Latin American Experience

We conducted phase III randomized double-blind, placebo-controlled trials in several Latin American locations that are characterized by low-transmission conditions (unstable malaria), few asymptomatic cases of parasite infection, and different ethnic groups.

In La Tola, Colombia, a trial involving 1,548 volunteers over 1 year of age was conducted by the Instituto de Inmunología (31). Of the volunteers, 738 received three doses of vaccine, and 810 received three doses of placebo. At 30 days after the third dose, 33% of the vaccinees and only 6.6% of the placebo-treated group had antibody titers against SPf66 of ≥1:200. During the year after the third dose, there were 168 cases of *P. falciparum* malaria in the vaccinee group and 297 cases in the placebo group. The protective efficacy against the first *P. falciparum* malaria episode was estimated as 33.6% (5% CI [18.8 to 45.7%]). In this trial, the vaccine efficacy was different in different age groups, being highest in children aged 1 to 4 years (77%) and adults older than 45 years (67%). The protective efficacy against the second episode of malaria was 50.5% (95% CI [12.9 to 71.9%]).

In La Te, Ecuador, a trial was performed by the research group of Fernando Sempértegui and the Ministry of Health of the Republic of Ecuador (28). A total of 537 subjects were randomized to receive either SPf66 malaria vaccine or placebo in three doses. Case detection surveillance was implemented by parasitological cross-sectional surveys every 2 months and by monthly household visits to each participant during 1 year of follow-up. At 30 days after the third dose, the prevalence of anti-SPf66 antibodies was 57% in the vaccinees and 8.8% in the placebo group.

Immune response was independent of age. Vaccine efficacy was calculated on the basis of a person's time of exposure. The protective effect when any malaria episode was considered was 66.8% (95% CI [−2.7 to 89.3%]) and that when only one episode per individual was considered was 60.2% (95% CI [−26 to 87.5%]).

Along the Rosario River (30) (on the Pacific coast of Colombia), 1,257 randomized individuals received the three vaccine doses. Both vaccinated and control groups were homogeneously distributed by age and sex. Active case detection by monthly visits allowed us to update the denominator of the incidence of population density required for the person–time-at-risk analysis. Passive case detection was performed by community volunteers and field workers. Only individuals with a previous confirmed diagnosis were treated. Cross-sectional urine tests were done for antimalarial agents. Indirect immunofluorescence titers against the whole *P. falciparum* parasite in this area showed that 31% of the population had IgG antibody titers of >1:80 against the whole parasite and 5% had IgG titers of >1:128 against SPf66. After the third dose, 71% of the vaccinees had titers of >1:128 against SPf66. In 22 months of follow-up, 131 *P. falciparum* malaria episodes were diagnosed (51 in the vaccinated group; 80 in the placebo group), yielding an attack rate of 5.26 cases per 100 person-years of follow-up in the vaccine group and 8.34/100 person-years in the placebo group. The estimated vaccine protective efficacy was 36.9% (95% CI [10.3 to 55%]; $P = 0.01$). In this area of low endemicity, the natural immune response expressed by indirect immunofluorescence titers increased with age and the parasitemia densities tended to decrease. This malaria profile suggests that the individuals in low-endemicity areas are able to create a natural immunity that is reflected in parasite densities but not in the clinical disease.

The results of all of these studies show that the chemically synthesized SPf66 malaria vaccine is safe, immunogenic, and protective. It has a protective efficacy against *P. falciparum* that ranges between 33.6 and 60.2%. We must emphasize, however, that in Latin American countries, the implementation of a well-done epidemiological follow-up by itself constitutes a secondary intervention that could diminish the incidence of malaria to a degree that could complicate the evaluation of the vaccine efficacy.

Several aspects of the way these trials have been designed that result from our previous experience in field trials deserve to be mentioned.

Determination of the study population

Initially, several candidate sites in which to conduct the studies were selected according to criteria such as the behavior of the disease during the past few years, malariometric indices (annual parasitological indices for *P. falciparum* and *P. vivax*, which must be high enough to produce enough cases to allow detection of a significant difference between vaccinated and control groups), infrastructure of the locality, access, willingness of the community to participate in a trial, etc. For many remote areas in South American countries, malaria cases are known to be grossly underregistered.

From the leaders of the communities, we determined the importance that town dwellers gave to the disease. Colombia is a leader in the WHO-sponsored Expanded Programs of Immunization, and vaccination is thus seen by most communities in a very positive light.

Once a site was selected, social workers, in close contact with the leaders of the community, achieved the participation of the entire community in the trial. The importance of these field workers and their interaction with the community cannot be overstated. These individuals had gone through extensive training in both malaria and the socioeconomic, cultural, and anthropological characteristics of the population to be studied, and in many cases, they were inhabitants of these communities. Trial designs for the different localities were sometimes modified according to the recommendations of these individuals.

A census of the total population was an important subsequent step, as it was the basis of the randomization procedure. The locality was dissected according to demographic (racial) characteristics, occupation, malariometric and medical histories, physical examination, etc. Individuals with no history of allergy or debilitating disease and with a normal physical examination were included in the study population. Pregnant women were excluded from the study. To establish whether a woman was pregnant, we performed blood or urine tests on every female aged 15 to 45 years. The tests were performed with blood samples and, if an ambiguous result was obtained, with urine samples. Family heads and every person over 18 years old were asked to sign a consent form to show that they accepted the facts that this was a double-blind placebo trial and that they could retire from the study whenever they chose to do so.

Individuals were assigned a code that greatly simplified the logistics of the study. The code consisted of a single letter corresponding to the locality followed by a two-digit number corresponding to the house number and a last letter corresponding to the individual's place within the family. A house number is assigned to each house by the malaria eradication-control institutions of each country and is widely used by most communities. This code simplified the task of tracking down single individuals when necessary.

Randomization

On the basis of the census of the universal population, each individual was classified in two different groupings: sex and age (1 to 4, 5 to 14, 15 to 45, and 45 years and older); in some studies, occupation was also used. The groups resulting from all possible combinations of these categories were randomized. Because detailed records of each individual were kept, the impact of some biases introduced into the study were minimized. Two different methods of randomizing the population were used. In one method, a computer at the headquarters of the research group randomized the coded individuals, and in the other method, the randomization procedure was done at the site of vaccination.

Prior to the beginning of each study, the Ethics Committee was designated by the health ministry of the respective country to monitor the field trial and assign letter codes to the vaccine and placebo vials. These vials were then dispatched to the vaccination post.

Vaccination

Early on in the design of our first trials, our group decided to use tetanus toxoid instead of plain aluminum hydroxide as the placebo, although we were well aware of the possible biases introduced into the study by doing so. The main factor that led us to this decision was an ethical one. We believed it was important for the individuals vaccinated with a placebo to receive at least minimal benefits for participating in this study in addition to the close follow-up. Besides, the physical aspect of both tetanus toxoid and the SPf66 formulations and the reactions they cause are very similar. To minimize the possibility that tetanus toxoid, an unspecific T-cell activator, would interfere with the results of the trial, it was applied only in the first dose of the vaccine.

For logistical purposes, the vaccination post was divided into the following four areas. (i) Registration site. Randomization is controlled at this point. The personnel at this site assign sequential letter codes to the members of each randomized groups in order of arrival (in the case of field randomization) or dictate the assigned code to the individual (in the case of randomization at headquarters). They check that the prerequisites for vaccination have been completely fulfilled (signature on a consent form, complete medical and census data, etc.). The individual then passes on to the vaccination site. (ii) Vaccination site. This site is unique for each vaccination post. The responsibility of vaccination with the correct letter-coded vial (as assigned by the registrar) lies with two persons in order to minimize the possibility of human error. It is important to select responsible people for these key positions. After immunization, the vaccinee passes on to the third site within the vaccination post. (iii) Observation site. The vaccinee remains at this site for 30 to 45 min. This site offers a good opportunity to participate with the community in health education and is a place where questions regarding general health matters are answered. The site is staffed by two doctors. (iv) Intensive care unit. Any individual who suffers any complications (local or systemic) during or after the 30-min observation period is taken immediately to this site. The basic equipment in this unit consists of oxygen tanks, respiratory support, intravenous liquids, drugs, and an ambulance where available. Critical-care doctors staff this unit, and the general practitioners that assist at the site had received proper training.

At 24 to 48 h after each vaccination, a team of doctors performed a medical check-up on the vaccinees to ensure that none were suffering delayed reactions. Apart from this, an independent Immediate Reaction Group (Grupo de Reacción Inmediata) composed of doctors belonging to governmental health institutions listened to and assessed complaints of any possible complication that might be attributed to the vaccine.

Epidemiological monitoring

A field station staffed by doctors, nurses, and personnel from the malaria control-eradication governmental institutions from that country was set up in each locality. The facilities installed at each station included computers for the collection and primary analysis of statistical data, microscopes, vehicles, boats, and communication systems. Both active and passive epidemiological monitoring was done for most trials. Active monitoring is an intensive, house-to-house search by a group of doctors for any malaria cases. Passive monitoring is performed by persons in the community, usually the leaders, who are trained in the reading of thick smears. This latter type of monitoring turned out to be rather important, for it increases the coverage of the monitoring procedure. However, it also reduces the power to detect the efficacy of the vaccine: early treatment decreases the possibility of contracting malaria by reducing the number of gametocytes. Diagnosis performed by these individuals was double-checked by the field microscopist, for whom we always chose a person with great experience, and later by the central laboratories of the malaria eradication-control institutions of each country.

African Experience

To test the vaccine under conditions of intense malaria transmission (stable malaria), a study in Tanzania was undertaken. The African study showed results similar to those described previously (1). The decrease in the frequency of episodes detected in children vaccinated with SPf66 in the Tanzanian trial could result from a decrease in the intensity of the symptoms rather than the number of attacks, thus preventing or reducing malaria-related morbidity and mortality.

Defining clinical malaria as an end point of the study is easier in Latin America than in Africa, because subclinical parasitemias are not easily observed. Infections with no symptoms occur at a high frequency in sub-Saharan Africa. For that reason, in the Tanzanian study, a case of malaria was defined as a parasitemia of 20,000 or more and an axillary temperature of 37.5°C or more (6). It is possible that the vaccine was able to reduce malaria mortality in a higher proportion of the population than the 31% recorded for the clinical episodes that were detected. These episodes could include most of the potentially severe cases.

Recently, details of a randomized trial with SPf66 in The Gambia in a group of children aged 6 to 11 months were published (11). The vaccine had no significant protective effect against clinical malaria episodes, but results under trial conditions are not yet conclusive. An analysis of the strong, active case detection and a long follow-up period during the next rainy season will be necessary to increase our knowledge of this important risk group. In the near future, testing of the vaccine against new end points such as severe malaria in infants, the high-risk group, will be carried out. Trials with differing epidemiological conditions such as seasonal malaria or in regions such as eastern Asia, where drug resistance is very high, will also be important, as will be evaluating how SPf66 could be delivered using the Extended Program of Immunization infrastructure.

CONCLUSIONS

Although knowledge of the malaria parasite's biology is still incomplete, SPf66 research has provided insight into some of the mechanisms that the parasite uses to evade host immunity. There now exists a basis for the adoption of an "antigenic cocktail" approach to obtaining a synthetic or recombinant subunit vaccine. The synthetic Colombian malaria vaccine, SPf66, is an example of this cocktail methodology. The basic research and the clinical and field stages of vaccine development are slow. Several human trials to test safety, immunogenicity, protection, and the formulation schedule of SPf66 were done with increasing number of volunteers (4, 19, 20, 24, 27). A vaccine requires a minimum of 10 years in the developmental process before being implemented in the field.

During this time, progress in malaria vaccine development has been dramatic. SPf66 vaccine became the only vaccine identified from such trials, and its development has involved many disciplines. Different methods for epidemiological evaluation have been designed and updated in order to be useful under clinical and field conditions (5, 10). The use of randomized double-blind and placebo-controlled trials plays a central role in the evaluation of the protective efficacy and side effects of new malaria vaccines (2, 7, 33).

We emphasize that in order to get to the human trial stage, one has to screen not only a range of vaccines but also a range of adjuvants and vaccine-adjuvant combinations. This need demonstrates the importance of bench research and the use of animal models before one proceeds to extremely expensive trials in humans. Before engaging in costly phase III trials, it is necessary to conduct phase I and II trials for evaluation of the safety, immunogenicity, and trends in protective efficacy of a designated group under controlled conditions.

The most critical step in the development of new products involves the phase II trial. Safe, adequate methods for experimental challenge must be implemented. To test a vaccine with epitopes from different parasite stages, it is necessary to set challenges through mosquito bites and/or intravenous infected erythrocytes. The volunteers have to be monitored very closely, and the strain of *P. falciparum* has to be a low-virulence native strain that is able to produce moderate to high parasitemias with a low clinical profile.

The knowledge required in a given situation to move from controlled trials to routine evaluation of a vaccine will certainly be different from one geographical area to another. It will be different from country to country, considering the different effects, areas, strains, vectors, populations or risk groups, magnitude of the problem, and public health impact. Together, all of these facts will lead to an independent but coordinated approach to assessing the potential of the vaccine in different regions of the world. In order to try and cover the whole clinical-malaria spectrum under unstable and stable malaria epidemiological conditions and with different ethnic groups, we performed studies of SPf66 in Latin America and Africa. Still more knowledge about the effect of the vaccine in different age groups, exposure levels, and immune status interaction is required in order to define an adequate scenario for potential public health implementation of the vaccine.

The importance of encouraging the necessary basic research and clinical and field trial strategies must be stressed if we expect to look forward to a new generation of vaccines (32).

REFERENCES

1. **Alonso, P. L., T. Smith, J. R. M. Armstrong Schellenberg, H. Masanja, S. Mwankusye, H. Urassa, I. Bastos de Azevedo, J. Chongela, S. Kobero, C. Menendez, N. Hurt, M.**

C. Thomas, E. Lyimo, N. A. Weiss, R. Hayes, A. Y. Kitua, M. C. Lopez, W. L. Kilama, T. Teuscher, and M. Tanner. 1994. Randomised trial of efficacy of SPf66 vaccine against *Plasmodium falciparum* malaria in children in southern Tanzania. *Lancet* **344:**1175–1181.

2. **Alonso, P. L., M. Tanner, T. Smith, R. J. Hayes, J. R. M. Schellenberg, M. C. Lopez, I. Bastos de Azevedo, C. Menendez, E. Lyimo, N. Weiss, W. L. Kilama, and T. Teuscher.** 1994. A trial of the synthetic malaria vaccine SPf66 in Tanzania: rationale and design. *Vaccine* **12:**181–186.
3. **Amador, R., A. Moreno, L. A. Murillo, O. Sierra, D. Saavedra, M. Rojas, A. L. Mora, C. L. Rocha, F. Alvarado, J. C. Falla, M. Orozco, C. Coronell, N. Ortega, A. Molano, J. F. Velásquez, M. V. Valero, L. Franco, F. Guzmán, L. M. Salazar, F. Espejo, E. Mora, R. Farfán, N. Zapata, J. Rosas, J. C. Calvo, J. Castro, T. Quinones, F. Nunez, and M. E. Patarroyo.** 1992. Safety and immunogenicity of the synthetic malaria vaccine Spf66 in a large field trial. *J. Infect. Dis.* **166:**139–144.
4. **Amador, R., A. Moreno, V. Valero, L. A. Murillo, A. Mora, M. Rojas, C. Rocha, M. Salcedo, F. Guzman, F. Espejo, F. Nunez, and M. E. Patarroyo.** 1992. The first field trials of the chemically synthesized malaria vaccine SPf66: safety, immunogenicity and protectivity. *Vaccine* **10:**179–184.
5. **Anonymous.** 1985. Principles of malaria vaccine trials: report of a joint meeting of the scientific working groups on immunology of malaria and on applied field research in malaria. *TDR/IMMAL/VAC* **85.3:**1–66.
6. **Armstrong-Schellenberg, J., T. Smith, P. Alonso, and R. Hayes.** 1994. What is clinical malaria? Finding case definitions for field research in highly endemic areas. *Parasitol. Today* **10:**439–442.
7. **Ballou, W. R.** 1995. Clinical trials of plasmodium falciparum erythrocytic stage vaccines. *Am. J. Trop. Med. Hyg.* **50:**59–65.
8. **Clyde, D. F., V. C. McCarthy, and W. E. Woodward.** 1975. Immunization of man against falciparum and vivax malaria by use of attenuated sporozoites. *Am. J. Med. Hyg.* **24:**397–401.
9. **Collins, W. E., H. McClure, E. Strobert, J. C. Skinner, B. B. Richardson, J. M. Roberts, G. G. Galland, J. Sullivan, C. L. Morris, and S. R. Adams.** 1993. Experimental infection of Anopheles gambiae s.s., Anopheles freeborni and Anopheles stephensi with Plasmodium malariae and Plasmodium brasilianum. *J. Am. Mosquito Control Assoc.* **9:**68–71.
10. **Committee for the Study on Malaria Prevention and Control: Status Review and Alternative Strategies.** 1991. Vaccines, p. 169–256. *In* S. C. Oaks, V. S. Mitchell, G. W. Pearson, and C. C. J. Carpenter (ed.), *Malaria: Obstacles and Opportunities*. National Academy Press, Washington, D.C.
11. **D'Alessandro, U., A. Leach, C. Drakeley, S. Bennett, B. Olaleye, G. Fegan, M. Jawara, P. Langerock, M. George, G. Targett, and B. Greenwood.** 1995. Efficacy trial of malaria vaccine SPf66 in Gambian infants. *Lancet* **346:**462–467.
12. **Herrington, D. A., D. F. Clyde, J. R. Davis, S. Baqar, J. R. Murphy, J. F. Cortese, R. S. Bank, E. Nardin, D. DiJohn, and R. S. Nussenzweig.** 1990. Human studies with synthetic peptide sporozoite vaccine (NANP)3-TT and immunization with irradiated sporozoites. *Bull. W.H.O.* **68**(Suppl.)**:**33–37.
13. **Marsh, K., and R. J. Howard.** 1986. Antigens induced on erythrocytes by P. falciparum: expression of diverse and conserved determinants. *Science* **231:**150–153.
14. **Merrifield, R.** 1963. Solid-phase peptide synthesis. I. Synthesis of a tetrapeptide. *J. Am. Chem. Soc.* **85:**2149–2154.
15. **Moreno, A., and M. E. Patarroyo.** 1989. Development of an asexual blood stage malaria vaccine. *Blood* **74:**537–546.
16. **Murillo, L. A., C. L. Rocha, A. L. Mora, J. Kalil, A. K. Goldemberg, and M. E. Patar-**

royo. 1992. Molecular analysis of each HLA DR4 B-1 gene in malaria vaccines. Typing and subtyping by PCR technique and oligonucleotides. *Parasite Immunol.* **13:**201–210.

17. **Murillo, L. A., F. Tenjo, O. Clavijo, M. Orozco, S. Sampaio, J. Kalil, and M. E. Patarroyo.** 1992. A specific T cell-receptor genotype preference in the immune response to a synthetic *Plasmodium falciparum* malaria vaccine. *Parasite Immunol.* **14:**87–94.
18. **Noya, G. O., Y. Gabaldon Berti, B. Alarcón de Noya, R. E. Borges, N. Zerpa, J. D. Urbáez, A. Madonna, E. Garrido, M. A. Jimenez, P. Garcia, I. Reyes, W. Prieto, C. Colmenares, R. Pabón, T. Barraez, L. G. de Caceres, N. Godoy, and R. Sifontes.** 1993. A population-based clinical trial with the SPf66 synthetic *Plasmodium falciparum* malaria vaccine in Venezuela. *J. Infect. Dis.* **170:**396–402.
19. **Patarroyo, G., L. Franco, R. Amador, L. A. Murillo, C. L. Rocha, M. Rojas, and M. E. Patarroyo.** 1992. Study of the safety and immunogenicity of the synthetic malaria SPf66 vaccine in children aged 1–14 years. *Vaccine* **10:**175–178.
20. **Patarroyo, M. E., R. Amador, P. Clavijo, A. Moreno, F. Guzman, P. Romero, R. Tascon, A. Franco, L. A. Murillo, G. Ponton, and G. Trujillo.** 1988. A synthetic vaccine protects humans against challenge with asexual blood stages of *Plasmodium falciparum* malaria. *Nature* (London) **332:**158–161.
21. **Patarroyo, M. E., P. Romero, M. L. Torres, P. Clavijo, D. Andrew, D. Lozada, L. Sanchez, P. del Portillo, C. Pinilla, A. Moreno, A. Alegria, and R. Houghten.** 1987. Protective synthetic peptides against experimental *P. falciparum* induced malaria, p. 117–124. *In* R. M. Chanock, R. A. Lerner, F. Brown, and H. Ginsberg (ed.), *Vaccines 87.* Cold Spring Harbor Laboratory Press, Cold Spring Harbor, N.Y.
22. **Patarroyo, M. E., P. Romero, M. L. Torres, P. Clavijo, A. Moreno, A. Martinez, R. Rodriguez, F. Guzman, and E. Cabezas.** 1987. Induction of protective immunity against experimental infection with malaria using synthetic peptides. *Nature* (London) **328:**629–632.
23. **Patarroyo, M. E., J. Vinasco, and R. Amador.** 1991. Genetic control of the immune response to a synthetic vaccine against *Plasmodium falciparum. Parasite Immunol.* **13:**509–516.
24. **Rocha, C. L., L. A. Murillo, A. L. Mora, M. Rojas, L. Franco, J. Cote, M. V. Valero, A. Moreno, R. Amador, F. Nunez, C. Coronell, and M. E. Patarroyo.** 1992. Determination of the immunization schedule for field trials with the synthetic malaria vaccine SPf66. *Parasite Immunol.* **14:**95–109.
25. **Rodriguez, R., A. Moreno, F. Guzman, M. Calvo, and M. E. Patarroyo.** 1990. Studies in owl monkeys leading to the development of a synthetic vaccine against the asexual blood stages of *Plasmodium falciparum. Am. J. Trop. Med. Hyg.* **43:**39–354.
26. **Romero, P.** 1992. Malaria vaccines. *Curr. Biol.* **4:**432–441.
27. **Salcedo, M., L. Barreto, M. Rojas, R. Moya, J. Cote, and M. E. Patarroyo.** 1991. Studies on the humoral immune response to a synthetic vaccine against *Plasmodium falciparum* malaria. *Clin. Exp. Immunol.* **84:**122–128.
28. **Sampértegui, F., B. Estrella, J. Moscoso, L. Piedrahita, D. Hernández, J. Gaybor, P. Naranjo, O. Mancero, S. Arias, R. Bernal, M. E. Códova, J. Suárez, and F. Zicker.** 1994. Safety, immunogenicity and protective effect of the SPf66 malaria synthetic vaccine against *Plasmodium falciparum* infection in randomized double-blind placebo-controlled field trials in an endemic area of Ecuador. *Vaccine* **12:**337–342.
29. **Teuscher, T., J. R. M. Armstrong Schellenberg, I. B. de Azevedo, N. Hurt, T. Smith, R. Hayes, H. Mansanja, Y. Silva, M. C. Lopez, A. Kitua, W. Kilama, M. Tanner, and P. L. Alonso.** 1994. Spf66, a chemically synthesized subunit malaria vaccine, is safe and immunogenic in Tanzanians exposed to intense malaria transmission. *Vaccine* **12:**328–336.
30. **Valero, M., R. Amador, J. Aponte, A. Narvaez, C. Galindo, Y. Silva, J. Rosas, F. Guz-**

man, and M. E. Patarroyo. Evaluation of SPf66 malaria vaccine during a 22-month follow-up field trial in the Pacific coast of Colombia. Submitted for publication.

31. **Valero, M. V., R. Amador, C. Galindo, J. Figueroa, M. S. Bello, L. A. Murillo, A. L. Mora, G. Patarroyo, C. L. Rocha, M. Rojas, J. J. Aponte, L. E. Sarmiento, D. Lozada, C. G. Coronell, N. Ortega, J. E. Rosas, P. L. Alonso, and M. E. Patarroyo.** 1993. Vaccination with Spf66, a chemically synthesised vaccine, against *Plasmodium falciparum* malaria in Colombia. *Lancet* **341:**705–710.
32. **WHO Division of Control of Tropical Diseases.** 1992. Outline for WHO's plan of work for malaria control 1993–2000. *CTD/MCM* **92.6:**1–21.
33. **WHO Malaria Action Programme.** 1989. Guidelines for the evaluation of *Plasmodium falciparum* asexual blood-stage vaccines in populations exposed to natural infection. *TDR/IMMAL/VAC* **85.3:**1–34.
34. **World Health Organization.** 1995. Twelfth Programme Report of the UNDP/World Bank/WHO special programme for research and training in Tropical diseases (TDR). *Bull. W.H.O.* **12:**64–76.

Malaria Vaccine Development: A Multi-Immune Response Approach
Edited by Stephen L. Hoffman

Chapter 10

Predicting the Effects of Malaria Vaccines on the Population Dynamics of Infection and Disease

Sunetra Gupta and Roy M. Anderson

INTRODUCTION

In the coming decade, various candidate preparations are likely to undergo extensive trials as potential vaccines to protect against malaria. Predicting the effect of these vaccines on the incidence of malaria infection and disease is not a straightforward exercise because of the many complex processes that influence the transmission success of the parasite in human and insect vector populations. A number of mathematical model systems (3, 10, 21) have been developed in an attempt to predict how malaria vaccines might interfere with the complex dynamics of malaria infection. These models investigate various population-level aspects of mass immunization by building upon the framework first developed by Ross (32) and later expanded upon by Macdonald (26). Models of malaria vaccines can provide qualitative insights into how the outcome of a vaccination program will vary with increasing vaccine efficacy and coverage and thus can provide valuable information both in the planning stage of a vaccine trial and in the interpretation of results. Models that include details of the age structure of the host population and of seasonality in transmission can address more complex questions such as the ideal age distribution of vaccine coverage or the optimal annual regimen for vaccination. Mathematical models can also provide some information on the likely quantitative effects of a vaccine of known efficacy in areas of differing malaria transmission intensities, which is also particularly helpful in selecting suitable sites for vaccination trials.

Mathematical models of malaria transmission are often founded upon crude assumptions regarding the interaction of the parasite with the human host and the mosquito vector. The crudity of these assumptions has to do partly with the need for a simple framework for examining the population dynamics of the infection. To create a highly reduced system capable of producing useful insights requires a distillation of the core processes involved in determining the typical courses of infection within the human host and mosquito vector and of those involved in the control of parasite transmission. The translation of these processes into a mathematical framework can promote an understanding of the fundamental principles of malaria transmission and the possibilities for disease control (2). It is also true, however, that certain model assumptions are bound to be crude, simply because of the paucity of information on many basic processes, such as the relationship of parasite density to infectiousness and disease, and the development of protective immunity. Information is also lacking on many essential parameters, such as the duration and intensity of infectiousness within the human host. Mathematical models are useful in this respect, since they can be used to explore the effects of a range of realistic parameter values. However, in order to make specific predictions regarding the outcome of a vaccine program, it is essential that we have accurate measurements of these parameters of transmission and of the degree of variation in these values within and between different regions. A major need in the assessment of vaccine trials through mathematical methods is therefore better data on many of the fundamental aspects of malaria infection.

The impact of any vaccination program depends upon the level of coverage and the duration of protection offered by the vaccine as well as upon the more complex effects of diminishing exposure to infection on the development and maintenance of natural immunity (2). In the case of malaria, the precise impact of a vaccine will also be determined by its influence on the genetic structure of the parasite population and on the overall transmission dynamics. This influence will depend on the position of the vaccine on two different axes of parasite diversity. One of these axes represents the change in antigen expression during the changing life cycle of the parasite (stage specificity), and the other represents the cross-sectional variation in antigenic properties within a particular life cycle stage, reflecting the genetic heterogeneity of the parasite (strain specificity). This chapter is thus organized to discuss the effects of sporozoite, blood-stage, and gametocyte vaccines separately. We also address, in a separate section, the issue of a strain-specific vaccine as opposed to an antigen "cocktail" that equally limits all parasite types. We use the word strain to reflect variation in immunogenicity rather than in strict association with a particular genotype. Thus, different genotypes that the immune system recognizes as antigenically identical are functionally regarded as the same strain.

The outcome of a vaccination program may be measured in several different ways. One method is to record reductions in the incidence of infections, with the limit being a complete blocking of transmission. However, as for many infectious diseases, malaria infection does not necessarily result in disease. The clinical severity of *Plasmodium falciparum* malaria varies widely both between and within age classes for reasons that are poorly understood. Typically, only a small proportion of infected hosts progress to disease, and an even smaller fraction exhibit severe disease symptoms (15). The low prevalence of severe illness may be explained by differences in host susceptibility to disease (23) or, equally, by differences in clinical virulence among parasite strains (18). The impact of vaccination on severe-disease incidence depends on what combination of host and parasite factors precipitates severe malaria. We conclude this chapter with a discussion on the potential effect of mass immunization on the incidence of severe malaria.

A SIMPLE MATHEMATICAL MODEL FOR MALARIA

Mathematical models of the transmission dynamics of infection provide a useful tool for assessing the relative effectiveness of different forms of intervention designed to control infection and/or disease. A simple mathematical model for a single malaria strain may be constructed by dividing the host population into susceptible and immune compartments, with the flow from the former to the latter being mediated by the process of infection. Upon becoming infected, an individual is (after a short delay) infectious to a susceptible mosquito for a certain length of time, which is assumed to be short relative to the duration of immunity. Thus, at any given time, a proportion of the effectively immune population is infectious. Infection, or the presence of parasites in the blood, is not necessarily synonymous with infectiousness; a person may remain infected long after ceasing to be infectious. For the purposes of modeling the dynamics of infection (at this very simple level, where immunity is not related to the duration or density of infection), it is necessary and sufficient to have information only on whether or not an individual is transmitting (infectious) and whether he or she is capable of receiving an infection (susceptible or immune). Figure 1 summarizes the interactions between the various components of host and vector populations, where the latter are simply structured as uninfected and infected (see Appendix for details of equations).

By definition, the number of infections in a community can grow only if every infected individual can infect at least one other person. In a totally susceptible population, this situation is described by a combination of parameters that influence parasite transmission, known as the basic reproductive rate (2). The basic reproductive rate of malaria may be given as

$$R_0 = m\, a^2\, b\, L_M\, H\, c \tag{1}$$

where m is the number of vector mosquitoes per human host, a is the mosquito biting rate (which is raised to the second power, since two biting events must take place to complete the cycle), b denotes the probability that an infective mosquito bite will produce an infection in the host,

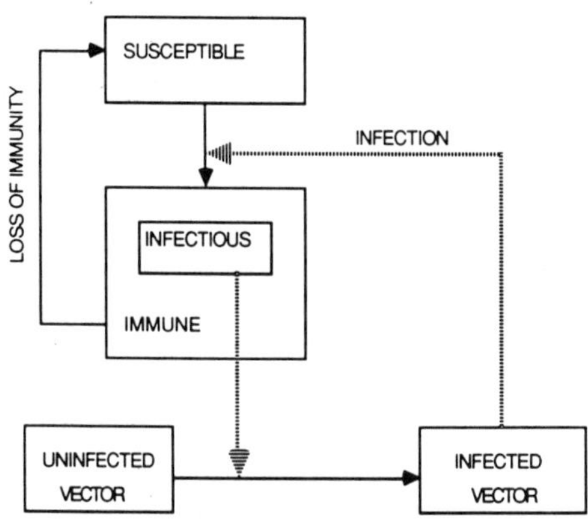

Figure 1. Schematic representation of a simple compartmental model for malaria transmission. Solid lines designate routes of flow between compartments, and broken lines represent the contact processes between host and vector that are responsible for these flows.

L_M is the average lifespan of the mosquito, H is the average duration of infectiousness in the human host, and c is the probability that an infectious feeding will lead to sporozoite production in the vector (4).

The basic reproductive rate defines the transmission potential of a parasite strain. The success of a vaccine will therefore depend on the R_0 of the infection (either with a given strain or as a composite measure of all strains circulating in a given community), and the outcome of intervention can often be conveniently expressed in terms of this basic property of the host-parasite system.

The number of hosts exhibiting (mild) disease symptoms in this simple model is related to the rate of infection and the rates of disease development and recovery and thus can be calculated as a proportion of the effectively immune population at any given time (see Appendix). However, the incidence of severe disease cannot be calculated simply as a fraction of the new infections arising in the population. The impact of vaccination on severe disease is therefore discussed separately.

SPOROZOITE VACCINE

A sporozoite vaccine, if entirely effective, will prevent infection. If not entirely effective, it may diminish the inoculation volume and consequently reduce the probability of disease development if the latter is related to inoculation dose. In this event, a "leaky" sporozoite vaccine would have the same effect as a merozoite vaccine. We thus restrict the discussion in this section to the effect of a perfect sporozoite vaccine. The mathematical model of Fig. 1 can easily be extended to incorporate an uninfectable compartment of vaccinated individuals, as shown in Fig. 2.

This simple model can be used to determine the proportion of the community that must be vaccinated yearly in order to interrupt transmission (i.e., achieve eradication). In the limit where immunity is lifelong (the best-case scenario), the critical proportion g (in units per year) is given by

$$g > \frac{1 - e^{-(R_0-1)} \left(\dfrac{1}{L} + \dfrac{1}{D}\right)}{h} \tag{2}$$

where vaccine efficacy h is measured as the fraction of those vaccinated that are protected against infection; D is the duration of protection offered by the vaccine; and L is the average life span of the human host (3). Figure 3 expresses the critical rate of vaccination required to achieve eradication (proportion to be treated each year) in relation to vaccine efficacy (h) and duration of protection (D). In areas of high transmission (Fig. 3A;

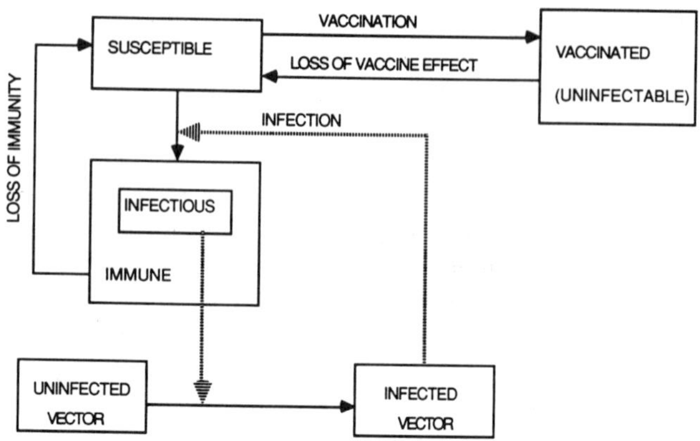

Figure 2. An extension of the scheme in Fig. 1 to include a compartment of vaccinated (uninfectable) individuals, thus providing a model for a sporozoite vaccine.

high R_0), eradication is possible only if the vaccine offers long duration of protection and is highly efficacious. In areas of low transmission (Fig. 3B; low R_0), the parasite can be eliminated at much lower levels of vaccine efficacy and shorter durations of protection. The relationship between transmission rates (as captured by the value of the basic reproductive rate, R_0) and the critical vaccination rate required for eradication is demonstrated in Fig. 4, where the latter is plotted against the former for different durations of protection against infection, assuming that the vaccine has an efficacy of 50% (i.e., half the individuals receiving the vaccine are protected against infection). Figure 4 confirms that eradicating malaria is difficult in regions where R_0 is high.

In the event that eradication is not feasible, vaccination strategies must be assessed in terms of their impact on levels of infection (or the incidence of new infections). Figure 5 records the proportional reduction in infection in relation to the proportion vaccinated and the duration of protection in an area of extremely high transmission ($R_0 = 50$). The relationships between reduction in infection, vaccine efficacy, and coverage are nonlinear but monotonic (i.e., higher coverage will always give a greater reduction in infection, though not in proportion to the increase in coverage). The monotonic relationship between the levels of control and of infection may not hold if the development of immunity is significantly influenced by exposure to infection. The simple model described in Fig. 2 can be modified to include two classes of exposed individuals (as shown in Fig. 6), when the semi-immune individual can be infected (at a lower probability than the nonimmune individuals) to become completely immune. Immune individuals cannot revert to a state of susceptibility. A drop in the level of infection as a result of vaccination will selectively impede the flow from semi-immune to immune individuals, since infection of semi-immune individuals has a lower probability than the infection of nonimmune individuals. As a consequence, a large proportion of the unvaccinated population may remain in a semi-immune state. Since immunity is lost at a faster rate in the semi-immune population, the rate of flow into the susceptible population will increase. If the reduction in overall probability of infection does not balance the increase in the size of the

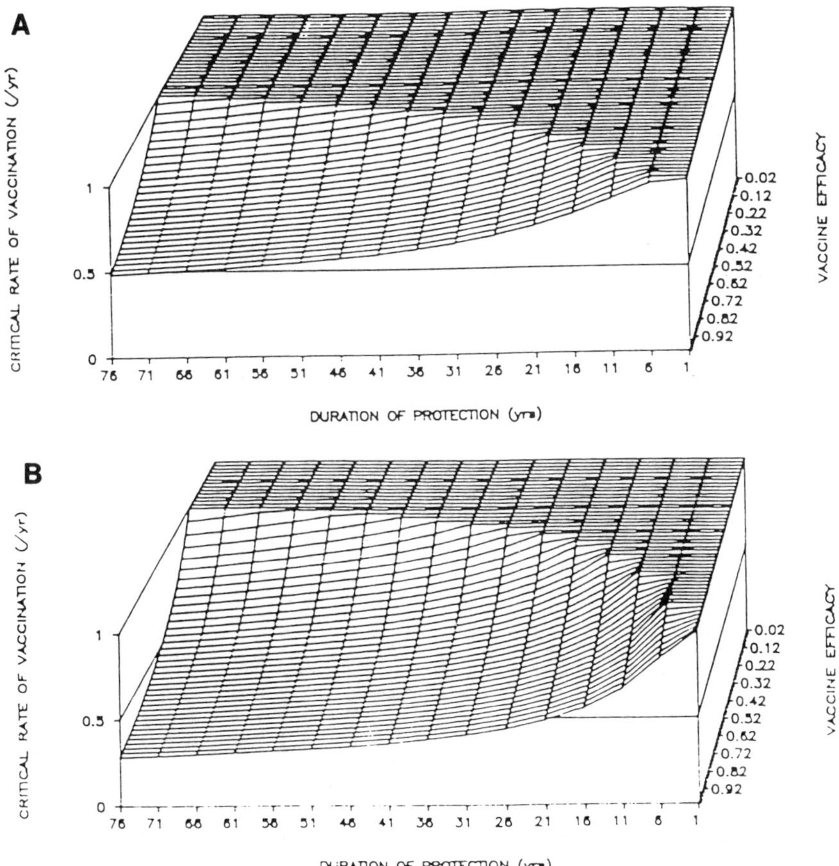

Figure 3. Plot of the critical rate of vaccination (proportion to be treated each year) required to achieve eradication as a function of vaccine efficacy (h) and the duration of protection provided (D). (A) $R_0 = 20$, $L = 50$ years; (B) $R_0 = 10$; $L = 50$ years.

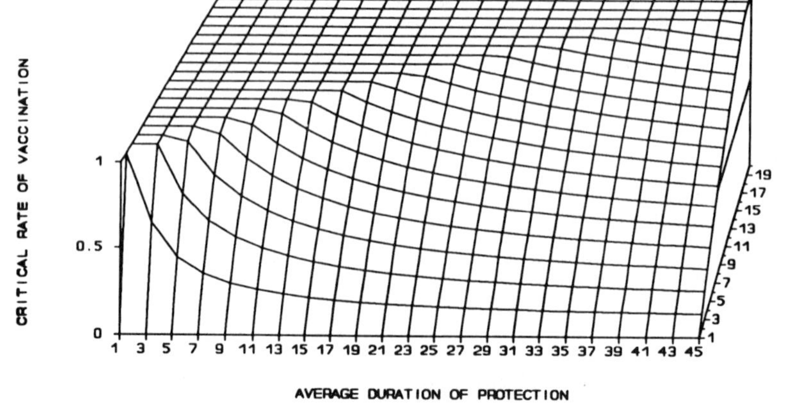

Figure 4. Critical vaccination rate plotted against the average duration of vaccine protection for a range of values of R_0.

susceptible population, then vaccination may result in higher equilibrium levels of infection. If infection in nonimmune individuals is more likely to manifest itself as disease than as infection among semi-immune individuals, then disease incidence may increase even when levels of infection are reduced as a result of the change in the distribution of new infections between nonimmune and semi-immune individuals (i.e., more

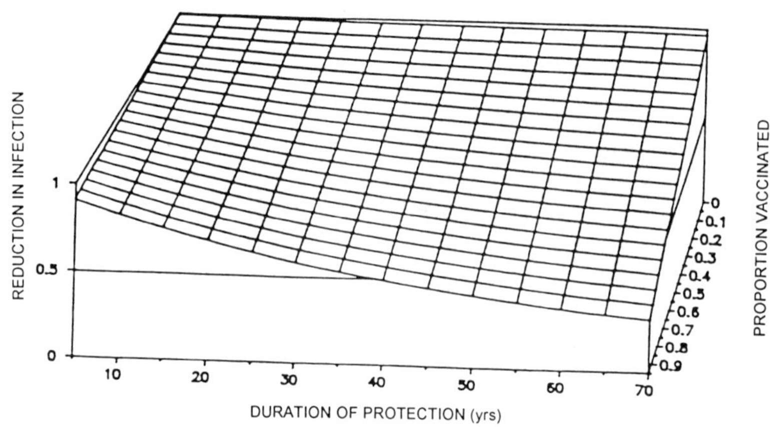

Figure 5. Proportional reduction in infection in relation to the proportion vaccinated and the duration of protection in an area of extremely high transmission ($R_0 = 50$).

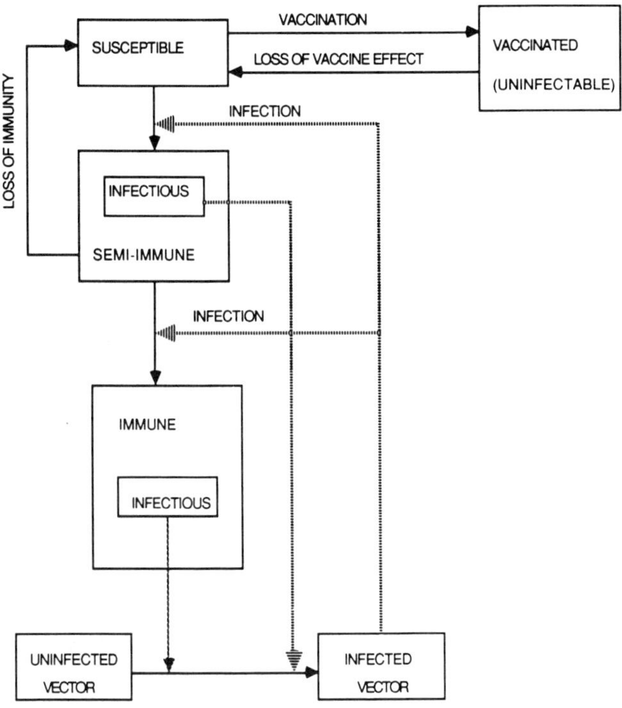

Figure 6. An extension of the simple model in Fig. 2 to include two tiers of immune individuals, where the semi-immune individual can be infected (at a lower probability than the nonimmune individuals) to become completely immune.

infections will occur among nonimmune than among semi-immune individuals, even if the total number of new infections is lower). The effect of the vaccine would thus be to compromise the development of natural immunity among the unvaccinated by constricting the transmission system, thereby forcing the "unvaccinated to bear the burden of maintaining transmission, possibly with increased disease" (21).

ASEXUAL-BLOOD-STAGE VACCINE

A vaccine that acts against the asexual blood stages of the malaria parasite will effectively block infection (and hence disease) if it completely limits parasite proliferation. Thus, a very effective blood-stage vaccine will have the same impact as a nonleaky sporozoite vaccine. However, if the vaccine acts to reduce parasite numbers rather than to eliminate the parasite, then the main effect of the vaccine is likely to be a reduction in the incidence of disease, because there appears to be some correlation between parasite burden and the development of clinical malaria (15). If the production of gametocytes is related to the density of the asexual blood stages, then a further effect may be a reduction in transmission. On the other hand, transmission-blocking immune responses may not be adequately induced at low parasite densities, so that theoretically, the duration of infectiousness could be extended by a blood-stage vaccine. The dynamics of infection will also be determined by the relationship between the development of protective immunity and the density of asexual blood stages. Figure 7 schematizes the disease-reducing effect of a blood-stage vaccine and also incorporates the assumption that density of asexual parasitemia can significantly affect the development of immunity. The system differs from that of the sporozoite vaccine in that the vaccinated population is no longer uninfectable but will, upon sustaining an infectious bite, enter (with the same probability as unvaccinated individuals) a class of infected (and subsequently immune) vaccinated individuals but develop low-density infections compared to their unvaccinated counterparts. The rate of (mild) disease among the vaccinated infected population is assumed to be lower than the disease rate in the unvaccinated population. If none of the parameters associated with transmission are affected by asexual-blood-stage parasite densities, then the equilibrium prevalence of disease will always be less than prevaccination levels, since

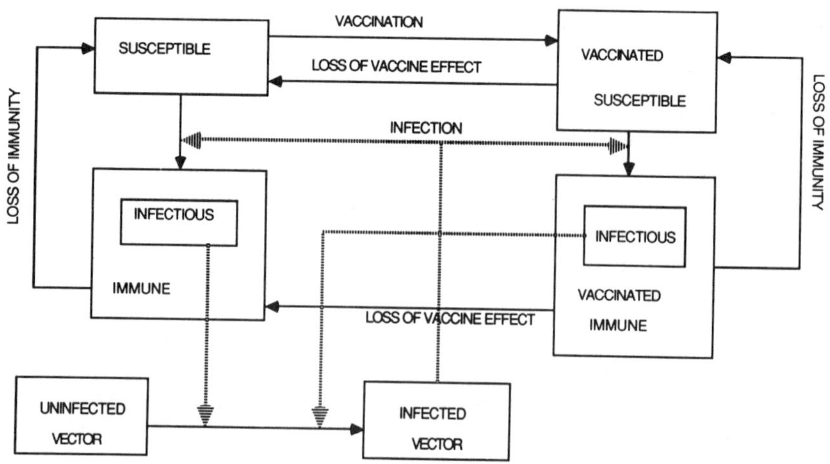

Figure 7. Structure of a compartmental model for an asexual-blood-stage vaccine that does not prevent infection but reduces parasite densities, thus limiting disease and also possibly the development of protective immunity.

while levels of infection remain unaffected by vaccination, there will be fewer clinical cases among the proportion of the infected population that is vaccinated.

The effect of reduced parasite densities on the development of immunity to further infection may be modeled by assigning a higher rate of immunity loss to the vaccinated immune individuals. Infection levels will thus increase within the community as a result of vaccination. However, the overall incidence of disease may still fall if the reduction in the proportion developing disease outweighs the increase in the prevalence of infection. The outcome of the tension between these opposing effects of a blood-stage vaccine will be determined by the precise relationship of parasite density with progression to disease and the development of protective immunity.

Parasite transmission may also be affected by blood-stage vaccines if the density of asexual parasites serves as a trigger for gametocyte production. This assumption may be explored by selectively reducing the infectivity of the vaccinated infectious population. Essentially, such a mechanism will have the same effect as a transmission-blocking vaccine and thus will reduce infection levels. However, the ultimate consequence of a reduction in gametocyte production can be assessed only in the context of the multiplicity of effects of a reduction in asexual-blood-stage density.

TRANSMISSION-BLOCKING VACCINE

Figure 8 summarizes the interactions between the various host and parasite components of infection that might occur with a transmission-blocking (gametocyte) vaccine that induces the function of an antibody that prevents fertilization in the mosquito. In this situation, vaccinated individuals can be infected and suffer disease but be unable to transmit the infection. Infection of vaccinated individuals confers natural immunity which will inhibit parasite transmission in addition to transmission blockage by vaccination. The effectiveness of such a vaccine in comparison to that of an infection-blocking (sporozoite or highly effective merozoite) vaccine depends on the duration of the transmission-blocking effect of the vaccine in relation to that of natural immunity. If vaccine protection is long-lasting, then the threshold criteria for eradication are identical for all types of vaccine, since in all cases, vaccination effectively reduces the size of the susceptible population. If, however, the effect of vaccination is short-lived, then a transmission-blocking vaccine may have an advantage over an infection-blocking vaccine, since the former does not prevent the development of immunity in the vaccinated population. The significance

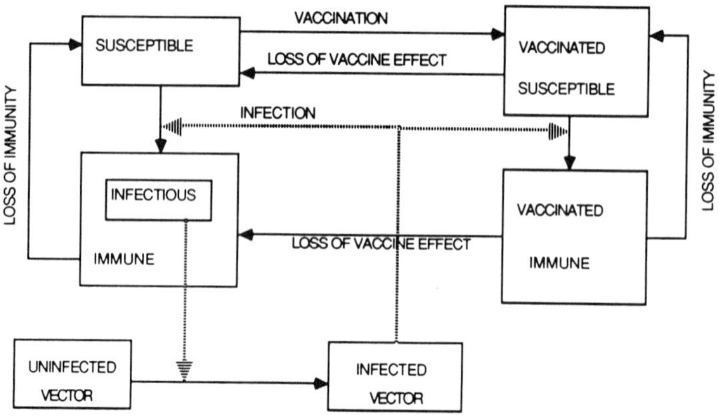

Figure 8. Structure of a compartmental model for a transmission-blocking vaccine, which prevents neither infection nor disease but does prevent parasite transmission.

of this effect is determined by the duration of natural immunity; if natural immunity is brief, then the effect is minimal.

The fact that a transmission-blocking vaccine does not prevent infection may facilitate eradication, but in situations where transmission cannot be interrupted, it will lead to less of a reduction in the prevalence of infection than an infection-blocking vaccine would provide. Thus, while a transmission-blocking vaccine compares favorably with infection-blocking vaccines in eliminating the parasite, it performs relatively poorly in reducing infection and mild disease under circumstances in which eradication is impossible to achieve. Figure 9 shows the nonlinear relationship between the proportional reduction in infection effected by mass administration of a transmission-blocking vaccine and the proportion vaccinated in an area of moderately high transmission ($R_0 = 10$). Only at very high levels of coverage, close to the threshold for eradication, is the vaccine able to generate a substantial reduction in numbers infected.

The effect of a transmission-blocking vaccine will be enhanced if infected mosquitoes taking a second blood meal on an immunized host suffer a degree of inhibition of infectivity (33). This phenomenon will have little effect on the equilibrium prevalence of infection unless the vaccination rate is high.

COHORT VACCINATION OF INFANTS

An alternative vaccination strategy to community-wide vaccination across all ages is to immunize a cohort of infants soon after birth. The critical proportion (for eradication) (p_s) of each yearly cohort of infants to be immunized by a sporozoite vaccine with an efficacy h and a duration of protection D is given by

$$p_s > \left(1 - \frac{1}{R_0}\right) \frac{\left(1 + \dfrac{L}{D}\right)}{h} \tag{3}$$

If natural immunity is lifelong, then the criterion for eradication for a transmission-blocking vaccine is less stringent than that for an infection-blocking vaccine, since the former has to be effective only over the average duration of infectiousness rather than over an average lifetime. The critical proportion (p_g) of infant cohort immunization associated with a transmission-blocking vaccine is given by

$$p_g = \left(1 - \frac{1}{R_0}\right) \frac{\left(1 + \dfrac{H}{D}\right)}{h} \tag{4}$$

where H is the average duration of infectiousness (3).

It is clear from this expression that the outcome is more sensitive to the duration of infectiousness (H) when the duration of protection (D) (strictly, vaccine effect rather than protection in this case) is short. When

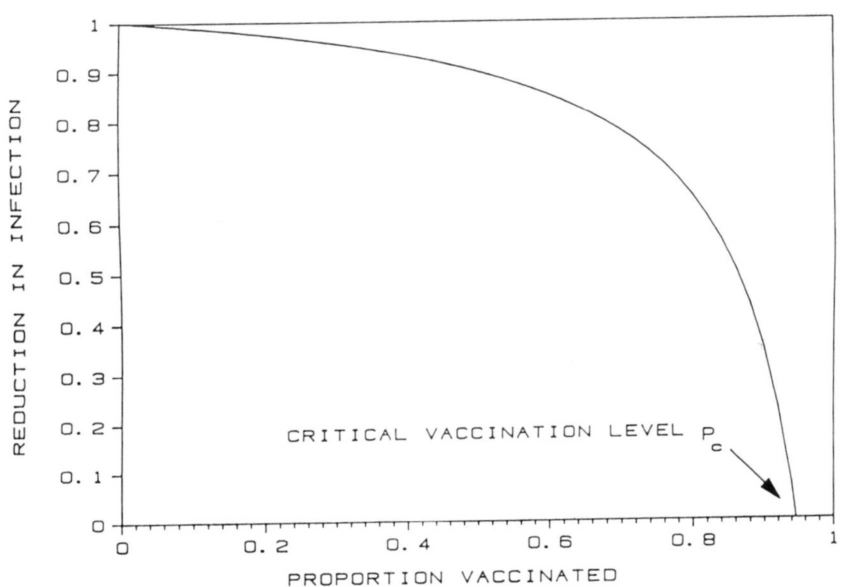

Figure 9. Relationship between the proportional reduction in infection effected by mass administration of a transmission-blocking vaccine and the proportion of the population vaccinated ($R_0 = 10$).

the duration of protection is long in comparison to the duration of infectiousness ($D>>H$), then the outcome is indifferent to the duration of infectiousness. Figure 10 compares the critical proportions to be vaccinated (in order to achieve eradication) in relation to duration of infectiousness (H) and basic reproductive rate (R_0) for a short duration (Fig. 10A; D = 6 months) and long duration (Fig. 10B; D = 10 years) of vaccine protection. If the average duration of vaccine effect is long (Fig. 10B), then the out-

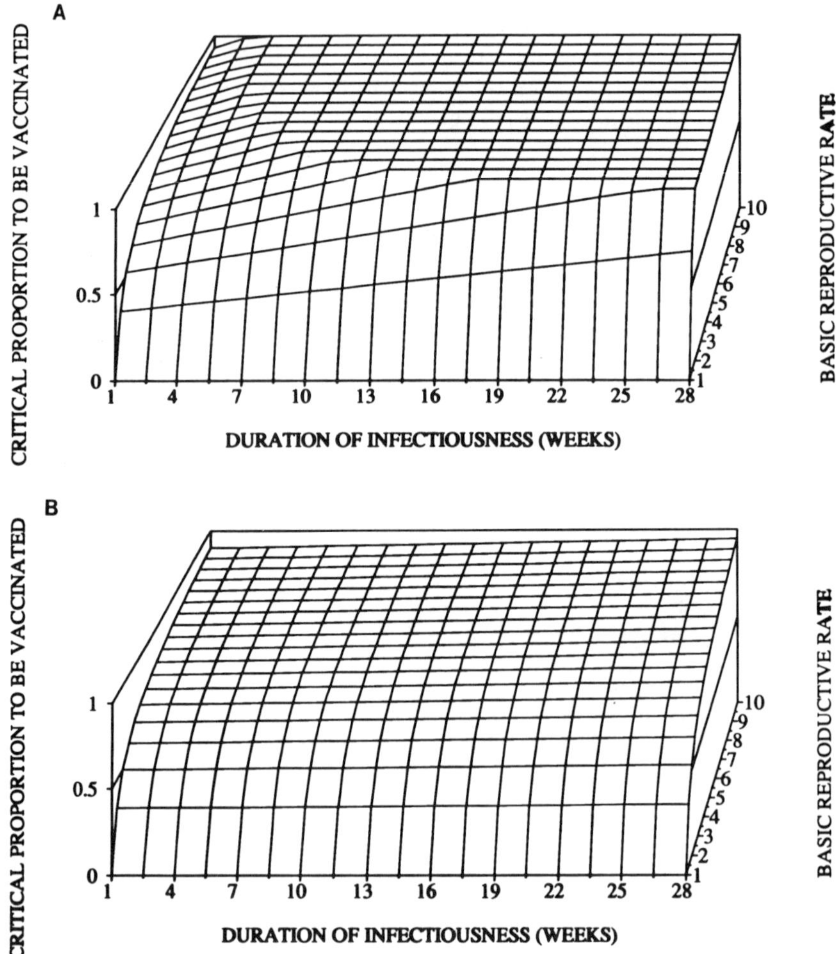

Figure 10. Critical proportions to be vaccinated (in order to achieve eradication) in relation to duration of infectiousness (H) and basic reproductive rate (R_0). (A) Average duration of protection by the vaccine (D) is 6 months; (B) vaccine endures for an average of 10 years.

come of a vaccine of a given efficacy is dependent only upon the basic reproductive rate, R_0. When D is small (Fig. 10A), the critical proportion to be vaccinated depends linearly on the duration of infectiousness. No satisfactory measures of this parameter are currently available. While *P. falciparum* infection can last for up to several months in an individual, the duration of infectiousness may be on the order of only a few weeks. Infectiousness is clearly related to patterns of gametocyte density, which several sources have described as wavelike (5, 7, 12, 24), decreasing in density and period during the course of the infection. Others (6) maintain that the intermittent occurrence of gametocytemia is not a universal phenomenon, reporting that in many cases, only one or two resurgences might occur. Furthermore, "the association between gametocytaemia and infectiousness is not perfect" (30), because certain criteria must be fulfilled before gametocytes can successfully infect a mosquito and be fertilized within it. One such requirement is that both male and female gametocytes must be present to ensure transmission. Sex ratios are severely biased toward the macrogametes at the tail end of a wave (5), suggesting that transmission is limited to the earlier section of the cycle. Also, if the quality of gametocytes deteriorates as the immune system begins to respond to the infection, then subsequent waves of gametocytemia may be noninfectious. In view of the uncertainty surrounding the issue of infectiousness, it is perhaps best to assume that the average duration of infectiousness H is of the order of a few weeks, in contrast to several months of parasitemia and a few months of intermittent noninfectious gametocytemia.

Figure 11 compares the effects of cohort vaccination with sporozoite and transmission-blocking vaccines. As expected, the transmission-blocking vaccine is immeasurably more effective than the sporozoite vaccine. Figure 11A indicates that a 100% efficient sporozoite vaccine is able to eradicate malaria only when the average duration of protection is more than 30 years and then only in regions of low transmission (low R_0). In contrast, a transmission-blocking vaccine is always able to achieve eradication over an identical range of parameter values (Fig. 11B). Figure 12 demonstrates the sensitivity of the outcome of a transmission-blocking vaccine to duration of protection when the latter is of the order of a few years rather than few decades; eradication may not be possible even in areas of fairly low transmission if the vaccine is short-acting. When eradication is not an achievable goal, a transmission-blocking vaccine performs poorly in comparison with the sporozoite vaccine in reducing infection levels, as in community-wide vaccination.

Cohort immunization of infants may result in a shift in the age distribution of infection. Unless the duration of protection against infection

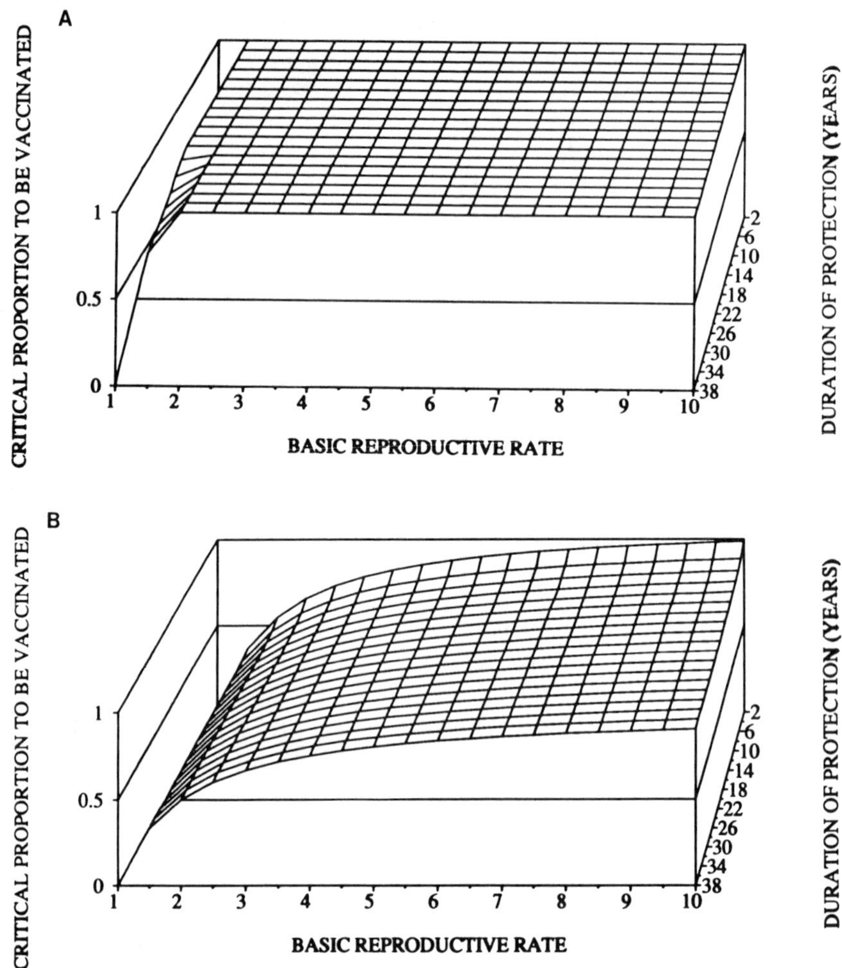

Figure 11. Critical proportions of infants to be vaccinated using sporozoite (A) and transmission-blocking (B) vaccines for a given range of values of R_0 and an average duration of vaccine protection.

is lifelong, the decay of a sporozoite vaccine will increase the proportion of nonimmune adults or older children in a population. Thus, although there will be fewer infections, more of them will occur among older individuals. This result may have serious consequences if naive older individuals are more likely than young infants to experience morbidity or serious disease in response to new infections. Such a trend, if it is there, will

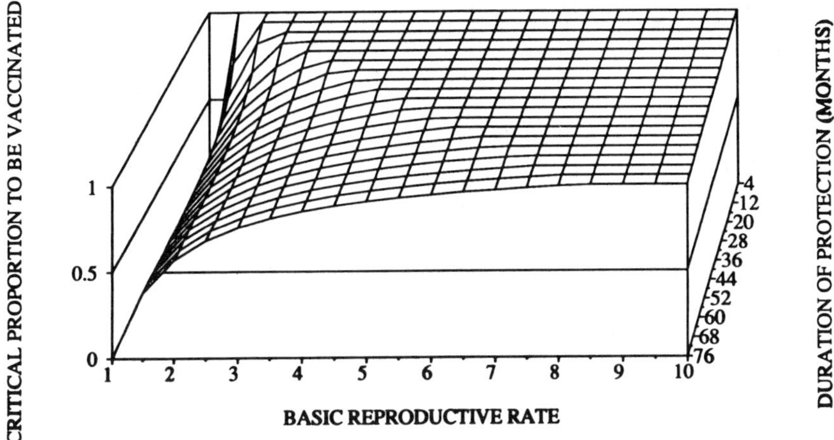

Figure 12. Critical proportion of infants to be vaccinated in order to achieve eradication as a function of R_0 when a transmission-blocking vaccine, for which the duration of protection is in the range of months rather than years, as recorded on the axis, is used.

have been masked in areas of endemicity by the progressive development of immunity with exposure. Thus, how the incidence of disease might be affected by a sporozoite vaccine is not clear. Since a transmission-blocking vaccine does not prevent the development of natural immunity, no change is likely to occur in the age distribution of infection unless levels of immunity are strongly correlated with exposure. Halloran et al. (22) explored the consequences of cohort immunization on the age distribution of infection in nonimmune as opposed to semi-immune individuals (i.e., when immunity is related to degree of exposure) and concluded that sporozoite and merozoite vaccines will reduce the prevalence of infection in both nonimmune and semi-immune individuals across all age classes. A transmission-blocking vaccine, however, while reducing the overall prevalence of infection, may increase the total number of infections occurring among the nonimmune and thus possibly the prevalence of disease in adults.

WHAT IS THE BASIC REPRODUCTIVE RATE OF MALARIA?

With any type of vaccine, the success of a control program clearly and critically depends on the basic reproductive rate (R_0) of malaria, which has been thought to range between unity and 5,000 (11, 25) in different areas. The rise in exposure to a particular disease agent with age can provide an indirect method of measuring its transmissibility or R_0. For many common viral and bacterial infections, such age-exposure profiles may be constructed from serological surveys for the presence or absence of specific antibodies (2). It can be shown that if these infections induce lifelong immunity, then the average age (A) at which an individual typically acquires an infection is inversely related to the basic reproductive rate (R_0) or transmissibility of the disease by the simple relationship

$$R_0 = L/A \tag{5}$$

where L is the average life span of the host (2). As is clearly indicated by equation 5, the average age at infection is low (1 to 2 years) for highly transmissible diseases such as measles and higher (9 to 10 years) for less transmissible diseases such as rubella (28, 31).

Age-exposure profiles for *P. falciparum* malaria as obtained by serological methods (29) generally indicate that the proportion of infants positive for *P. falciparum* antibodies rises sharply with age after an initial decline associated with the loss of maternal antibodies, so that in many areas in which *P. falciparum* is endemic, more than 80% of children are exposed to *P. falciparum* by the age of 1 year. A direct extension of the relationship between R_0 and the average age at infection as defined in equation 5 would therefore lead to an extremely high estimate (of the order of 60 to 80) for the R_0 of *P. falciparum* malaria. Malaria, however, does not induce lifelong immunity upon exposure. Young children in zones of hyperendemic malaria may experience anywhere between one and five clinical attacks of malaria per year (15). Older subjects appear to develop a functional but nonsterilizing immunity that manifests itself in a reduction of the number of clinical episodes. However, a substantial reduction in parasite rates (proportion of individuals with asexual malaria parasites) is observed only in adults. Parasite rates tend to peak in children of around 10 years of age and slowly decline to low but significant levels in adults (16, 27). As a result, the mean duration of protective immunity across all age classes (D_I) may be much lower than the average life span of an individual (L). This difference in the establishment of immunity can significantly alter the age-exposure profile of malaria (20) such that the average age at first infection (or exposure) is given, to a good approximation, by

$$A_I = D_I/R_0 \tag{6}$$

The steepness of age-exposure profiles commonly encountered for malaria is thus not necessarily indicative of a high basic reproductive rate but may instead be a consequence of the low average duration of immunity in the population. To say that the average duration of immunity in the population is low does not imply that there is necessarily any decay of immunity. In the case of malaria, a low average duration of immunity in a population is a simple expression of the fact that a large proportion of individuals do not appear to be immune at all, despite having experienced infection. Among those who are immune, immunity may be long-lived; a recent study (9) implies that protection developed in childhood can persist without boosting for at least 30 years.

The long period required to develop protective immunity to malaria can be explained by the existence of several antigenic types or strains, each of which may induce lifelong strain-specific immunity (8). Under these circumstances, immunity to each strain may have average duration L, while the population's average duration of immunity D_I to "malaria" (effectively a collection of these strains) may be significantly lower than L. Gupta et al. (20) calculated the basic reproductive rate of malaria to be around 6 to 7 according to data on patterns of seroconversion to five antigenically distinct isolates of *P. falciparum* in an area of holoendemic malaria (Madang region of Papua New Guinea). At these low values of R_0, malaria will be easier to control than previously thought.

EFFECT OF VACCINATION ON THE DYNAMICS OF DISEASE IN AREAS OF VERY LOW TRANSMISSION

Instabilities in the transmission dynamics of malaria are higher in areas of low transmission (low R_0), particularly if the duration of infectiousness is short and immunity is long-lived (19). Figure 13 demonstrates the effects of a low-efficacy transmission-blocking vaccine on the temporal trends in disease incidence in an area of low transmission. The vaccine appears to increase the interepidemic period, particularly between the first and second peaks. This effect increases with the proportion of the population vaccinated per unit time. Identical dynamics will be induced by a highly efficacious vaccine if coverage is very low. When the interepidemic period is very long, we may assume that eradication has effectively occurred, since only in a very large population would the proportion of lingering infections in the interepidemic period translate into real numbers of individuals. At intermediate values of extension of the interepidemic period, the population may be large enough to support such a decline in proportion infected, and a second epidemic may occur after the disease appears to have been eradicated. Conversely, at a particular time, vaccine intervention may appear to have increased the prevalence of infection or disease, even though the system will eventually settle to a lower equilibrium. Transient instabilities may thus confound the

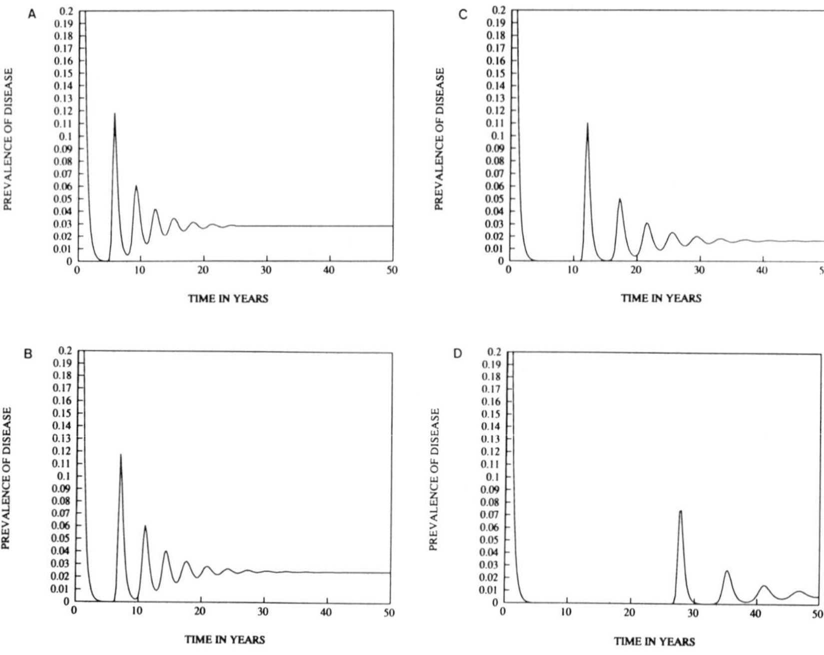

Figure 13. Temporal trends in disease incidence upon application of a transmission-blocking vaccine in an area of low transmission ($R_0 = 2.2$) for different levels of coverage with a low-efficacy vaccine (10% efficacy). Coverage: (A) no vaccination; (B) 10%; (C) 30%; (D) 50%.

assessment of a vaccination trial in areas of low transmission. More important, however, these instabilities may facilitate eradication. Seasonality in transmission may have complex effects on the inherent instabilities of a transmission system resulting from the interaction of two oscillatory phenomena. Since transmission is highly seasonal in many malarious areas, further research on the dynamics induced by seasonal exposure to infection is required.

EFFECTS OF STRAIN SPECIFICITY

Periodicity in malaria transmission has been documented by longitudinal village-based studies of the transmission dynamics of the FC27 serotype in the Madang region of Papua New Guinea (14). Mathematical models of parasite heterogeneity in malaria reveal that sustained oscillations can be precipitated by strain heterogeneity as a result of complex interactions between the immune responses to the different strains (19).

Thus, by reducing parasite heterogeneity, a strain-specific vaccine may alter the dynamics of infection. This alteration may be manifested as an increase in the stability of the system, since single-strain models tend to exhibit damped rather than sustained oscillations and are also unstable only over a short range of basic reproductive rates. As mentioned in the previous section, inherent instabilities tend to be amplified by vaccines and therefore enhance prospects for eradication. Reducing strain diversity may thus diminish the probability of eradication by causing the system to settle to a stable robust equilibrium.

A more direct consequence of strain specificity is the selection of strains resistant to a particular vaccine preparation. This eventuality is of particular importance in the context of differences in strain virulence. If strains of differing virulence are in polymorphic equilibrium and the effect of a vaccine is to select for the more virulent types, then there could be a concomitant increase in disease incidence.

From their analysis of seroconversion patterns to antigenically distinct isolates of *P. falciparum*, Gupta et al. (20) concluded that malaria can be seen as a composite of mildly transmissible strains, each of which may induce lifelong strain-specific immunity. In order to exploit the low transmissibility of these strains (and collectively of malaria), it is therefore necessary to invest in the design of a vaccine that circumvents the antigenic diversity of the parasite.

IMPACT ON INCIDENCE OF SEVERE DISEASE

Halloran et al. (22) distinguished between serious disease as infection in nonimmune individuals and mild disease as infection in semi-immune individuals, in whom immunity is exposure dependent. Since semi-immune individuals are less likely than nonimmune individuals to experience infection, the incidence of severe disease in this model may exceed that of mild disease. The low prevalence of severe disease as observed in the field refutes this association between exposure-dependent immunity and disease. Furthermore, the Halloran et al. (22) model predicts that an increase in the nonimmune population with vaccination (as a result of diminishing infection levels) will have a disproportionate effect on the incidence of serious disease. Fortunately, there is little to suggest that severe disease is likely to be more severe in the semi-immune than in the nonimmune individual or, the converse, that prior sensitization leads to the severe immunopathological reactions described for some other infections (e.g., dengue fever).

Several host and parasite factors may precipitate severe illness. There is increasing evidence of genetically determined differences in host susceptibility to disease. A high prevalence of erythrocyte abnormalities such as sickle cell trait, α-thalassemia, β-thalassemia, and possibly glucose-6-phosphate dehydrogenase deficiency has been found in malaria-endemic areas (1, 13, 34). Certain HLA types may also be associated with protection against severe malaria (23). The development of disease may also be an exposure-dependent phenomenon, as suggested by the observation that the use of bed nets can reduce the incidence of severe disease without inducing any significant change in prevalence of infection (15). However, this reduction may also be interpreted as the consequence of the differential impact of reduction in transmission (by bed net use) on more virulent and less transmissible parasite strains (17). The effect of a vaccine on the prevalence of severe disease will depend on the combination of host and parasite factors responsible.

Both infection-blocking and transmission-blocking vaccines are likely to have a proportionately greater impact on the incidence of severe disease if the disease is caused by a distinct, highly virulent parasite strain constituting an independent low-transmission (low-R_0) system (18). This effect is analogous to the reduction in disease caused by bed net use. Severe disease can manifest itself either as cerebral malaria or as severe malarial anemia. If the former is caused by a particularly "sticky" parasite strain while the latter is an extreme aggravation of mild malaria, then we may expect a greater reduction in cerebral malaria cases but not observe this effect in the incidence of severe malarial anemia. If host factors outweigh parasite factors as the cause of severe disease, then vaccination may have little impact on the case rate of severe disease, since the probability of covering the small proportion of the population predisposed to severe disease is low. In particular, with a transmission-blocking vaccine (which does not prevent disease), changes in the incidence of severe disease are likely to be imperceptible (unless complete eradication is achieved) if the latter is precipitated by host factors alone. Thus, a vaccine might provide information on the relative importance of host and parasite heterogeneity as a factor in severe disease.

If severe disease is associated with a virulent parasite strain, the incidence of severe malaria may be selectively diminished by a vaccine specific to that strain. Conversely, the situation may be aggravated by a vaccine that selectively removes less virulent strains from the population.

The relationship between the density of asexual blood stages and severe disease is unclear. Certainly, severe malarial anemia could be a consequence of a massive depletion of erythrocytes by high parasite densities. Cerebral malaria could also be precipitated by saturation of cytoadherence sites, resulting in sequestration in the brain. Blood-stage vaccines may thus have a significant impact on the incidence of severe disease without having much effect on the prevalence of infection and mild disease if they act to reduce parasite densities below the critical threshold for severe disease. Conversely, a general reduction in parasite densities may increase the relative importance of any transmissibility advantage associated with higher levels or specific types of cytoadherence and thereby select for more virulent parasite strains by creating a bottleneck in the system.

SUMMARY AND CONCLUSIONS

A malaria vaccine may not always be capable of achieving eradication but may cause significant changes in patterns of infection and disease and consequently in levels of malaria mortality and morbidity. Malaria vaccines may be categorized as those that block infection (sporozoite and highly effective asexual-blood-stage vaccines), those that block transmission without preventing infection (antiookinete), and those that reduce disease (leaky sporozoite and asexual-blood-stage vaccines). The impact of a stage-specific vaccine will be determined, however, not merely by its infection- and transmission-blocking or disease-reducing properties but also by its relationship to the development of natural immunity. In the absence of natural immunity, transmission-blocking vaccines are equivalent to infection-blocking vaccines in their capacity to eliminate malaria. Since transmission-blocking vaccines do not compromise the development of natural immunity, they have a greater likelihood of interrupting the malaria cycle than infection-blocking vaccines have but are less effective at reducing infection and disease when eradication is not an achievable goal. The effect of diminishing exposure to infection on the development and maintenance of natural immunity may interfere with the success of vaccination, especially if clinical malaria is less likely to occur in individuals with some degree of immunity. Similarly, the disease-reducing effect of an asexual-blood-stage vaccine may be diminished by the effect of parasite densities on the development of natural immunity.

The effect of a vaccine will vary between areas of high and low transmission. The critical proportion to be vaccinated over a certain time interval increases nonlinearly with the basic reproductive rate (R_0) of malaria. A recent analysis (20) of exposure patterns to antigenically distinct isolates suggests that R_0 values for malaria have been grossly overestimated. Even in areas of high endemicity, malaria may have low R_0 values and yet generate high exposure rates as a consequence of the delay in development of protective immunity (20). This situation implies that the control of malaria through mass vaccination may be easier than previously thought. In areas of low transmission (low R_0), eradication may also be facilitated by the amplification of the intrinsic instabilities of the system by a vaccine. These instabilities may be intensified by seasonalities in transmission. Thus, the effect of a vaccine may also vary between areas with high and low amplitudes in seasonal transmission intensities. Transient instabilities induced by a vaccine may, however, confound the interpretation of the results of a vaccine trial.

The impact of vaccination on the incidence of severe malaria will be determined by the relative importance of host and parasite heterogeneity as factors in severe disease (18, 23). If severe disease is associated with a virulent parasite strain of low transmissibility, vaccination will result in a greater reduction in the incidence of severe malaria than in the incidence of infection and mild malaria. If, however, severe malaria is precipitated by host factors, then at best, a proportionate reduction may be expected in the case rate of severe disease.

Parasite heterogeneity is likely to create major complications in the design of vaccination strategies. Parasite strains resistant to any particular vaccine may be selected in a manner analogous to the development of drug resistance. This effect will be particularly deleterious if the property of resistance is linked to parasite virulence. Furthermore, a reduction in parasite diversity may cause the transmission system to switch to a more stable dynamical regime, which may reduce the probability of eradication.

Essentially, a broad range of complex outcomes is associated with each type of stage-specific vaccine. The impact of vaccination will be determined both by standard parameters such as the duration of protection, vaccine efficacy, and vaccination rates and by the dynamical consequences of the relationships between natural immunity and exposure to infection, virulence and transmissibility, parasite density, and infectiousness. Vaccine trials must be designed in the context of these uncertainties until more information regarding the basic processes and parameters of malaria transmission is available.

Acknowledgments. We thank Brian Greenwood for his invaluable participation in the development of this chapter.

S.G. thanks the Leverhulme Trust and the British Society for Parasitology, and R.M.A. thanks the Wellcome Trust.

APPENDIX

A simple mathematical model for malaria transmission, as schematically presented in Fig. 1, may be described by a pair of differential equations describing the changes over time in the proportion of the host population immune to infection (x) and the proportion of the vector population infectious to the host (y):

$$\frac{dx}{dt} = mab(1 - x)y \quad d_I x \tag{A1}$$

$$\frac{dy}{dt} = ma^2bHc(1 - x)y(1 - y)\left(\frac{1}{L_M}\right)y \tag{A2}$$

where m is the number of vector mosquitoes per human host, a is the mosquito biting rate, b denotes the probability that an infective mosquito bite will produce an infection in the host, L_M is the average life span of the mosquito, H is the average duration of infectiousness in the human host, c is the probability that an infectious feeding will lead to sporozoite production in the vector, and d_I is the average duration of immunity.

The dynamics of the proportion of the host population infected with malaria (x_P) (i.e., with asexual parasites in the blood) may be expressed as

$$\frac{dx_P}{dt} = mab(1 - x)y \quad r_P x_P \tag{A3}$$

where r_P is the rate of recovery from infection (clearance of parasitemia).

Similarly, if a proportion γ of new infections results in clinical malaria, then changes in the proportion of the host population with mild malaria (x_D) may be described by the following equation:

$$\frac{dx_D}{dt} = \gamma mab(1 - x)y \quad r_D x_D \tag{A4}$$

where r_D is the rate of recovery from mild disease.

This basic framework may be extended to explore the effects of different stage-specific vaccines by adding compartments of vaccinated individuals as appropriate to the particular type of vaccine.

REFERENCES

1. **Allison, A. C.** 1964. Polymorphism and natural selection in human populations. *Cold Spring Harbor Symp. Quant. Biol.* **29:**137–149.
2. **Anderson, R. M., and R. M. May.** 1991. *Infectious Diseases of Humans: Dynamics and Control.* Oxford University Press, Oxford.
3. **Anderson, R. M., R. M. May, and S. Gupta.** 1988. Non-linear phenomena in host-parasite interactions. *Parasitology* **99:**S59–S79.
4. **Aron, J. L., and R. M. May.** 1982. The population dynamics of malaria, p. 139–179. *In* R. M. Anderson (ed.), *Population Dynamics of Infectious Diseases: Theory and Applications.* Chapman & Hall, Ltd., London.

5. **Boyd, M. F.** 1941. Epidemiology of malaria: factors related to the intermediate host, p. 551–607. *In* M. F. Boyd (ed.), *Malariology*. The W. B. Saunders Co., Philadelphia.
6. **Bruce-Chwatt, L. J.** 1963. A longitudinal survey of natural malaria infection in a group of West African adults. *W. Afr. Med. J.* **12:**141–173, 199–217.
7. **Brumpt, E.** 1941. The human parasites of the genus Plasmodium, p. 65–121. *In* M. F. Boyd (ed.), *Malariology*. The W. B. Saunders Co., Philadelphia.
8. **Day, K. P., and K. Marsh.** 1990. Naturally acquired immunity to *Plasmodium falciparum*. *Parasitol. Today* **7:**A68–A70.
9. **Deloron, P., and C. Chougnet.** 1992. Is immunity to malaria really short-lived? *Parasitol. Today* **8:**375–378.
10. **de Zoysa, A. P. K., C. Mendis, A. C. Gamage-Mendis, S. Weerasinghe, P. R. Herath, and K. N. Mendis.** 1991. A mathematical model for *Plasmodium vivax* malaria transmission: estimation of the impact of transmission-blocking immunity in an endemic area. *Bull. W.H.O.* **69:**725–734.
11. **Dietz, K.** 1988. Mathematical models for the transmission and control of malaria, p. 1091–1134. *In* W. H. Wernsdorfer and I. McGregor (ed.), *Principles and Practice of Malariology*. Churchill Livingstone, London.
12. **Eyles, D. E., and M. D. Young.** 1951. The duration of untreated or inadequately treated *Plasmodium falciparum* infections in the human host. *J. Natl. Malar. Soc.* **10:**327–336.
13. **Flint, J., A. V. S. Hill, D. K. Bowden, S. J. Oppenheimer, P. R. Sill, S. W. Serjeantson, J. Bana-Koiri, K. Bhatia, M. Alpers, A. J. Boyce, D. J. Weatherall, and J. B. Clegg.** 1986. High frequencies of α thalassaemia are the result of natural selection by malaria. *Nature* (London) **321:**744–749.
14. **Forsyth, K. P., R. F. Anders, J. Cattani, and M. A. Alpers.** 1989. Small area variation in prevalence of an S-antigen serotype of *Plasmodium falciparum* in villages of Madang, Papua New Guinea. *Am. J. Trop. Med. Hyg.* **40:**344–350.
15. **Greenwood, B. M., K. Marsh, and R. W. Snow.** 1991. Why do some children develop severe malaria? *Parasitol. Today* **7:**277–281.
16. **Gupta, S., and K. P. Day.** 1994. A theoretical framework for the immunoepidemiology of *Plasmodium falciparum* malaria. *Parasite Immunol.* **16:**361–370.
17. **Gupta, S., and A. V. S. Hill.** 1995. Dynamic interactions in malaria: host heterogeneity meets parasite polymorphism. *Proc. R. Soc. Ser. B* **261:**271–277.
18. **Gupta, S., A. V. S. Hill, D. Kwiatkowski, A. M. Greenwood, B. M. Greenwood, and K. P. Day.** 1994. Parasite virulence and disease patterns in *P. falciparum* malaria. *Proc. Natl. Acad. Sci. USA* **91:**3715–3719.
19. **Gupta, S., J. Swinton, and R. M. Anderson.** 1994. Theoretical studies of the effect of genetic heterogeneity in the parasite population on the transmission dynamics of malaria. *Proc. R. Soc. Ser. B* **256:**231–238.
20. **Gupta, S., K. Trenholme, R. M. Anderson, and K. P. Day.** 1994. Antigenic diversity and the transmission dynamics of *Plasmodium falciparum*. *Science* **263:**961–963.
21. **Halloran, M. E., and C. J. Struchiner.** 1992. Modelling transmission dynamics of stage-specific malaria vaccines. *Parasitol. Today* **8:**77–85.
22. **Halloran, M. E., C. J. Struchiner, and A. Spielman.** 1989. Modelling malaria vaccines. II. Population effects of stage-specific malaria vaccines dependent on natural boosting. *Math. Biosci.* **94:**115–149.
23. **Hill, A. V. S., C. E. Allsopp, D. Kwiatkowski, N. M. Anstey, P. Twumasi, P. A. Rowe, S. Bennett, D. Brewster, A. J. McMichael, and B. M. Greenwood.** 1991. Common West African HLA antigens are associated with protection from severe malaria. *Nature* (London) **352:**495–500.

24. **Jeffrey, G. M., and D. E. Eyles.** 1955. Infectivity to mosquitoes of *Plasmodium falciparum* as related to gametocyte density and duration of infection. *Am. J. Trop. Med. Hyg.* **4:**781–789.
25. **Macdonald, G.** 1956. Theory of eradication of malaria. *Bull. W.H.O.* **15:**369–387.
26. **Macdonald, G.** 1957. *The Epidemiology and Control of Malaria.* Oxford University Press, London.
27. **Marsh, K.** 1992. Malaria—a neglected disease? *Parasitol. Today* **104:**553–569.
28. **McLean, A. R., and R. M. Anderson.** 1988. Measles in developing countries. I. Epidemiological parameters and patterns. *Epidemiol. Infect.* **100:**111–133.
29. **Molineaux, L., and G. Gramiccia.** 1980. *The Garki Project.* World Health Organisation, Geneva.
30. **Nedelman, J.** 1984. Inoculation and recovery rates in the malaria model of Dietz, Molineaux and Thomas. *Math. Biosci.* **69:**209–233.
31. **Nokes, D. J., and R. M. Anderson.** 1987. Rubella vaccination policy: a note of caution. *Lancet* **i:**1441–1442.
32. **Ross, R.** 1911. *The Prevention of Malaria,* 2nd ed. Murray, London.
33. **Sinden, R. E.** 1983. Sexual development of malaria parasites. *Adv. Parasitol.* **22:**154–216.
34. **Yenchitsomanus, P., K. M. Summers, K. Bhatia, J. Cattani, and P. G. Board.** 1985. Extremely high frequencies of α-globin gene deletions in Madang and on Kar Kar Island, Papua New Guinea. *Am. J. Hum. Genet.* **37:**778–784.

Malaria Vaccine Development: A Multi-Immune Response Approach
Edited by Stephen L. Hoffman

Chapter 11

What Can Be Expected from Malaria Vaccines?

Brian Greenwood

INTRODUCTION

The wait for an effective malaria vaccine has been a long one, much longer than might have been anticipated 20 years ago, when it was first demonstrated that humans could be protected against malaria by vaccination with irradiated sporozoites (19). Ten years ago, there was widespread optimism that exciting developments in immunology and molecular biology would rapidly provide the key to the development of a subunit malaria vaccine (45). Unfortunately, this has not happened, because identifying the parasite molecules that are most important in the induction of protective immunity against malaria infection has proved to be more difficult than expected. Nevertheless, some progress has been made. Several vaccines based on the circumsporozoite protein have been developed and shown to give high antibody levels in most immunized subjects and to protect a few from challenge infections (4, 27, 39, 60). When tried under conditions of natural challenge, these vaccines were relatively ineffective, but they were safe (15, 38, 59, 65). In South America, a synthetic peptide vaccine based on three blood-stage antigens of *Plasmodium falciparum* gave a moderate level of protection against clinical malaria (70). This vaccine also gave modest protection in Tanzania (3) but no protection in The Gambia (22) (see chapter 9). Several other sporozoite, blood-stage, and transmission-blocking vaccines are approaching clinical trials. In view of this slow but steady progress, it is not unreasonable to imagine that a partially effective malaria vaccine will be available for widespread use in malaria-endemic areas within the next decade. Therefore, it may not be too early to consider some of the possible overall effects of the introduction of such a vaccine into a malaria-endemic area. Some of these issues are addressed in this chapter.

THE OBJECTIVES OF MALARIA VACCINATION

From the global point of view, the major requirement of a malaria vaccine is that it prevent deaths from malaria and abort the severe infections that may leave permanent neurological sequelae (14). To achieve this objective, it may not be necessary for a vaccine to prevent malaria infection, and doing so could even be a disadvantage (see below). A vaccine that could give young, susceptible children a level of immunity equivalent to that acquired by their parents as a result of prolonged natural exposure to infection would go a long way toward solving the malaria problem of most areas of high endemicity.

Unfortunately, such a vaccine would have limited value for nonimmune visitors to malaria-endemic areas, for whom even a mild malaria infection of the kind experienced by semi-immune adults would be unacceptable. For this group of individuals, a vaccine that gives a very high degree of protection against infection is required.

For both humanitarian and economic reasons, eradication of malaria must be the ultimate aim of malaria vaccine development. Malaria has several of the desirable characteristics of an infection that is a candidate for eradication. For example, there is no significant animal reservoir, and in the circumstances that are likely to prevail toward the end of an eradication campaign, most infections will be symptomatic and therefore accessible to an active case detection and treatment program. The success achieved in previous limited eradication campaigns based on vector control indicates that eradication of malaria is possible and should be a legitimate objective of malaria vaccine development.

TYPES OF MALARIA VACCINES AND THEIR LIKELY BIOLOGICAL EFFECTS

Three main groups of malaria vaccines are recognized: pre-erythrocytic, asexual-blood-stage, and transmission-blocking vaccines. However, subgroups of vaccines are being developed within each of these three main categories, and each subgroup has some unique characteristics. In this section, the likely biological consequences of immunization with the three main groups of malaria vaccines are considered. Any type of malaria vaccine that could induce an immune response that was 100% effective at killing parasites or blocking transmission would result in eradication of malaria in the community in which it was given, provided that a sufficiently high level of vaccine coverage was achieved. Such a paragon of a vaccine is an unlikely prospect, and the question of interest to the epidemiologist and the public health physician is what effects partially effective vaccines are likely to achieve (Table 1).

Pre-Erythrocytic-Stage Vaccines

It may be possible to develop vaccines that induce an immune response that can destroy sporozoites before they invade a hepatocyte. Because of the short time that sporozoites are in the circulation, this "Star Wars" approach will be difficult to achieve. Pre-erythrocytic vaccines,

Table 1. Likely effects of partially effective pre-erythrocytic, erythrocytic, and transmission-blocking vaccines

Type of vaccine	Reduction in[a]:			
	Death and severe disease	Mild disease	Infection	Transmission
Pre-erythrocytic	++	+	+	+
Erythrocytic	++	+	−	−
Transmission blocking	+	+	+	+

[a]++, great reduction; +, reduction; −, no reduction.

even those based on sporozoite antigens, are more likely to be effective when they induce an immune response that acts against liver-stage parasites, perhaps through the mediation of cytotoxic T lymphocytes. There is no evidence that the development of malaria parasites within the liver has any direct harmful effects on the host. Thus, the clinical consequences of anti-sporozoite and anti-liver-stage vaccines are likely to be similar.

A pre-erythrocytic vaccine that could induce an immune response sufficiently powerful to prevent the release of any merozoites from liver-stage parasites, regardless of the dose of sporozoites inoculated or of their antigenic characteristics, would provide 100% protection against clinical infections of all grades of severity and would interrupt transmission. Unfortunately, the first generation of pre-erythrocytic vaccines is likely to have a more limited ability to prevent the development of liver stages. What clinical effects are such vaccines likely to achieve?

It has been argued that because just one liver-stage schizont can release into the blood 10,000 to 30,000 merozoites, which double in numbers every 2 to 3 days, a pre-erythrocytic vaccine will be clinically useless unless it can achieve a very high level of protection against the development of liver-stage parasites. This argument is flawed on two grounds.

First, it appears that despite the very large numbers of sporozoites present in the salivary glands of infected mosquitoes, only a few sporozoites are inoculated during the majority of mosquito bites. The number of sporozoites found in the salivary glands of an infected mosquito varies from less than 100 to more than 100,000. Thus, in a recent study undertaken in western Kenya, the geometric mean number of sporozoites found in naturally infected *Anopheles gambiae* was 962 (9). Larger numbers have been found in experimentally infected mosquitoes of this species (7) and in earlier studies (58). Prevention of the infection of one hepatocyte in the face of an attack by 1,000 sporozoites would require an immune response of extraordinary efficiency. Fortunately, it appears that only a small proportion (2 to 3%) of the sporozoites present in the salivary glands is usually inoculated when an infected mosquito bites a susceptible host. This area is difficult to investigate, but by using a variety of ingenious in vitro techniques, a number of recent studies suggested that the number of inoculated sporozoites averages around 5 to 20 and is independent of the number of sporozoites present in the salivary glands (8, 57, 61). In some instances, no sporozoites were inoculated despite moderately heavy salivary gland infections. Simple calculations (Table 2) show that if the average number of sporozoites inoculated is 100 or more, then a pre-erythrocytic vaccine that is 90% effective would have a negligible effect on preventing infection. However, if the average number of sporozoites in-

Table 2. Efficacy of a pre-erythrocytic vaccine in preventing clinical malaria infection in relation to size of sporozoite inoculum[a]

No. of sporozoites inoculated	% of patients protected against disease with vaccine efficacy of:		
	80%	90%	95%
1	83	91	95
5	40	62	78
10	16	39	61
20	3	15	38
100	0	0	1

[a]This table assumes that the number of sporozoites inoculated follows a Poisson distribution and that one viable sporozoite reaching the liver gives rise to a clinical infection.

oculated is only 10, a more realistic estimate, then a vaccine that is 90% effective at destroying sporozoites or liver-stage parasites would prevent approximately 40% of subjects from progressing to a blood-stage infection.

Second, the effects of the immune response of the host must be taken into account. From the time that parasites first appear in the blood, the host will mount an immune response to blood-stage antigens, and this response may occur at an even earlier stage of the infection in the case of blood-stage antigens, such as MSA1, that are also expressed in the pre-erythrocytic stages of the parasite. This immune response is likely to be brisk in semi-immune individuals but will also take place in those experiencing their first malaria infection. Recent studies demonstrated that a substantial number of subjects who have not been previously exposed to malaria have CD4 cells that respond to malaria antigens, perhaps because of prior exposure to cross-reactive antigens (29, 73). In addition to these specific immune responses, some nonspecific protective immune mechanisms, such as enhanced production of C-reactive protein and orosomucoid, will come into play (26, 56). If the initial blood inoculum comes from only 1 liver schizont rather than from 100, then the time during which the host is exposed to blood-stage antigens before symptomatic levels of parasitemia are reached will, in the case of *P. falciparum*, be delayed by 2 to 4 days, i.e., one or two blood-stage cycles. This grace period could provide the additional time needed for the host to develop an immune response that could constrain parasite multiplication to a degree sufficient to prevent potentially fatal levels of parasitemia being achieved (Fig. 1).

Thus, in addition to preventing blood-stage infections in a proportion of subjects, a partially effective pre-erythrocytic vaccine could ame-

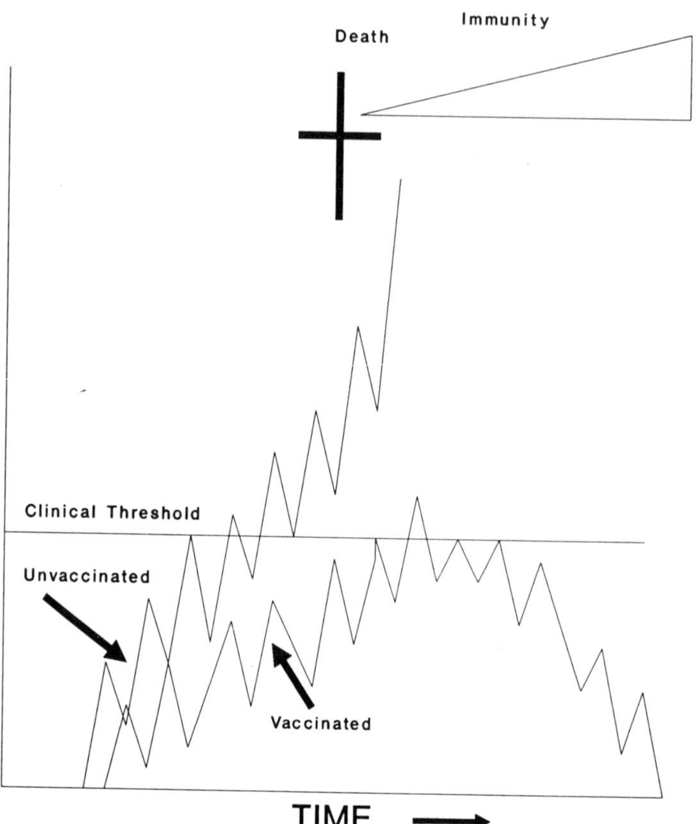

Figure 1. Effect of a partially effective pre-erythrocytic vaccine on parasite density and clinical outcome.

liorate the clinical features of malaria in another group of patients. That this might occur is suggested by the results of trials of insecticide-impregnated bed nets. Surprisingly, several studies have shown that impregnated bed nets are more effective at preventing deaths from malaria or clinical malaria than at preventing infection (1) and more effective at preventing high-density infections than low-density ones (16, 66). The way in which insecticide-treated bed nets achieve their protective effect is not known, but they may interfere with mosquito feeding in such a way as to reduce the sporozoite inoculum. If this is the case, their mode of action may be similar to that of a pre-erythrocytic vaccine.

Blood-Stage Vaccines

A vaccine that is 100% effective at destroying merozoites would prevent all clinical infections, prevent the development of gametocytes, and, if given to a high enough proportion of the population, stop transmission. However, it is very unlikely that the first blood-stage malaria vaccines will be so effective. The occurrence of strain variability is likely to pose a particular problem for blood-stage vaccines, since many of the antigens under consideration as vaccine candidates are known to vary from isolate to isolate.

The relationship between levels of parasitemia and the clinical features of malaria is a complex one, but in general, there is a correlation between high parasitemia and disease severity (24). Thus, if a blood-stage vaccine could induce an immune response that contained, although it did not prevent, parasite multiplication, then that vaccine would probably prevent deaths from malaria and severe clinical disease while having little effect on the overall incidence of infection.

The means by which *P. falciparum* produces the clinical features of severe malaria are still not understood completely, but at least three main mechanisms are thought to be involved (Table 3). Severe anemia is a major cause of admission to the hospital and of malaria deaths, and the incidence of this complication of malaria is increasing in Africa as parasites become increasingly resistant to chloroquine. While the pathogenesis of malarial anemia is complex (25), the most important mechanism is likely to be direct destruction of erythrocytes by parasites. Thus, if a malaria vaccine can induce an immune response that constrains parasite multi-

Table 3. Subgroups of blood-stage malaria vaccines and their possible effects

Subgroup	Mode of action	Effect on[a]:		
		Malaria fever	Cerebral malaria	Severe anemia
Anti-infection	Interference with reinvasion; removal of parasitized erythrocytes	+	+	+++
Antitoxin	Neutralization of "toxins" that stimulate cytokine release	+++	++	+
Antiadherence	Blocking of cytoadherence receptor on parasitized erythrocytes	−	+++	−

[a]+++, marked reduction; ++, moderate reduction; +, slight reduction; −, no reduction.

plication even though it does not prevent infection, then it would be likely to prevent severe malarial anemia.

Two mechanisms are believed to play central roles in the pathogenesis of cerebral malaria, the most feared complication of *P. falciparum* infections. These mechanisms are adherence of parasitized erythrocytes to the endothelial cells of small cerebral vessels and production of toxic levels of cytokines such as tumor necrosis factor. Good progress is being made in understanding the mechanisms of both phenomena (18, 41, 64), thus opening the way to the development of vaccines that could induce immune responses that prevent these processes. Therefore, developing a vaccine that induces the formation of antibodies to the surface receptor(s) of *P. falciparum* that is involved in cytoadherence may be possible. Alternatively, developing a vaccine that stimulates the production of antibodies to the phospholipid molecules that are believed to play an important part in cytokine induction may be possible (6). Such vaccines, although protecting against cerebral malaria, might not have any effect on parasite multiplication and so would be unlikely to protect against severe malarial anemia, although there is some evidence that tumor necrosis factor might play some part in its pathogenesis (17, 50). Therefore, "antitoxin" vaccines are unlikely to be used alone, but they might be included as a component of a vaccine cocktail.

Transmission-Blocking Vaccines

Transmission of malaria may be prevented by immune responses that destroy gametocytes, interfere with fertilization, or prevent the development of the malaria parasite within the mosquito. Vaccines to achieve each of these objectives are being developed. However, the biological effects of each of these vaccines will be similar: they will prevent or reduce malaria transmission. Mathematical modeling (see below) indicates that in areas of high malaria endemicity, very high levels of vaccine efficacy and vaccine coverage will be required to interrupt transmission.

Sometimes it is argued that transmission-blocking vaccines will be difficult to promote because even if they are highly effective, they will not provide the individual recipient with any kind of protection against malaria infection. This is the case if a subject is vaccinated in isolation. However, malaria transmission is frequently focal, so that if a large proportion of the members of a household or village are immunized with an effective transmission-blocking vaccine, then each individual in these communities will obtain some personal protection. However, in contrast to blood-stage vaccines, a transmission-blocking vaccine that is not sufficiently immunogenic to induce a very high level of immunity will have no beneficial clinical effect.

QUANTITATING THE BIOLOGICAL EFFECTS OF MALARIA VACCINES

A knowledge of the biology of the malaria parasite and the characteristics of the immune response that it induces allows us to make informed guesses as to the likely qualitative effects of the three main types of malaria vaccine. The modeling approach allows this guesswork to be taken a stage further by providing some indication of the likely quantitative effects of each vaccine when applied under various conditions. The kinds of helpful information that can be provided by modeling are described in detail in another chapter in this book. Vaccine modeling requires the use of terms that describe various biological characteristics of malaria infection. Unfortunately, it is frequently necessary to make major assumptions when giving these terms a quantitative value. For example, equations describing the effects of a transmission-blocking vaccine may use a term describing the duration of infectiousness following a malaria infection. However, there is almost no information on the duration of infectiousness following malaria infections in semi-immune subjects resident in areas of endemicity, so in this case, a major assumption has to be made. Two of the indirect benefits of the modeling approach have been the highlighting of areas of ignorance about the biology of naturally acquired malaria and the challenge to epidemiologists to provide the data that can fill some of these gaps.

THE CLINICAL SPECTRUM OF MALARIA AND ITS POSSIBLE MODULATION BY VACCINATION

The consequences of inoculation with sporozoites through the bite of an infected mosquito range from an aborted infection to an illness so severe that it kills within a day of the first appearance of symptoms (Fig. 2). The factors that determine the outcome of any individual infection are complex (37). One important determinant is the level of immunity of the host, which in turn is dependent to a large extent upon the frequency with which he or she has been infected with an antigenically similar malaria parasite in the past. Thus, in areas where malaria infections are infrequent, a large proportion of infected subjects become ill, and many have severe disease. In contrast, in areas of high malaria endemicity, only a small proportion of infected subjects, perhaps as few as 2%, develop severe disease. What is special about the latter group of subjects is uncertain. One possibility is that they possess certain genetic characteristics that make them susceptible to severe disease (40). Another possibility is that only a limited number of parasites possess the characteristics needed to produce severe complica-

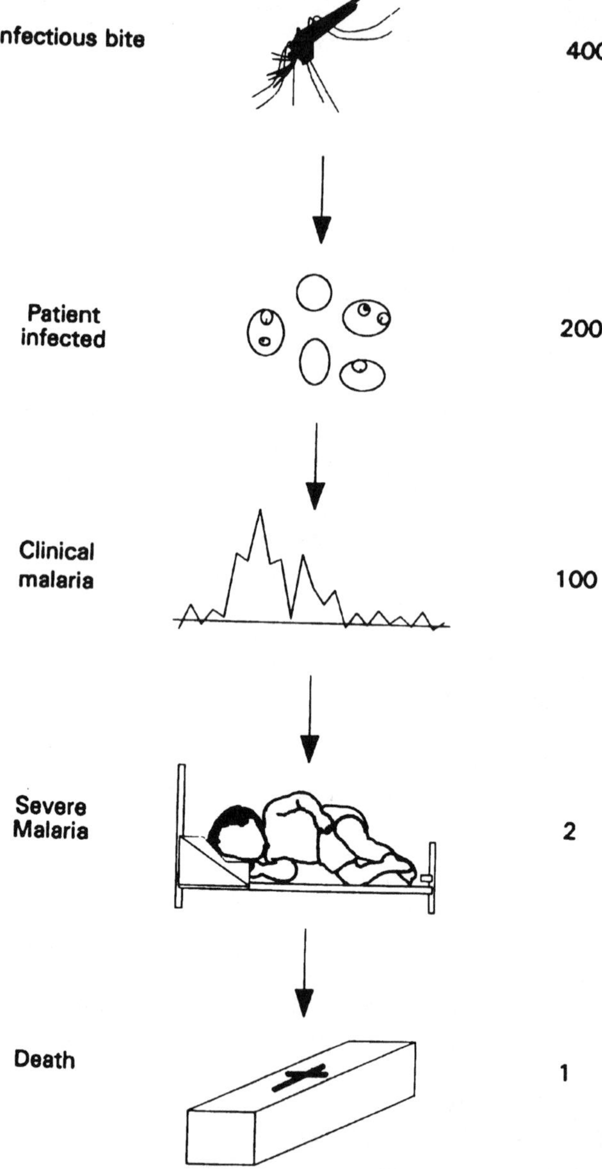

Figure 2. Clinical spectrum of *P. falciparum* malaria in African children. Numbers on the right-hand side indicate the approximate number of individuals in each category per death from malaria in an area of moderate malaria transmission. Reprinted from reference 37 with permission.

tions. Environmental factors such as the availability of treatment may play some part in determining the outcome of an individual infection, but these do not appear to be as important as might have been expected (45a).

Because the nature of the factors that determine the balance between severe and mild infections in semi-immune populations is not clearly understood, the possibility exists that under certain circumstances, vaccination will enhance rather than decrease mortality from malaria. Thus, by substantially reducing the overall prevalence of infection in a malaria-endemic community, a pre-erythrocytic or blood-stage vaccine could interfere with the development of naturally acquired immunity in a high proportion of the population in such a way as to upset the previous balance between severe and mild disease. Disturbance of the established equilibrium could be expressed in several ways. First, it is possible that the proportion of severe to mild infections among the groups previously most at risk, for example, young children in Africa, will be increased. Second, loss of established immunity as a result of lack of exposure might result in severe disease developing in adult subjects who had previously experienced only mild or asymptomatic infections. Finally, reduction of infection might change a situation of stable malaria transmission into an unstable one, where large epidemics would be possible. This last possibility would be particularly likely in the case of a vaccine that gave only a limited period of protection and was not boosted by natural infection. In each of these situations, it is possible to envisage a situation in which malaria mortality over a period of years could be higher after vaccination than it would have been without (Table 4). That this risk is not just theoretical is indicated by experience in Madagascar and São Tome. On each of these

Table 4. Possible effects of two different malaria vaccines on mortality from malaria in a malaria-endemic area

Parameter	Result with:		
	No vaccine	Vaccine A	Vaccine B
Vaccine efficacy		70%	90%
Effect on natural immunity		None	Marked
Population of children at risk	1,000	1,000	1,000
No. of clinical cases of malaria[a]	1,000	300	100
No. of severe cases	20[b]	6[b]	30[c]
No. of deaths[d]	10	3	15

[a]Based on the assumption of one clinical attack per year.
[b]Based on a 2% incidence of severe disease (semi-immune).
[c]Based on a 30% incidence of severe disease (nonimmune).
[d]Based on a 50% mortality among severe cases.

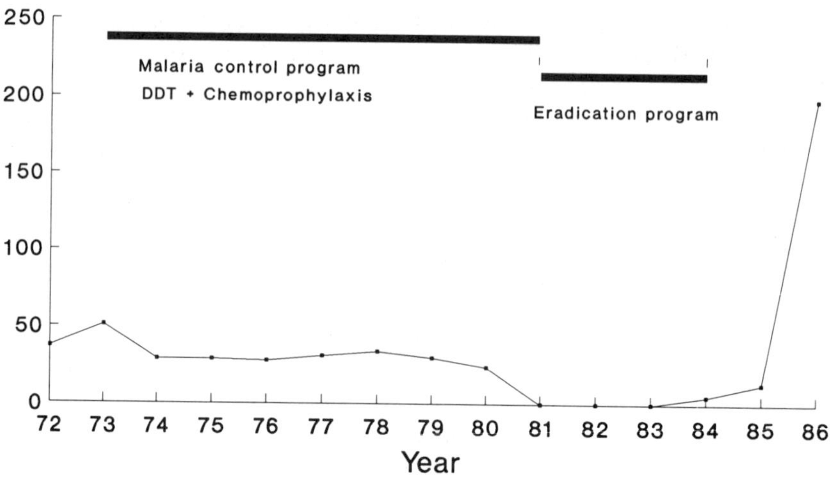

Figure 3. Deaths from malaria (●) on the island of Saõ Tome during and after a malaria eradication program. Reprinted from reference 23 with permission. © 1987 American Institute of Biological Sciences.

islands, a period of successful malaria control was followed by major epidemics (23). On São Tome, as many deaths occurred during one annual epidemic as had occurred during about 10 years in the period preceding the introduction of malaria control (Fig. 3).

Fortunately, experience with chemoprophylaxis for the population of malaria-endemic areas indicates that it is possible to reduce mortality and morbidity from malaria and to significantly reduce the incidence of malaria infection without leading to substantial loss of naturally acquired immunity (32). Thus, concerns about this possible outcome of the introduction of a partially effective vaccine into an area of endemic malaria may be unwarranted, but once large-scale vaccination programs are launched, careful long-term surveillance will be essential.

THE PUBLIC HEALTH IMPACT OF AN EFFECTIVE MALARIA VACCINE

In this section, the likely impact of a malaria vaccine that could prevent nearly all clinical attacks of malaria and could interrupt transmission is considered.

Mortality

Recently, several attempts to estimate the number of deaths caused each year by malaria have been made. This task is impossible to accomplish with any degree of accuracy, but most attempts have come up with a figure of between 0.5 and 2 million deaths from malaria each year (33, 72). Nearly all of these deaths occur in children in sub-Saharan Africa.

It is reasonable to expect that an effective pre-erythrocytic or blood-stage vaccine would prevent death from malaria in vaccine recipients, even if it did not provide complete protection against infection. Thus, provided that the vaccine could be delivered to the majority of the "at risk" population, then a major proportion of the 1 million or so deaths directly attributed to malaria would be prevented.

Two factors complicate considerations of the effects of a malaria vaccine on mortality: the concept of replacement mortality and the indirect effect of malaria on other diseases (20, 52). In deprived communities, a proportion of children may have a high risk of dying because of factors such as a low birth weight, malnutrition, or extreme resource deprivation for their family; if such children are prevented from dying from one condition by a specific intervention, such as vaccination, then they may succumb to another. Direct evidence to support this concept is meager. The consequences of an epidemic of measles in the village of Keneba in The Gambia are sometimes cited to support this concept. McGregor (47) reported that mortality during the rainy season after a dry-season epidemic of measles with a high mortality was lower than usual but that the effect was not marked: four deaths compared to six and seven during the preceding 2 years. More striking was an increase in the number of deaths during the subsequent two rainy seasons, perhaps partly due to the late effects of measles on mortality (42) but due also to an increase in the birth rate following the measles epidemic. The results of the large malaria intervention trial undertaken in Garki in northern Nigeria have also been used to support the concept of replacement mortality. In this study, the reduction in infant mortality rate in villages where insecticide and chemoprophylaxis were used was less than the reduction in the daily infant conversion rate from negative to positive blood film, and the suggestion was made that this was because infants saved from death from malaria died from something else (53). This explanation for the finding is plausible, but there is no reason that the level of reduction in malaria mortality should be quantitatively directly related to the level of reduction of infection; many factors can influence the balance between death and infection, as discussed above. Sickle cell disease may represent a better example of replacement mortality. In rural areas of Africa, the life expectancy of a child with sickle cell disease is poor, and many such children die from malaria. In such communities, it is likely that a child with sickle cell disease who is protected against malaria by vaccination will die from another infection, such as pneumococcal septicemia, unless the overall level of health care provided in the community is improved. Nevertheless, any reduction in the efficacy of a malaria vaccine in reducing child mortality as a result of replacement mortality is likely to be small and more than offset by the indirect effects of the vaccine in preventing deaths from other causes.

Acute malaria impairs the functions of several components of the immune response, and consequently, malaria increases susceptibility to several other infections, particularly to nontyphoidal salmonella septicemia, which has a high mortality in children (46). In addition, malaria has an adverse effect on nutrition (48, 62, 67). Thus, control of malaria might lead to a reduction in deaths from other infections and from malnutrition. There is good evidence that this is the case. When malaria was controlled successfully in Ceylon in the late 1940s, overall mortality fell to a much greater extent than did malaria mortality. A part of this fall may have been related to other improvements in the health services, for example, the introduction of penicillin, but reduction of deaths from other causes by controlling malaria was probably a contributory factor (30). When malaria was controlled successfully in Guyana, mortality from renal disease was reduced, probably as a result of control of *Plasmodium malariae* infections, but the number of deaths from pneumonia was also reduced (28). (Fig. 4). In Nigeria, protection of

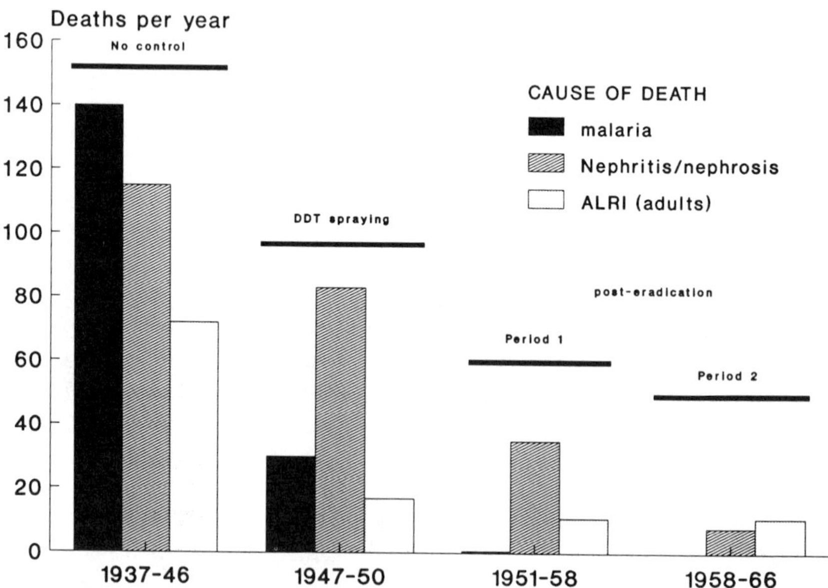

Figure 4. Deaths from malaria, chronic nephritis, and acute lower respiratory infections (ALRI) in Guyana before and after a malaria eradication program. Reprinted from reference 28 with permission.

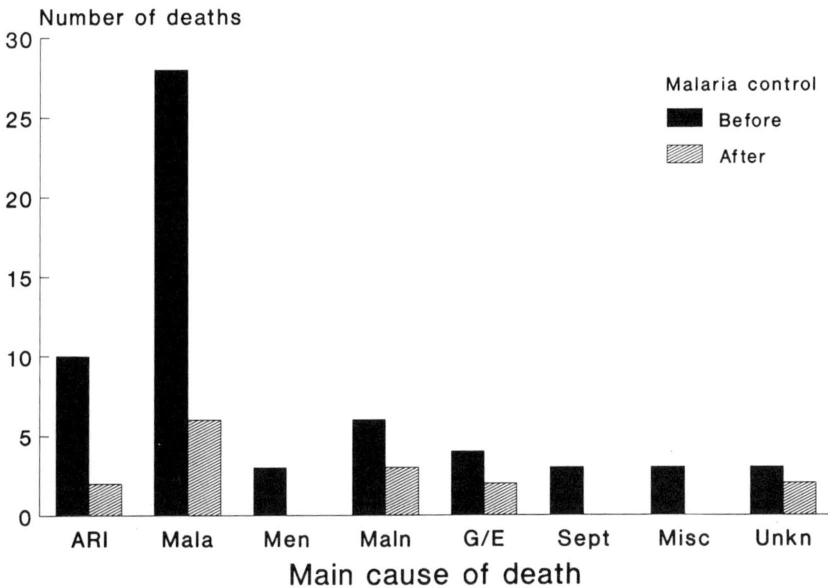

Figure 5. Deaths in Gambian children under the age of 6 years before and after the introduction of a malaria control program employing insecticide-treated bed nets and chemoprophylaxis. ARI, acute respiratory infection; Mala, malaria; Men, meningitis; Maln, malnutrition; G/E, gastroenteritis; Sept, septicemia; Misc, miscellaneous; Unkn, unknown. From reference 2 with permission.

young children from malaria by chemoprophylaxis with chloroquine led to a reduction in deaths from malnutrition (11), and this finding was confirmed in a larger study subsequently undertaken in The Gambia (34). More recently, a similar situation was observed among Gambian children protected from malaria by chemoprophylaxis and insecticide-treated bed nets. Preintervention studies, using the postmortem-questionnaire technique, suggested that malaria was responsible for about 25% of deaths in children aged 1 to 4 years who were resident in the study area. However, control of malaria by chemoprophylaxis and insecticide-treated bed nets led to a reduction in mortality of over 50% in children in this age group (1, 2). Analysis of causes of death showed that not only deaths from malaria but also those from acute respiratory infections and malnutrition were prevented (Fig. 5). Part of this effect may have been due to failure of the postmortem-questionnaire approach to establish the correct cause of death, in particular to differentiate between deaths from acute malaria and those from pneumonia, but this successful malaria control campaign probably did lead to a reduction in deaths from causes not related directly to malaria infection. Effective control of malaria by vaccination should have the same effect.

Thus, despite the possibility of replacement mortality, the limited data available suggest that an effective malaria vaccine would lead to a reduction in childhood mortality by an amount greater than the 1 million deaths currently attributed to the infection.

Morbidity

Estimating the number of clinical attacks of malaria that occur worldwide each year is at least as difficult as estimating the level of mortality caused by the infection. Many attacks are mild and either treated at home or left untreated, so official statistics reflect only a small proportion of the morbidity caused by malaria. For example, the figure of 100 million attacks per year cited by the World Health Organization (72) is probably a substantial underestimate. Stürchler's estimate (68) of nearly 500 million cases a year, with 300 million of them in Africa, is probably a better estimate. Recently, a number of attempts have been made to determine the incidence of malaria in different parts of Africa more accurately by using active surveillance. Figures in the range of one to five attacks per year were found for children in rural areas (33, 69, 71) but attack rates are much less in large urban areas, where about 30% of the population of Africa now lives. Thus, an overall rate of one or two clinical attacks per year in children under the age of 10 years is probably a reasonable estimate, giving 150 to 300 million clinical attacks in African children each year, to which must be added a substantial number of mild attacks in adults. Thus, it is no surprise that fever, often due to malaria, is the most frequent cause of attendance at clinics and dispensaries in many parts of the tropics.

There is ample evidence of the importance of malaria as a cause of admission to hospital in many parts of Africa. A recent survey showed that in a group of mission hospitals in East Africa, the proportion of hospital admissions due to malaria is increasing, as is mortality from the infection (55). At the large Mamayemo hospital in Kinshasa, Zaire, malaria accounted for 38% of all admissions to the pediatric wards during 1985 through 1986 and caused 13% of pediatric deaths (31). At the central government hospital in The Gambia (The Royal Victoria Hospital), malaria is also the most frequent cause of admission to the pediatric department, accounting for 24% of pediatric admissions and 24% of pediatric deaths during a 3-year period (1988 to 1991) (13). Mortality among children admitted to this hospital with cerebral malaria is in the vicinity of 20%, and 12% of survivors are left with permanent neurological sequelae (14).

Thus, an effective vaccine that prevented clinical illness due to malaria infection would, in many parts of Africa, lead to a reduction in outpatient and clinical attendances of as much as 25% and to a reduction in pediatric admissions by a similar figure.

The adverse effects of malaria on nutrition have been demonstrated clearly (48). Malaria probably exerts this effect by decreasing food intake and exerting metabolic effects (67). Three studies have shown that children protected from malaria with chemoprophylaxis grow better than control children living in the same environment (12, 35, 49) (Fig. 6). Although the effect was not very marked in any of these studies, an effective malaria vaccine is likely to lead to a significant improvement in the overall nutritional state of the community in which it was used, unless the supply of food became a limiting factor.

The Outcome of Pregnancy

Malaria has a number of adverse effects on pregnancy, especially in primigravidae (10). In nonimmune subjects, acute malaria may lead to abortion, stillbirth, or premature delivery. In semi-immune subjects, malaria rarely results in an acute febrile illness, but it may cause severe anemia, especially in primigravidae. Infection of the placenta impairs fe-

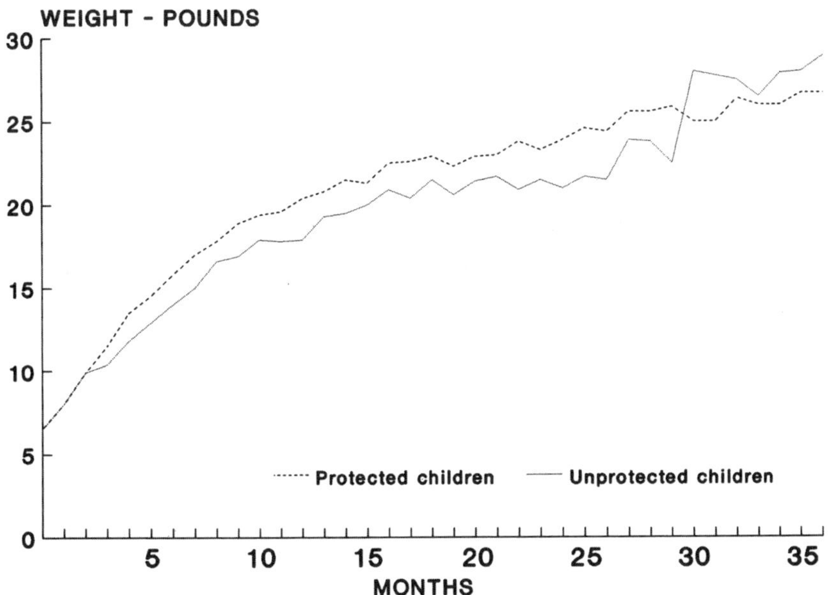

Figure 6. Effect of the regular administration of chemoprophylaxis on the weight of young Gambian children. From reference 49 with permission.

tal development and is associated with low birth weight. Some, but not all, studies have shown that these serious complications of pregnancy can be prevented by chemoprophylaxis (21, 36, 54). Thus, pregnant women are likely to be a high-priority target for vaccination. A vaccine that could prevent a high proportion of infections would improve the outcome of pregnancy in women exposed to malaria, especially primigravidae. Whether a vaccine that was able to contain levels of parasitemia but not prevent infection would have any effect is less certain. It might reduce the prevalence of severe anemia, but whether it could prevent infection of the placenta and consequent low birth weight is more doubtful.

THE ECONOMIC EFFECTS OF AN EFFECTIVE MALARIA VACCINE

Malaria can cause economic loss in several ways (Table 5). It can result in loss of work among adults, for example, among migrant workers in Amazonia and gem miners in Cambodia and Thailand; reduce work efficiency; or cause affected communities to adopt easier but less productive agricultural practices (5). Among schoolchildren and students, it may lead to absenteeism. Each clinical attack of malaria that requires treatment results in direct costs in relation to the costs of medical care, including drugs and the time of the staff who diagnose and care for the patient. In addition, there may be costs to the family in bringing a patient to a treatment center and in loss of earnings by those who look after the patient. If the patient dies, the family will be faced with the direct costs of premature funeral expenses and loss of time at work for the mourning period, and there will be indirect costs related to the loss of years of productive life. Quantitating these costs is difficult, but recently, a number

Table 5. Economic costs of malaria

Type of cost	Actual cost of:	
	Clinical attack of malaria	Death from malaria
Direct	Treatment costs	Costs of treatment prior to death
	Drugs	Premature funeral expenses
	Time of medical staff	
	Building overheads	
	Transport costs	
Indirect	Loss of work time by patient	Loss of work time by family during illness and mourning period
	Loss of work time by caregiver	Productive years of life lost

of attempts have been made to do this in various communities. However, most of these estimates have been based on assumptions of the likely effects of malaria on a family; more accurate assessments based on direct observations are needed. Several recent studies estimated the costs of malaria in Africa (63). Estimates of the cost per case have ranged from $6 to $13, with an average cost combining direct and indirect costs of approximately $10. Shepard et al. (63) estimated that this figure would increase to $16 by 1995. Each case study has shown that indirect costs contribute a larger share to the total costs than do direct costs. If it is accepted that 300 million cases of malaria that are severe enough to interrupt the activities of the affected household occur in Africa each year, then the economic costs of malaria in Africa alone are likely to be at least $1 billion per year. To this must be added the substantial costs of malaria in areas of Asia and South America, where loss of productivity among adult workers may be high.

The likely costs of the first commercially available malaria vaccines are difficult to guess, because these costs will be influenced by whether a single antigen will suffice or whether a vaccine cocktail will be required and by the nature of the production methods finally chosen. However, once development costs have been met, perhaps through sales to tourists and visitors to areas of endemicity, and large sales of vaccine are guaranteed, it is reasonable to hope that a course of two or three doses of malaria vaccine might cost something in the range of $10 to $20 excluding delivery costs. Provided that the vaccine can be given within the framework of the existing Expanded Programs of Immunization, now established very effectively in many developing countries, delivery costs will be low. However, delivery costs will be substantially higher if specific malaria vaccine programs are required. In tropical Africa, a malaria vaccine that cost $10 and provided lifelong protection against death from malaria and 5 years of protection against clinical attacks of malaria vaccination would cost $2 per clinical attack averted and $250 per life saved, figures that compare favorably with those achieved by other successful intervention strategies such as vector control and chemotherapy (43, 51).

The vaccine costs of routine immunization of all children in Africa with a malaria vaccine that cost $10 a course would be something on the order of $200 million a year. Although such a program would almost certainly be cost-effective, some of the savings, for example, in years of healthy life saved, would not generate cash for the health authorities responsible for buying the vaccine. Thus, if the vaccine was to be used widely in the areas where it was most needed, then some international financial assistance would be required. The prospects of being asked for indefinite support for the purchase of malaria vaccines could provide a powerful stimulus to the donor agencies to consider the possibility of an eradication campaign. If eradication could be achieved by vaccination of all at-risk subjects, then its costs would probably be covered within a few years by abolition of the need to vaccinate all children in malaria-endemic areas and all visitors to such communities.

CONCLUSION

An effective malaria vaccine would have tremendous public health and economic consequences. It would reduce childhood mortality in Africa, would substantially reduce the costs presently spent on treatment, would improve nutrition, and could bring direct economic benefits by reducing absenteeism and improving work efficiency. Widespread deployment of a highly effective malaria vaccine is probably some years away, but the possibility of some sudden and unexpected breakthrough cannot be excluded. Therefore, it is appropriate to give some thought to the possible long-term consequences of the deployment of such a vaccine.

Most thoughtful people consider that the greatest challenge facing the developing world today is the uncontrolled growth of population. This growth is most marked in tropical Africa, the area that has the fewest resources and in which the impact of a malaria vaccine is likely to be most dramatic. At present, the population growth rate of much of Africa is in the region of 3%, which means a doubling of the population every 25 years. How might this situation be changed by an effective malaria vaccine? In Fig. 7, I show the likely effect of an effective malaria vaccine on the population of The Gambia. This model assumes that the vaccine is 90% effective in preventing death from malaria, that the infant mortality rate is 100/1,000, that malaria accounts for 10% of deaths in infants, and that the child mortality rate is 25/1,000, half of which are due to malaria. In this model, 10 years after the introduction of the vaccine, the population would be about 6% greater than would otherwise have been predicted.

Recently, King drew attention to the doubtful prospects that face the ever-increasing number of children and young people in the developing world (44). Many face a life of severe economic hardship characterized by poor housing, poor nutrition, lack of education, and lack of job opportunities. It would be a tragedy if one of the main consequences of the successful development of an effective malaria vaccine, a major scientific achievement, was to make this situation worse. If the gains promised by a malaria vaccine are to be realized then its introduction must be accompanied by measures that ensure that its success will not be accompanied by a further increase in the numbers of children surviving to face a dismal future.

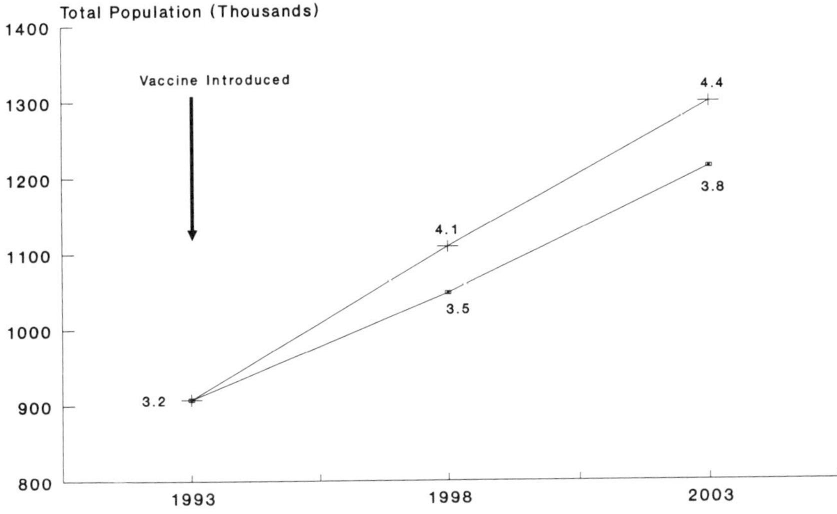

Figure 7. Predicted effect of an effective malaria vaccine on the population of The Gambia. Total population and percentage growth rates are shown. Estimates are based on the assumptions that (i) malaria accounts for 10% of deaths in infants and 50% of deaths in children aged 1 to 4 years, (ii) the vaccine is 90% effective at preventing death from malaria, and (iii) there is no change in fertility rate. Symbol: ■, no vaccination; +, vaccination. (From reference 23a.)

REFERENCES

1. **Alonso, P. L., S. W. Lindsay, J. R. M. Armstrong, M. Conteh, A. G. Hill, P. H. David, G. Fegan, A. De Francisco, A. J. Hall, F. C. Shenton, K. Cham, and B. M. Greenwood.** 1991. The effect of insecticide-treated bednets on mortality of Gambian children. *Lancet* **337:**1499–1502.
2. **Alonso, P. L., S. W. Lindsay, J. R. M. Armstrong, K. Keita, P. Gomez, F. C. Shenton, A. G. Hill, P. H. David, G. Fegan, M. K. Cham, and B. M. Greenwood.** 1992. A malaria control trial using insecticide-treated bednets and targeted chemoprophylaxis in a rural area of The Gambia, West Africa. The impact of the interventions on mortality and morbidity from malaria. *Trans. R. Soc. Trop. Med. Hyg.* **87**(Suppl. 2):37–44.
3. **Alonso, P. L., T. Smith, J. R. M. Armstrong Schellenberg, H. Masanja, S. Mwankusye, H. Urasse, I. Bastos de Azeveda, J. Chongela, S. Kobero, C. Menendez, N. Hurt, M. C. Thomas, E. Lyimo, R. Hayes, A. Y. Kitua, M. C. Lopez, W. L. Kilama, T. Teuscher, and M. Tanner.** 1994. Randomised trial of efficacy of Spf66 vaccine against *Plasmodium falciparum* malaria in children in Southern Tanzania. *Lancet* **344:**1175–1181.
4. **Ballou, W. R., S. L. Hoffman, J. A. Sherwood, M. R. Hollingdale, F. A. Neva, W. T. Hockmeyer, D. M. Gordon, I. Schneider, R. A. Wirtz, J. F. Young, G. F. Wasserman, P. Reeve, C. L. Diggs, and J. D. Chulay.** 1987. Safety and efficacy of a recombinant DNA *Plasmodium falciparum* sporozoite vaccine. *Lancet* **i:**1277–1281.
5. **Barlow, R., and L. M. Grobar.** 1986. *Costs and Benefits of Controlling Parasitic Diseases.* PHN technical notes 85-17. World Bank, Washington D.C.

6. **Bate, C. A., J. Taverne, E. Roman, C. Moreno, and J. H. Playfair.** 1992. Tumor necrosis factor induction by malaria exoantigens depends upon phospholipid. *Immunology* **75:**129–135.
7. **Beier, J. C., J. R. Davis, J. A. Vaughan, B. H. Noden, and M. S. Beier.** 1991. Quantitation of *Plasmodium falciparum* sporozoites transmitted *in vitro* by experimentally infected *Anopheles gambiae* and *Anopheles stephensi. Am. J. Trop. Med. Hyg.* **44:**564–570.
8. **Beier, J. C., F. K. Oyango, J. K. Koros, M. Ramadhan, R. Ogwang, R. A. Wirtz, D. K. Koech, and C. R. Roberts.** 1991. Quantitation of malaria sporozoites transmitted *in vitro* during salivation by wild Afrotropical *Anopheles. Med. Vet. Entomol.* **5:**71–79.
9. **Beier, J. C., F. K. Onyango, R. Ramadhan, J. K. Koros, C. M. Asiago, R. A. Wirtz, D. K. Koech, and C. R. Roberts.** 1991. Quantitation of malaria sporozoites in the salivary glands of wild Afrotropical *Anopheles. Med. Vet. Entomol.* **5:**63–70.
10. **Brabin, B. J.** 1983. An analysis of malaria in pregnancy in Africa. *Bull. W.H.O.* **61:**1005–1016.
11. **Bradley-Moore, A. M., B. M. Greenwood, A. K. Bradley, A. Bartlett, D. E. Bidwell, A. Voller, B. R. Kirkwood, and H. M. Gilles.** 1985. Malaria chemoprophylaxis with chloroquine in young Nigerian children. 1. Its effect on mortality, morbidity and the prevalence of malaria. *Ann. Trop. Med. Parasitol.* **79:**549–562.
12. **Bradley-Moore, A. M., B. M. Greenwood, A. K. Bradley, B. R. Kirkwood, and H. M. Gilles.** 1985. Malaria chemoprophylaxis with chloroquine in young Nigerian children. III. Its effect on nutrition. *Ann. Trop. Med. Parasitol.* **79:**575–584.
13. **Brewster, D. R., and B. M. Greenwood.** 1993. Seasonality of paediatric diseases in The Gambia, West Africa. *Ann. Trop. Paediatr.* **13:**133–146.
14. **Brewster, D. R., D. Kwiatkowski, and N. J. White.** 1990. Neurological sequelae of cerebral malaria in children. *Lancet* **336:**1039–1043.
15. **Brown, A. E., P. Singharaj, H. K. Webster, J. Pipithkul, D. M. Gordon, J. W. Boslego, K. Krinchai, P. Suarchawaratana, C. Wongsrichanalai, W. R. Ballou, B. Permpanich, K. C. Kain, M. R. Hollingdale, J. Wittes, J. U. Que, M. Gross, S. J. Cryz, and J.C. Sadoff.** 1994. Safety, immunogenicity, and limited efficacy study of a recombinant Plasmodium falciparum circumsporozoite vaccine in Thai soldiers. *Vaccine* **12:**102–108.
16. **Carnevale, P., V. Robert, C. Boudin, J.-M. Halna, L. Pazart, P. Gazin, A. Richard, and J. Mouchet.** 1988. La lutte contre le paludisme par des moustiquaires imprégnées de pyréthrinoides au Burkina Faso. *Bull. Soc. Pathol. Exot.* **81:**832–846.
17. **Clark, I. A., and G. Chaudhri.** 1988. Tumour necrosis factor may contribute to the anaemia of malaria by causing dyserythropoiesis and erythrophagocytosis. *Br. J. Haematol.* **70:**99–103.
18. **Clark, I. A., K. A. Rockett, and W. B. Cowden.** 1992. TNF in malaria, p. 303–328. *In* B. Beutler (ed.), *Tumour Necrosis Factors: the Molecules and Their Emerging Role in Medicine.* Raven Press, New York.
19. **Clyde, D. F., H. Most, V. C. McCarthy, and J. P. Vanderberg.** 1973. Immunization of man against sporozoite-induced falciparum malaria. *Am. J. Med. Sci.* **266:**169–177.
20. **Cohen, J. E.** 1988. Estimating the effects of successful malaria control programmes on mortality. *Popul. Bull. U.N.* **25:**6–26.
21. **Cot, M., A. Roisin, D. Barro, A. Yada, J.-P. Verhave, P. Carnevale, and G. Breart.** 1992. Effects of chloroquine chemoprophylaxis during pregnancy on birth weight: results of a randomized trial. *Am. J. Trop. Med. Hyg.* **46:**21–27.
22. **D'Allesandro, U., A. Leach, C. J. Drakely, S. Bennett, B. O. Olaleye, G. W. Fegan, M. Jawara, P. Langerock, M. O. George, G. A. Targett, and B. M. Greenwood.** 1995. Efficacy trial of malaria vaccine SPf66 in Gambian infants. *Lancet* **346:**462–467.
23. **de Ceita, J. G. V.** 1987. Malaria in Sao Tome and Principe, p. 142–155. *In* A. A. Buck (ed.), *Proceedings of the Conference on Malaria in Africa.* American Institute of Biological Sciences, Washington, D.C.

23a. **Fegan, G. W., and B. M. Greenwood.** Unpublished data.

24. **Field, J. W., and J. C. Niven.** 1937. A note on prognosis in relation to parasite counts in acute subtertian malaria. *Trans. R. Soc. Trop. Med. Hyg.* **30:**569–574.

25. **Fleming, A. F.** 1981. Haematological manifestations of malaria and other parasitic diseases. *Clin. Haematol.* **10:**983–1011.

26. **Friedman, M. J.** 1983. Control of malaria virulence by α_1-acid glycoprotein (orosomucoid), an acute-phase (inflammatory) reactant. *Proc. Natl. Acad. Sci. USA* **80:**5421–5424.

27. **Fries, L. F., D. M. Gordon, R. L. Richards, J. E. Egan, M. R. Hollingdale, M. Gross, C. Silverman, and C. R. Alving.** 1992. Liposomal malaria vaccine in humans: a safe and potent adjuvant strategy. *Proc. Natl. Acad. Sci. USA* **89:**358–362.

28. **Gigliogi, G.** 1972. Changes in the pattern of mortality following the eradication of hyperendemic malaria from a highly susceptible community. *Bull. W.H.O.* **46:**181–202.

29. **Goodier, M., P. Fey, K. Eichmann, and J. Langhorne.** 1992. Human peripheral blood gamma delta T cells respond to antigens of *Plasmodium falciparum*. *Int. Immunol.* **4:**33–41.

30. **Gray, R. H.** 1974. The decline of mortality in Ceylon and the demographic effects of malaria control. *Popul. Stud.* **28:**205–229.

31. **Greenberg, A. E., M. Ntumbanzondo, N. Ntula, L. Mawa, J. Howell, and F. Davachi.** 1989. Hospital-based surveillance of malaria-related paediatric morbidity and mortality in Kinshasa, Zaire. *Bull. W.H.O.* **67:**189–196.

32. **Greenwood, B. M., P. H. David, L. N. Otoo-Forbes, S. J. Allen, P. L. Alonso, J. R. Armstrong Schellenberg, P. Byass, M. Hurwitz, A. Menon, and R. W. Snow.** 1995. Mortality and morbidity from malaria after stopping malaria chemoprophylaxis. *Trans. R. Soc. Trop. Med. Hyg.* **89:**629–633.

33. **Greenwood, B. M., A. M. Greenwood, A. K. Bradley, P. Byass, K. Jammeh, K. Marsh, S. Tulloch, F. S. J. Oldfield, and R. J. Hayes.** 1987. Mortality and morbidity from malaria among children in a rural area of The Gambia, West Africa. *Trans. R. Soc. Trop. Med. Hyg.* **81:**478–486.

34. **Greenwood, B. M., A. M. Greenwood, A. K. Bradley, R. W. Snow, P. Byass, R. J. Hayes, and A. B. N'Jie.** 1988. A comparison of two strategies for control of malaria within a Primary Health Care Programme in The Gambia. *Lancet* **i:**1121–1127.

35. **Greenwood, B. M., A. M. Greenwood, A. W. Smith, A. Menon, A. K. Bradley, R. W. Snow, F. Sisay, S. Bennett, W. M. Watkins, and A. B. H. N'Jie.** 1989. A comparative study of Lapudrine® (chlorproguanil) and Maloprim® (pyrimethamine and dapsone) as chemoprophylactics against malaria in Gambian children. *Trans. R. Soc. Trop. Med. Hyg.* **83:**182–188.

36. **Greenwood, B. M., A. M. Greenwood, R. W. Snow, P. Byass, S. Bennett, and A. B. Hatib-N'Jie.** 1989. The effects of malaria chemoprophylaxis given by traditional birth attendants on the course and outcome of pregnancy. *Trans. R. Soc. Trop. Med. Hyg.* **83:**589–594.

37. **Greenwood, B., K. Marsh, and R. Snow.** 1991. Why do some children develop severe malaria. *Parasitol. Today* **10:**277–281.

38. **Guiguemdé, T. R., D. Sturchler, J. B. Ouédraogo, M. Drabo, H. Etlinger, C. Douchet, A. R. Gbary, L. Haller, S. Kambou, and M. Fernex.** 1990. Vaccination contre le paludisme: premier essai avec un vaccin sporozoite, le $(NANP)_3$-TT (RO 40-2361) en Afrique (Bobo-Dioulasso, Burkina Faso). *Bull. Soc. Pathol. Exp.* **83:**217–227.

39. **Herrington, D. A., D. F. Clyde, G. Losonsky, M. Cortesia, J. R. Murphy, J. Davis, S. Baqar, A. M. Felix, E. P. Heimer, D. Gillessen, E. Nardin, R. S. Nussenzweig, V. Nussenzweig, M. R. Hollingdale, and M. Levine.** 1987. Safety and immunogenicity in man of a synthetic peptide malaria vaccine against *Plasmodium falciparum* sporozoite. *Nature* (London) **328:**257–259.

40. **Hill, A. V. S., C. E. M. Allsopp, D. Kwiatkowski, N. M. Anstey, P. Twumasi, P. A.**

Rowe, S. Bennett, D. Brewster, A. J. McMichael, and B. M. Greenwood. 1991. Common West African HLA antigens are associated with protection from severe malaria. *Nature* (London) **352:**595–600.

41. **Howard, R. J., and A. D. Gilladoga.** 1989. Molecular studies related to the pathogenesis of cerebral malaria. *Blood* **74:**2603–2618.
42. **Hull, H. F., P. J. Williams, and F. Oldfield.** 1983. Measles mortality and vaccine efficacy in rural West Africa. *Lancet* **i:**972–975.
43. **Institute of Medicine.** 1991. Economics of malaria control, p. 237–256. *In* S. C. Oaks, V. S. Mitchell, G. W. Pearson, and C. C. J. Carpenter (ed.), *Malaria: Obstacles and Opportunities.* National Academy Press, Washington D.C.
44. **King, M.** 1990. Health is a sustainable state. *Lancet* **336:**664–667.
45. **Kolata, G.** 1984. The search for a malaria vaccine. *Science* **226:**679–682.

45a. **Koram, K. A., S. Bennett, J. H. Adiamah, and B. M. Greenwood.** 1995. Socio-economic determinants are not major risk factors for severe malaria in Gambian children. *Trans. R. Soc. Trop. Med. Hyg.* **89:**151–154.

46. **Mabey, D. C. W., A. Brown, and B. M. Greenwood.** 1987. *Plasmodium falciparum* malaria and *Salmonella* infections in Gambian children. *J. Infect. Dis.* **155:**1319–1321.
47. **McGregor, I. A.** 1964. Measles and child mortality in The Gambia. *W. Afr. Med. J.* **13:**251–257.
48. **McGregor, I. A.** 1982. Malaria: nutritional implications. *Rev. Infect. Dis.* **4:**798–804.
49. **McGregor, I. A., H. M. Gilles, J. H. Walters, A. H. Davies, and F. A. Pearson.** 1956. Effects of heavy and repeated malaria infection on Gambian infants and children. *Br. Med. J.* **ii:**686–692.
50. **Miller, K. L., P. H. Silverman, B. Kullgren, and L. J. Mahlmann.** 1989. Tumour necrosis factor alpha and the anemia associated with murine malaria. *Infect. Immun.* **57:**1542–1546.
51. **Mills, A.** 1991. The economics of malaria control, p. 141–165. *In* G. A. T. Targett (ed.), *Malaria: Waiting for the Vaccine.* John Wiley and Sons Ltd., Chichester, England.
52. **Molineaux, L.** 1985. The impact of parasitic diseases and their control, with an emphasis on malaria and Africa, p. 13–44. *In* J. Vallin and A. D. Lopez (ed.), *Health Policy, Social Policy and Mortality Prospects.* Ordina, Liege, Belgium.
53. **Molineaux, L., and G. Gramiccia.** 1980. *The Garki Project,* p. 231–249. World Health Organization, Geneva.
54. **Morley, D., M. Woodland, and W. F. J. Cuthbertson.** 1964. Controlled trial of pyrimethamine in pregnant women in an African village. *Br. Med. J.* **i:**667–668.
55. **Petit, P. L. C., and J. K. S. van Ginneken.** 1992. Mortality and morbidity from infectious diseases in rural hospitals in West and East Africa over the past 16 years, p. 174. *Abstr. 5th Int. Congr. Infect. Dis. Nairobi, 1992.*
56. **Pied, S., A. Nussler, M. Pontet, F. Miltgen, H. Matile, P.-H. Lambert, and D. Mazier.** 1989. C-reactive protein protects against preerythrocytic stages of malaria. *Infect. Immun.* **57:**278–282.
57. **Ponnudurai, T., A. H. Lensen, J. van Gemert, M. G. Bolmer, and J. H. Meuwissen.** 1991. Feeding behaviour and sporozoite ejection by infected *Anopheles stephensi. Trans. R. Soc. Trop. Med. Hyg.* **85:**175–180.
58. **Pringle, G.** 1966. A quantitative study of naturally-acquired malaria infections in *Anopheles gambiae* and *Anopheles funestus* in a highly malarious area of East Africa. *Trans. R. Soc. Trop. Med. Hyg.* **60:**626–632.
59. **Reber-Liske, R., L. A. Salako, H. Matile, A. Sowunmi, and D. Stürchler.** 1995. $[NANP]_{19}$-51. A malaria vaccine field trial in Nigerian children. *Trop. Geograph. Med.* **47:**61–63.

60. **Rickman, L. S., D. M. Gordon, R. Wistar, U. Krzych, M. Gross, M. R. Hollingdale, J. E. Egan, J. D. Chulay, and S. L. Hoffman.** 1991. Use of adjuvant containing mycobacterial cell-wall skeleton, monophosphoryl lipid A, and squalene in malaria circumsporozoite protein vaccine. *Lancet* **337:**998–1001.
61. **Rosenberg, R., R. A. Wirtz, I. Schneider, and R. Burge.** 1990. An estimation of the number of sporozoites ejected by a feeding mosquito. *Trans. R. Soc. Trop. Med. Hyg.* **84:**209–212.
62. **Rowland, M. G. M., T. J. Cole, and R. Whitehead.** 1977. A quantitative study into the role of infection in determining nutritional status in Gambian village children. *Br. J. Nutr.* **37:**441–450.
63. **Shepard, D. S., M. B. Ettling, U. Brinkmann, and R. Sauerborn.** 1991. The economic cost of malaria in Africa. *Trop. Med. Parasitol.* **42:**199–203.
64. **Sherman, I. W., I. Crandall, and H. Smith.** 1992. Membrane proteins involved in the adherence of *Plasmodium falciparum*-infected erythrocytes to the endothelium. *Biol. Cell* **74:**161–178.
65. **Sherwood, J. A., R. S. Copeland, K. A. Taylor, K. Abok, T. K. Ruebush, H. A. O. Ondolo, J. B. O. Were, A. J. Oloo, J. I. Githure, C. J. Mason, R. A. Wirtz, I. P. Schneider, D. M. Gordon, W. R. Ballou, M. R. Hollingdale, M. Gross, J. S. Wittis, J. C. Sadoff, and C. R. Roberts.** 1991. Immunogenicity and efficacy of a *Plasmodium falciparum* circumsporozoite vaccine in a malaria endemic area of Kenya, p. 287. *Abstr. 40th Annu. Meet. Am. Soc. Trop. Med. Hyg. Boston, 1991.*
66. **Snow, R. W., S. W. Lindsay, R. J. Hayes, and B. M. Greenwood.** 1988. Permethrin-treated bed nets (mosquito nets) prevent malaria in Gambian children. *Trans. R. Soc. Trop. Med. Hyg.* **82:**838–842.
67. **Stettler, N., Y. Schutz, R. Whitehead, and E. Jéquier.** 1992. Effect of malaria and fever on energy metabolism in Gambian children. *Pediatr. Res.* **31:**102–106.
68. **Stürchler, D.** 1989. How much malaria is there worldwide? *Parasitol. Today* **5:**39–40.
69. **Trape, J. F., A. Zoulani, and M. C. Quinet.** 1987. Assessment of the incidence and prevalence of clinical malaria in semi-immune children exposed to intense and perennial transmission. *Am. J. Epidemiol.* **126:**193–201.
70. **Valero, M. V., L. R. Amador, C. Galindo, J. Figueroa, M. S. Bello, A. L. Murillo, A. L. Mora, G. Patarroyo, C. L. Rocha, M. Rojas, J. J. Aponte, L. E. Sarmiento, D. M. Lozada, C. G. Coronell, N. M. Ortega, J. E. Rosas, P. L. Alonso, and M. E. Patarroyo.** 1993. Vaccination with SPf66, a chemically synthesized vaccine, against *Plasmodium falciparum* malaria in Colombia. *Lancet* **341:**705–710.
71. **Velema, J. P., E. M. Alihonou, J.-P. Chippaux, Y. van Boxel, E. Gbedji, and R. Adegbini.** 1991. Malaria morbidity and mortality in children under three years of age on the coast of Benin, West Africa. *Trans. R. Soc. Trop. Med. Hyg.* **85:**430–435.
72. **World Health Organization.** 1990. World malaria situation, 1988. *Rapport Trimest. Sanit. Mond.* **43:**68–79.
73. **Zevering, Y., F. Amante, A. Smillie, J. Currier, G. Smith, R. A. Houghten, and M. F. Good.** 1992. High frequency of malaria-specific T cells in non-exposed humans. *Eur. J. Immunol.* **22:**689–696.

Index